Pathologist of the Mind

Pathologist of the Mind

Adolf Meyer and the Origins of American Psychiatry

S. D. LAMB

Johns Hopkins University Press

Baltimore

Johns Hopkins University Press
2715 North Charles Street
Baltimore, Maryland 21218-4363
www.press.jhu.edu

All illustrations courtesy of and reproduced by permission of the Alan Mason
Chesney Medical Archives of The Johns Hopkins Medical Institutions.

Library of Congress Cataloging-in-Publication Data
Lamb, S. D., 1971– author.
Pathologist of the mind : Adolf Meyer and the origins
of American psychiatry / S. D. Lamb.
p. ; cm.
Includes bibliographical references and index.
ISBN-13: 978-1-4214-1484-3 (hardcover : alk. paper)
ISBN-10: 1-4214-1484-8 (hardcover : alk. paper)
ISBN-13: 978-1-4214-1485-0
ISBN-10: 1-4214-1485-6
I. Title.
[DNLM: 1. Meyer, Adolf, 1866–1950.   2. Psychiatry—history—United States.
3. History, 19th Century—United States.   4. History, 20th Century—United
States.   5. Psychotherapy—history—United States.   WM 11 AA1]
RC454.4
616.89—dc23   2013048850

A catalog record for this book is available from the British Library.

Special discounts are available for bulk purchases of this book.
For more information, please contact Special Sales at 410-516-6936
or specialsales@press.jhu.edu.

Johns Hopkins University Press uses environmentally friendly book materials,
including recycled text paper that is composed of at least 30 percent
post-consumer waste, whenever possible.

In memory of my parents,
Stan and Carol Lamb

CONTENTS

Illustrations follow pages 98 and 160

ACKNOWLEDGMENTS

Psychobiology, the concept upon which all of Adolf Meyer's reforms, practices, and teaching were based, evolved from his extensive knowledge of complex scientific and philosophical principles. In numerous scientific publications, he discussed these complexities from several demanding disciplinary perspectives, using weighty terminology, obscure literary references (for which he assumed his readers needed no context), and a writing style that was painfully long-winded, dense, convoluted, and abstruse. In order to research and write this book, I worked to become fluent in the Meyerian syntax. I learned to identify and analyze the substantive content of Meyer's writing, and I searched for ways to reinterpret his meaning in intelligible terms for both myself and my readers—all of which regularly tested my intellectual and emotional limits. That I survived this decade-long ordeal owes much to the advice, comments, and support that I received from a number of people, whom I now have the pleasure of acknowledging and thanking.

Pathologist of the Mind was born and fostered in the Institute of the History of Medicine at Johns Hopkins University. The institute's faculty, under the chairmanship of Randall Packard, created an intellectually diverse and supportive environment that was ideal for scholarly development. My thesis supervisor, Daniel Todes, cultivated a working relationship that integrated pedagogy and collegiality seamlessly. He was (and remains) enthusiastic, generous, challenging, responsive, and shrewd; I am fortunate and forever grateful to reap the benefits of Dan's exceptional commitment to academic supervision and mentorship. Thank you to my dissertation committee members—Nathaniel Comfort, Ronald Walters, Jay Schulkin, and Ruth Leys—for their many keen insights and constructive comments. Special thanks to Ruth Leys, whose singular expertise on Meyer's ideology guided my ongoing evaluation of the abundant source materials. At an early stage in the research, Harry Marks suggested that I read Morton White's 1949 study *Social Thought in America: The Revolt Against Formalism*, assuring me that sound scholarship has no expiry date; without Harry's suggestion and White's thesis, I would still be scratching my head and wondering what on earth Meyer was up to. Nonetheless, there was plenty of confusion—and frustration and misery—assuaged by a phenomenal group of fellow graduate students at Hopkins without whom there would still be no book on Adolf Meyer: Cassidy Brown, Sandra Eder, Melissa Graffe, Alexa Green, Tulley Long, Massimo Patrozzi, Katie Reinhart, Pierce Salguero, Jimmy Schafer, Nick Spicher, and Mark Waddell. I am also indebted to the institute's curator of historical collections, Christine Ruggere, and its head administrator, Coraleeze Thompson, for unfailingly answering questions and resolving problems both big and small.

I owe a special debt to Nancy McCall, director of the Alan Mason Chesney Medical Archives of the Johns Hopkins Medical Institutions, who, in my first week of graduate school—after hearing me describe a passionate but still vague interest in clinical approaches to mind-body problems—encouraged me to look at the Hopkins psychiatrist Adolf Meyer. Our fruitful discussions strengthened and enhanced this research, as did the invaluable help I received from archivists Marjorie Kehoe, Andrew Harrison, and Phoebe Evans Letocha while working with the extraordinary archival materials in the Adolf Meyer Collection. Two of Meyer's successors, Raymond DePaulo, the current chief of psychiatry of Johns Hopkins Hospital, and his predecessor, Paul McHugh, have shown ongoing and enthusiastic support for this work. I am grateful to both for sponsoring my request to have the medical records of Phipps patients unsealed for my historical research. As a result, my thanks are also owed to Louise Thompson and the staff of the hospital's Medical Records Service for locating the microfilmed records, and for welcoming me into their office space, where I spent hundreds of hours collecting data. A warm thank you to Christina Ascher and Klaus Asher (the former chose the German spelling) for our fascinating discussions about their grandfather, Adolf Meyer, and for the splendid insights and materials they shared with me.

This research received generous financial support from the Social Sciences and Humanities Research Council of Canada, the Andrew W. Mellon Foundation, the American Council of Learned Societies, and the Canada Research Chair in the Social History of Medicine at McGill University.

Over the years my thinking about Meyer and the medical and cultural worlds he inhabited benefited from the incisiveness of innumerable scholars, especially the members of the American Association for the History of Medicine and the Canadian Society for the History of Medicine. Participants in various forums generated helpful commentary on my claims and findings: the Gender Workshop and the Colloquium in the History of Science, Medicine, and Technology at Johns Hopkins University; the Richardson Seminar at Cornell University; the Joint-Atlantic Seminars for the History of Medicine; the Workshop for Technique, Technology and Therapy in the Brain Sciences at Clarkson University; the Conference on Psychotherapeutics at the Centre for the History of Psychological Disciplines at University College London; and the Nursing History Research Unit at the University of Ottawa. In particular, Michael Bliss, John Burnham, Stephen Casper, Jackie Duffin, Matthew Gambino, Delia Gavrus, Christopher Green, Gerald Grob, Katja Guenther, Laura Hirshbein, Cheryl Logan, Nathan Moon, Hazel Morrison, Richard Noll, Isabelle Perreault, Thomas Schlick, Nick Whitfield, and David Wright shared much-appreciated knowledge, materials, and insights. Edward Shorter at the University of Toronto and Andrea Tone at McGill University created invaluable opportunities for professional and personal growth. Elisabeth Tutscheck kindly helped with translations. A sincere thank you to Robert Maunder, without whom this journey could not have been conceived by me, let alone undertaken and completed. The process of turning the thesis into a book manuscript was an incredibly positive experience thanks to the guidance and generosity of editor Jacqueline Wehmueller, and the judicious copy editing of Anne Whitmore at Johns Hopkins University Press. While I earnestly considered all the comments and advice that I received, the final decisions and interpretations in this study are my own and I bear full responsibility for them and any errors that remain.

Finally, I would like to thank my wonderful friends and family in Toronto for backing this academic (ad)venture, transitioning along with me without hesitation when I explained that stand-up comedy at Yuk Yuk's was out and a Ph.D. at Johns Hopkins was in. They were all cheerleaders; thank you to Alistair Hepburn, Karen Hepburn, Linda North, and Arlene Lott, who were also proofreaders of book chapters. In addition, Arlene Lott graciously prepared the digital images for publication. My sister, Teri Bowers, deserves special recognition. She proofread the dissertation by reading every word aloud. She ensured that I was installed in a safe apartment to begin graduate school in 2004. And, more important, she had the emotional fortitude to drive away and leave me standing alone on a curb in Baltimore City.

In the final stretch to complete my dissertation six years later, she organized a greeting card drive—what I refer to as the Hallmark Blitzkrieg of the Terrible Summer—that filled my mail box with encouraging words from home for weeks on end. In truth, by that time Baltimore had become my home, in part because of the camaraderie and conviviality I enjoyed with my fellow Charm City Roller Girls (it is remarkable how effectively the high-speed, full-contact sport of roller derby can relieve the stress of contemplating a complicated figure like Adolf Meyer).

While my mother, Carol Lamb, harbored reservations about my roller skating career in Baltimore, she took great pleasure and pride in my academic accomplishments. Her unexpected death marked the beginning of the Terrible Summer, but with the assistance and kindness of so many of the people already mentioned, I was able to fulfil her wishes (which she had restated periodically during my residency in the United States): finish your degree as quickly as possible, return home immediately thereafter, marry a Canadian. I miss her terribly and will always be sorry that she never got the chance to address me as Doctor Lamb with a twinkle in her eye, or to meet my new family. Thanks to the fantastic opportunity given to me by Andrea Tone to continue my research at McGill University in Montreal, I met and married a wonderful Canuck and full-time father. No sooner did Félix and Amélie welcome a stepmother into their lives, however, than I locked myself in my basement office to work on something called "the book" for the next three years—*merci mes amis*. Lastly, my amazing husband and fellow academic, Joey Paquet, sustained me, intellectually and emotionally, during the solitude of revising and expanding my analysis away from the supportive environment of graduate school. Joey spent hours discussing the principles of psychobiology, pragmatism, and psychoanalysis with me, and his keen observation that Adolf Meyer did not view his psychiatric patients the way Charles Darwin did his barnacles (as a medium for proving his theory) marked a key turning point in my revision process. The publication of *Pathologist of the Mind* is a direct result of his love, patience, and support, for which I am eternally thankful.

Pathologist of the Mind

Introduction

During the first half of the twentieth century, Adolf Meyer was the most recognizable, authoritative, and influential psychiatrist in the United States. When Johns Hopkins Hospital received a large endowment in 1908 to establish the country's first university clinic devoted to psychiatry—still a nascent medical specialty at that time—Meyer was selected to oversee the enterprise. He was a Swiss-German émigré, aged forty-two, with a reputation for introducing scientific methods into large, outdated American institutions known as lunatic asylums. The Henry Phipps Psychiatric Clinic, named after its benefactor, opened at Johns Hopkins in 1913, and Meyer remained psychiatrist-in-chief at the nation's preeminent medical school and hospital until 1941. From this powerful institutional position, he implemented his grand vision for American psychiatry: to establish the specialty as a clinical science operating like the other academic disciplines at the pioneering research university.

As it did for his fellow professors, Meyer's affiliation with Johns Hopkins bestowed national medical authority and assured him a leadership role within his profession. His opinions on scientific and social issues peppered newspaper and magazine articles. Americans also read about Meyer's famous patients at the Phipps Clinic and his expert testimony at sensationalized legal trials. He was a public face for a national movement—for which he suggested the name "mental hygiene"—to promote the need for healthy habits of mind as well as body. For three decades, every Johns Hopkins medical student was obliged to pass his course on psychobiology, Meyer's distinctive approach to psychiatry as a branch of biology. During World War I, the military trained scores of medical specialists to mobilize Meyerian psychiatry for the war effort. He was dubbed the "Dean of American Psychiatry," a moniker that acknowledged his de facto role as superintendent of the young and impressionable specialty. He held extraordinary sway over academic appointments in psychiatry, as well as state policies regarding the training and certification of psychiatrists. Based on Meyer's concept of the person as a biological organism, the Rockefeller Foundation funneled millions of dollars into psychosomatic research and, in 1952, the American Psychiatric Association produced the first edition of its *Diagnostic and Statistical Manual*, popularly known today as "the *DSM.*" Meyer trained hundreds of specialists who ultimately occupied leadership positions and made significant contributions in psychiatry, neurology, experimental psychology, social work, and public health. In a period of rapid growth for modern psychiatry in the 1930s and 1940s, a Meyerian diaspora transported his ideas and practices throughout the United States and beyond to Canada, Britain, and parts of Asia and Europe.

Direct the historical gaze toward the origins of American psychiatry— within medicine or popular culture—and all roads lead to Adolf Meyer. This is not hagiography; his wholesale and indelible influence on American psychiatry is the one thing about which historians agree when it comes to Meyer. Yet, Meyerian psychiatry has never been the focus of systematic historical analysis. His convoluted theory of psychobiology, and his notoriously ineffective attempts to explain it, leave most readers of his scientific papers baffled about its meaning and implications, about Meyer's objectives, and about the reasons for his great influence. The primary instrument of his influence, the Phipps Clinic at Johns Hopkins, has remained *terra incognita* for historians of medicine. In general histories of American psychiatry, Meyer is universally credited with making the discipline a part of science-based medicine—a founding father if ever there was one—but the name Adolf Meyer is not familiar to psychiatrists today. This state of affairs would have surprised his contemporaries, supporters and critics alike.

"We all are, in some ways, students of Adolf Meyer," the psychiatrist Franklin Ebaugh declared upon the death of his mentor in 1950, "and will remember him as one of the most illustrious of modern physicians." D. K. Henderson, the inaugural assistant director of the Phipps Clinic, added, "Meyer's biological viewpoint brought an interest and hopefulness into psychiatry which will always endure." Oskar Diethelm, head of psychiatry at Cornell Medical School, who trained at the Phipps Clinic, concluded in his obituary of his teacher that "through him more than anyone else, American psychiatry became what it is today." The value and originality of Meyer's contributions were incomprehensibly large, another obituarist reiterated in

1950, because so much of what Meyer conceived and implemented had become the discipline's standard procedures. In 1966, the *American Journal of Psychiatry* honored the centenary of his birth with a special issue teeming with testimony as to the essentially Meyerian character of American psychiatry. Less than two decades later, however, commentators began to note the unexpected disappearance of Adolf Meyer. "His influence pervades every aspect of our field," explained the Johns Hopkins psychiatrist and Meyer trainee Jerome Frank in 1980, but his teachings "slipped into the mainstream of psychiatry without an identifying tag." Paul McHugh, one of Meyer's successors as chief of psychiatry at Johns Hopkins (1975–2001), still reminds psychiatric residents that "you might think you've discovered something new and that usually means you just haven't read Meyer." Throughout the rise and decline of psychoanalysis after World War II and the emergence of new paradigms of brain chemistry and genetics, senior clinicians continued to attest to the endurance of Meyer's principles.[1]

Both historians and psychiatrists have overwhelmingly concluded that the basic infrastructure for American psychiatry evolved from the concepts and methods pioneered by Meyer at the turn of the twentieth century, yet those very concepts and methods appear utterly obscure to us today. Without understanding what they were, we can neither hope nor propose to understand the nature or impact of Meyer's influence. The principal aim of this book, then, is to rediscover Meyerian psychiatry.

This study examines Meyer's efforts between 1892 and 1917 to institute a clinical science of psychiatry in the United States that harmonized the practices and expectations of scientific medicine with his biological conception of mental illness. This period was a transformative interlude in the development of American psychiatry. It was spurred by a vanguard of psychiatrists, within which Meyer was a leading figure, that transcended the cynicism fostered by the deterioration of nineteenth-century "lunatic asylums." Inspired by an environmentalist and functional view of mental disorders, this cohort established the institutions and procedures with which American psychiatry has pursued successes and faced failures throughout the twentieth century. In his intellectual biography of Meyer's contemporary Sigmund Freud, Frank Sulloway argues that Freud's fundamental conceptions were biological by inspiration and implication, making Freud a biologist of the mind. In this same sense, I argue, the Meyerian mold for American psychiatry was the product of a pathologist of the mind.[2]

Meyer's elite medical training in Europe predisposed him to a clinical model of university medicine that combined work with hospital patients, laboratory research, and bedside teaching. Clinical medicine was based on pathology, the study of disease and diagnosis. Pathology used comparative and quantitative methods to identify otherwise indiscernible patterns or markers of distinct diseases. In the nineteenth century, anatomists and bacteriologists developed sophisticated methods for differentiating pathological mechanisms responsible for disease, and this new knowledge was utilized to cure, ameliorate, or prevent sickness. At the outset of his career, Meyer reoriented his own scientific interests away from brain research (what he called comparative neurology) and toward work with patients who had mental disturbances (something he distinguished as psychiatry). Inspired by American pragmatism, he adopted the view that the human mind had evolved according to its use as an instrument of biological adaptation. This instrumentalist view cast subjective experience and social interaction as functions of human biology. It followed, for Meyer, that miscarriages of these biological functions were the responsibility of scientific medicine. Meyer defined scientific medicine as the union of pathology and therapeutics. He insisted that it was the established methods of pathology that transformed the art of medicine into a scientific enterprise. In turn, he believed, pathology remained a handmaiden to practical medicine—it was not a basic laboratory science and it was always in pursuit of new therapeutic applications. Mental dysfunction, as much as brain disease, he held to be a medical condition resulting from pathological processes.

In the early 1890s, Meyer embarked on a crusade to establish a science of psychiatry based on the compatibility of pathology as method and mind as biology. At the center of this campaign was his concept of "psychobiology." He framed prevalent forms of mental illness not as distinct brain diseases, as did a great majority of his peers in that era, but as failed adaptation. Emerging from Meyer's steadfast allegiances—one methodological and the other conceptual—were two interlocking conclusions that guided his vision for American psychiatry. First, the pathological processes underlying mental disorder took place not exclusively at the level of brain tissues or metabolism but also at the level of adaptive behavior and individual experience. Second, despite its immateriality, a pathological experience could be distinguished from normal experiences in the same way the pathologist differentiated diseased and healthy tissue—by systematic and comparative analyses of large numbers of specimens. Psychobiology was Meyer's attempt to combine established clinical methods and an emergent concept, articulated by American pragmatists, that mental activity was a causal force in the human organism's dynamic interaction with its environment.

On the basis of the theory of natural selection, he rejected all forms of mind-body dualism. He also dismissed traditional mechanical views of the nervous system in which mental activity was a meaningless byproduct. He suggested that some mental disturbances were best understood as correctible maladjustment, not incurable diseases. Meyer called for a scientific and biological approach to mental disorders, insisting that no fundamental distinction existed between the natural sciences (characterized by so-called objective knowledge of physical phenomena and universal causal laws) and the human sciences (dealing with subjective interpretations of historical, behavioral, and social processes). He lobbied traditional asylums to adopt the procedures of modern hospitals but also emphasized the need for urban institutions that made voluntary medical help for psychiatric illness accessible and appealing to the public. He insisted, too, that the United States establish specialized institutes at medical universities devoted to research and teaching in psychiatry. Early in his career, he managed to secure advantageous platforms from which to broadcast his optimistic vision for twentieth-century psychiatry.

Meyer was hardly alone in any of these objectives, but his psychobiology—and, by extension, the practices he used and taught at Johns Hopkins—represented a distinctly American and Meyerian way to pursue them.

The Luminous Fog of Adolf Meyer's Self-Expression

One seeks in vain to understand Meyer's vision, claims, or practices by analyzing him as a thinker and plunging into his published articles. He never succeeded in completing a synthetic monograph and his prose is notoriously opaque and indirect. When his roughly 250 scientific papers were published in four handsome tomes after his death in 1950, an enthusiastic reviewer acclaimed their great historical significance even though "it is not possible to recommend these volumes for reading." Senior medical students at Johns Hopkins once staged annual burlesque shows that lampooned the eccentricities of their revered medical professors. In the 1920s, a foreign student took to the stage costumed as the chief of psychiatry and gave a lengthy recitation in Chinese, occasionally interposing the word *psychobiology*. The audience, filled with students who had endured Meyer's famously indecipherable lectures, roared with laughter. In 1940, someone plucked a phrase from an article Meyer published in the journal *Science* and nominated it to compete in the weekly Non-Stop Sentence Derby in *The New Yorker* magazine. It won. This amusing anecdotal evidence highlights the extent and notoriety of Meyer's inability to express himself clearly. For him, however, it was no laughing matter. He remained sensitive about these shortcomings throughout his career, eventually asking his followers to write the textbooks that he was continually under pressure to produce.[3]

A confluence of factors accounted for the difficulty of Meyer's formal prose. One contemporary recalled that he often qualified any core statement with a bewildering series of dependent clauses. These qualifiers frequently preceded the statement's subject (common in Meyer's native German, but confusing for English speakers). His tendency toward awkward phrasing left readers and listeners dizzy. "I feel rather skeptical toward certain claims," he wrote in reference to a colleague's work, "but certainly not too much so not to consider it unwise to disregard his findings." Contradictory intellectual impulses created the worst problems. One Phipps Clinic staffer recalled that Meyer was cautious in his pronouncements, never peremptory, which created "a very characteristic circumlocution." Meyer dreaded making an imprecise, incomplete, or abstract statement; yet he habitually skirted conclusions, declarations, and absolutes that might be construed as dogma. His compulsion to be exhaustive and precise, while maintaining a thoroughgoing relativism, produced mystifying displays of verbal gymnastics. To take a trivial example, he began one paper with the phrase *In long periods of historically accessible time.* Evidently, the more traditional *Throughout history* implied too strongly the invalid notion that the entirety of history was knowable. Above all, he abhorred the thought of reducing complex phenomena to simplified terms. While admirable, his caution made discussing such phenomena problematic.[4]

Meyer thought clearly and logically, the psychiatrist Leo Kanner insisted of his mentor. Kanner observed that, due to Meyer's status, their colleagues rarely criticized psychobiology openly—or constructively. Instead, criticism reached Meyer through back-channels. "Not relying fully on the true profundity of his ideas," Kanner concluded, "a note of defensive self-justification seemed to creep into his sayings and writings." While his explanations remained consistent throughout his published papers, Meyer routinely tailored his terminology and analogies to suit each new intended audience. His aim was probably to enhance clarity and avoid oversimplifications that he imagined might invalidate his claims in the minds of his listeners and readers, but the practice deprived his corpus of programmatic coherence and further weakened its explanatory power. The obfuscating impact of these idiosyncrasies was compounded by the difficulties of his subject matter and by the epidemic of neologisms generated by the theoretical chaos that characterized psychiatry and neurology at the end of the nineteenth century. It was a recipe for textual disaster.[5]

Despite the inadequacy of his formal communications, Meyer conveyed his ideas successfully in other forums, and he attracted devotees. His contemporaries speculated endlessly about the Delphic quality of his formal remarks in comparison to the lucidity of his casual conversations with colleagues and patients.[6] "There were always pithy remarks," the psychiatric epidemiologist Alexander Leighton testified, "that created a desire to hear and understand more."[7] The New York psychiatrist Joseph Wortis, a Phipps staffer who later founded the journal *Biological Psychiatry*, admitted that Meyer's cryptic formulations during daily staff meetings had tempted him, occasionally, to cry out that the emperor had no clothes. What prevented him, he claimed, was the chief's beneficent aura. "One could get the drift even when the details were obscure," Wortis recalled; "the fog was luminous."[8] William Alanson White, the director of St. Elizabeths State Hospital in Washington, wished to know who had formulated one principle that he found particularly useful. He received the answer from the Meyerian psychiatrist John MacCurdy at Cambridge University. It was at a national meeting in 1914, MacCurdy told White, when "poor old Southard gave one of his mazy papers" (a reference to E. E. Southard, director of the Boston Psychopathic Hospital). Southard's topic—the significance of the *un* in the term *unconscious*—had spawned acrimonious debate about how to classify mental disorders. When Meyer stood and suggested that definition in psychiatry "must express direction rather than delimitation," the debaters reportedly fell into a satisfied silence. "This priceless formulation always stuck in my mind," mused MacCurdy, "it was one of those pearls which he has ever been prone to drop when off his guard."[9] In conversation he was clear, even lapidary, according to these testimonies. Thus did Meyer manage to acquire loyal adherents eager to implement, transmit, and expand upon his ideas.

In the absence of any systematic study and clarification of Meyerian psychiatry, and without a critical biography about its originator, historians have confidently and assertively provided their own characterizations of Meyer's temperament, motives, intelligence, moral fiber, ideas, methodology,

and long-term impact on the discipline. Based on circumstantial evidence—
which is, indeed, plentiful—historians with strikingly different scholarly
approaches unanimously cite Meyer as *the* definitive influence on the development of American psychiatry. This consensus is especially intriguing
given that causal reductionism to any single dead white man has become
passé within the discipline of history.[10] A few scholars offer condescending
assertions. One historian contends that Meyer's proclamations about psychobiology were "devoid of substance" and not well-respected.[11] Another
declares him a "second-rate thinker" who arrested psychiatry's scientific advance with indiscriminate eclecticism.[12] The practices Meyer employed and
taught at the nation's leading medical school for three decades, claims a
third historian, were an "elusive charade" perpetrated by someone whose
confidence outstripped his competence.[13] None of these scholars claims to
have analyzed systematically the substance or reception of Meyer's ideas
and practices. The form and content of this book evolved from the assumption that any verdicts or indictments of Adolf Meyer are premature before
we rediscover Meyerian psychiatry.

Sources and Scope

Rather than relying exclusively on Meyer's abysmal attempts to explain his
views, this study explores the meaning of his psychobiology by examining
how he put it into practice at the Phipps Clinic. The breadth and richness of
two extraordinary sources of documentary evidence enabled me to eaves-
drop on Meyer when he was expressing himself most clearly; these sources
are his extensive professional and personal correspondence and the medical
records of Phipps patients. His papers are housed in the medical archives at
Johns Hopkins University in Baltimore. The documents and artifacts in the
Adolf Meyer Collection are diverse and exhaustive, spanning his life (1866–
1950) and offering insights into every aspect of his personal and professional
activities. In addition to the textured and candid exchanges in his written
correspondence, Meyer's private scientific notes, unpublished lectures, and
travel diaries provide a trove of information. I also made extensive use of
the architectural plans and photographs of the Phipps Clinic. I was privi-
leged to be the first scholar granted permission to use for historical research
the hospital records of patients admitted to the Phipps Clinic. These records
are unusually detailed—a hallmark of Meyerian psychiatry—and they allowed me to situate Meyer's ideas within the clinical situations to which he

applied them. Expressive, often evocative, notations by Meyer and his staff describe institutional procedures and day-to-day interactions with patients. These dossiers also include stenographic transcripts of interviews with patients as well as texts written by patients in the form of daily logs, letters of complaint, and dream journals. Patients' incoming and outgoing letters were screened by staff and, if deemed significant to the clinical enterprise, were transcribed into the record. To protect patients' privacy, I have substituted suitable alternatives for some biographical details while preserving gender, social status, and ethnicity.

Scholars seeking to understand Meyer face an embarrassment of archival riches and raw data, enough to yield stacks of diverse monographs from a dozen different disciplinary perspectives. I limited myself to what seemed a significant and manageable period, 1908 to 1917, the decade that encompassed the planning and early operation of the Phipps Clinic. The Progressive Era, with its promotion of scientific method, was in full swing at that time and proved vital to twentieth-century American culture and psychiatry. The United States' declaration of war on Germany and the discovery of the malarial cure for syphilitic insanity—psychiatry's first so-called magic bullet—made 1917 a critical turning point within American culture and medical research on mental disorders. Concentrating on this decade offered me the opportunity to explore my interests in the intersection of medical theory and medical practice, and in relationships among psychiatrists, patients, and kinship networks. This focus determined the set of case histories I chose to examine: 1,897 cases constructed from all admissions in 1913 through 1917, corresponding to 1,772 individuals (some patients were admitted more than once in this five-year period). Compiling demographic and medical information, I analyzed quantitatively the patient population in this period. Extensive biographical information, collected by the clinicians from each patient and from those closest to him or her, helped contextualize data. My approach to using clinical records to explore interactions among physicians, patients, and the illnesses that dominated their lives was guided by Chris Feudtner's masterful study of the Joslin Diabetes Clinic in early twentieth-century Boston.[14]

Pathologist of the Mind is not a comprehensive history of Meyerian psychiatry. Nor is it a critical assessment of psychobiology's internal logic, intellectual debts, or reception by Meyer's contemporaries. It is an historical interpretation of his early ideas and practices with an emphasis on exploration and explanation. Before we analyze what *we* think Meyer was saying

and doing we need a better working theory of what *he* thought he was saying and doing. With clearer context, one can more profitably pursue, for
instance, the intriguing hypothesis suggested by historian Ruth Leys that,
in Meyer's writings, psychosexual conflict took the form of a pathological,
revolutionary feminist, or Bolshevik desire, or consider whether the Phipps
Clinic represented a disciplinary economy that conditioned the conduct of
patients, as historian Eric Engstrom demonstrates for a similar institution.[15]
A fuller grasp of Meyer's meaning and the objectives that shaped his practices, moreover, will contextualize the critiques exchanged between Meyer
and his peers in the United States (for example, E. E. Southard, Stanley
Abbot, Earl Bond, and Edward Kempf) and his affiliations with and criticisms of the two other Europeans whose influence on American psychiatry
equaled his, Emil Kraepelin and Sigmund Freud. I therefore focus on clarify-
ing, comprehending, and contextualizing Meyer's conceptual formulations
and clinical practices between 1892 and 1917.[16]

My historical narrative revolves around the methods and concepts that
Meyer devised in order to study and treat abnormal mental phenomena, or
psychopathology. Throughout this study, I refer to a limited number of typical forms of psychopathology that he observed in his clinic and discussed
with his peers in the terms used at the beginning of the twentieth century:
affective disorders (in which depression or paranoia typically dominated),
manic-depressive insanity, organic disease (such as a brain tumor or psychosis caused by syphilis), schizophrenia and dementia praecox, functional
disorders, or "neuroses" (including hysteria, neurasthenia, psychasthenia),
and psychopathology related to auto-intoxication (for example, due to di-
etary deficiency or metabolic imbalance) or to the prolonged use of external
intoxicants such as alcohol or morphine. These mental disorders—which
was Meyer's preferred term—shared a wide variety of mild to severe symptoms, such as exhaustion, confusion, memory loss, hallucinations, delusions,
depression, paranoia, catatonia, anxiety, compulsions, irritability, and violent or impulsive behavior. He and his peers recognized that a single patient
often exhibited two or more typical symptom pictures, either simultaneously or progressively. For this reason, he emphasized repeatedly that the
majority of these disorders could not be defined or diagnosed categorically.
Nevertheless, in most cases he entered into the patient's medical record a
nonstandardized diagnostic description that was allied to one of the types
listed above. These descriptions indicate that between 1913 and 1917 the patient population at the Phipps Clinic was composed of relatively even num-

bers (15–18%) of each of the abovementioned categories, with slightly more affective disorders (20%) and significantly fewer cases of manic-depressive insanity (7%). Two percent of cases were unclassified, and 5 percent were designated "constitutional psychopathic personality," which implied a fundamental defect in the personality.

Exploring medical and cultural understandings of mental illness has been, historically, a controversial enterprise. At one extreme, there are scholars who insist that mental diseases are universal and timeless, independent of the names different societies assign to them. Other historians maintain that mental illnesses are cultural constructions and cannot be compared across time. I approach the issue from a perspective most recently reiterated by the historian Charles Rosenberg in a discussion of bipolar disorder and its history. Various forms of mental disorder have enduring idiosyncratic tendencies that can be traced historically over long periods, yet particular cultures and eras perceive and manage them in historically specific ways. While my purpose was not to analyze how mental illnesses were clinically or culturally construed in the Phipps Clinic, this study does shed some light on this important aspect of the history of American psychiatry and its patients.[17]

Adolf Meyer the Person

The following description of Meyer's appearance and disposition emerges from impressions recorded by his students and colleagues after his death and archival photographs. A man of small stature and dark complexion, he wore his beard in the Van Dyke style and dressed immaculately but informally in tweeds. He reportedly walked with a measured tread, spoke with a Swiss-German accent, and stroked his beard meditatively when he was listening to a patient or colleague. He looked like a stage psychiatrist and very much the European professor, as one colleague remarked. Apparently, he was often diplomatic and optimistic to a fault. Other recurring descriptors are: dignified, erudite, modest, demanding, meticulous, perceptive, and generous. "He would not tolerate levity," recalled Jerome Frank, "to the point that some members of the staff found him intimidating." Many writers affirmed that he demanded independent thinking, downplayed theoretical discussions, and was delighted to be asked questions. In informal settings, he was relaxed and displayed a lively sense of humor. Most Sundays, Meyer welcomed members of his staff to his home in the affluent Baltimore neighborhood of Roland Park, a tradition among Johns Hopkins professors popularized by

William Osler, the hospital's renowned first chief of medicine. "Dr. Meyer would regale us with a seemingly inexhaustible fund of amusing, often slightly barbed, reminiscences of the prominent persons he had known," recalled Frank. Once a year, the entire first-year class of Johns Hopkins medical students was invited to the chief of psychiatry's home. "They may not have been able to understand him fully," commented Oskar Diethelm, "but they recognized his wisdom and his deep sincerity as a teacher and loved and admired him for it." Several intimates commented on his eyes. "A man who looked at you straight in the eye with a gaze which pierced your innermost thoughts," was how D. K. Henderson described his mentor and friend. "It was his eyes which registered every variation of mood," Henderson mused, "joy, disappointment, pity, and sorrow, and more occasionally blazing anger, which was never long sustained." Leo Kanner testified that those eyes "could encourage or deter you, make you feel proud of yourself or annihilate you, take you into his confidence or keep you at a distance." What stood out most, another emphasized, was his remarkable energy and his zeal to teach.[18]

Several other aspects of Meyer's personality are important for understanding his professional activities in the prewar era. The first was his perseverance in implementing his vision for psychiatry. He was constantly in campaign mode. Early in his career when his boss, an influential superintendent of a large public asylum, was unable to read due to an eye injury, a young Meyer insisted upon sitting with him each evening, for weeks, to translate aloud the pioneering work of the German clinician Emil Kraepelin.[19] His activities within countless professional associations, social agencies, and government committees for half a century vastly enlarged his pool of potential captive audiences. In 1919, the psychiatrist E. E. Southard proclaimed that there was no greater power within psychiatry to change minds than the personality of Adolf Meyer—he was a ferment, an enzyme, a catalyzer. "I don't know that we could abide two of him," Southard confessed, "but in our present status we must be glad there was one."[20] His tenacious resolve was fundamental to how Meyer was able to commandeer American psychiatry in the opening decades of the twentieth century.

His outlook on social and professional relations and his experience of his own power are also noteworthy. Respectful of the existing social order, he rarely discussed politics or religion and recoiled from public controversy of any kind. He was fiercely democratic in his view of scientific inquiry and hesitant to interfere with the pursuits of his colleagues, including juniors. He eagerly utilized his authority to advance his ideology within psychiatry and society, but he exhibited indecision in difficult situations and often retreated, even when his influence might have resolved disputes or saved careers.[21] Some historians have interpreted this tendency as malevolent, insisting that Meyer had the power to suppress but refrained from curtailing therapeutic practices that appear repugnant to us today, such as indiscriminate surgeries on psychiatric patients.[22] "He was not unaware of the nimbus with which he was being surrounded," observed Leo Kanner, but "his love of unpretentious simplicity was not strong enough to break through it." Meyer's reluctance (or inability) to indoctrinate followers and form a consolidated school of Meyerian psychiatry, despite his faith in the validity of his own theory, appealed to some adherents and contemporaries, while others found it aggravating and insincere.[23]

Finally, his own mother's serious mental illness shaped Meyer's professional and intellectual trajectories. Just weeks after emigrating to the United States in 1892, he wrote to his younger brother, "You certainly must have noticed how in the last years she has declined mentally." He later wrote, "It had a great influence on my choice of career." He avowed, however, that he would not have left Zurich had he suspected that she would experience a "mental breakdown." His letters to his mother, Anna Walden Meyer, and his brother, Hermann, between 1892 and 1898 tell the story of a severe and protracted mental disturbance. His mother was institutionalized at the well-known psychiatric clinic and hospital where her son had trained, the Burghölzli at the University of Zurich. Around the turn of the century, she experienced a robust recovery that lasted several years, but she eventually became unwell again. In the spring of 1912, accompanied by his wife, Mary Potter Brooks, Meyer sailed to Europe to see his mother. She was in a "serious depression," confused and delusional, he noted. It was a heartbreaking visit for Meyer, but his mother improved slightly afterward. "I am happy that Mama is better and jolly," he told Hermann with the same optimism he injected into his correspondence with the families of his patients. "Step by step, the insecurity will make room for a strong and healthy confidence in reality." Shortly after, however, she suffered a stroke and died. The opening of the Phipps Clinic was five months away. Meyer remained convinced that strenuous life circumstances surrounding his emigration had acted as a precipitating factor in his mother's first breakdown and that the respite of her hospitalization had enabled her to regain her emotional balance. These deeply personal experiences of mental illness textured Meyer's thinking and practices throughout his career.[24]

Meyer emigrated in late 1892, from a small Germanic Swiss canton to Chicago, Illinois. Like any newcomer, he confronted strange cultural realities. His earliest letters to his mother and brother reveal that he was appalled by Americans' brashness, ignorance, and fixation with the "almighty dollar." When he attended a church service (to drum up business for his fledgling medical practice), he heard the preacher estimate the monetary value of eternal salvation to be above $25,000. "This is what the Americans understand!" Meyer editorialized. Other interactions were more amusing. When he filed for citizenship at the Chicago courthouse, the clerk informed him that immigrants must foreswear allegiance to all foreign rulers, including the Duke of Switzerland. Meyer explained politely that his native country had been a democratic republic since 1291 and was not a duchy. The bureaucrat was insistent that anyone with a "Dutch accent" must have a duke. In order to expedite his application, therefore, Meyer convincingly renounced his imaginary sovereign. Soon, though, he reinterpreted such audacity as a distinctly American blend of energy and enterprise. He felt that an empirical and practical outlook suffused American life—the upshot of common sense traditions, rooted in the Enlightenment philosophies that had prevailed throughout the young nation's history. This national personality resonated with his own intellectual stance at that time, the philosophy of naïve realism. His upbringing in a direct democracy, in which citizens called referendums to amend canton law and a simple majority carried the vote, also shaped his experience of the United States. In his new home, Meyer identified Swiss values that were dear to him: democracy, broadmindedness, practicality, and respect for both individual and collective judgment. He also encountered small groups of like-minded liberals interested in systematic solutions to the social problems of modern societies.[25]

Adolf Meyer's Cultural and Scientific Contexts

In the 1890s, a groundswell of diverse reform movements emerged in the United States. They were spurred by middle-class anxieties about the disorder generated by industrialism and capitalism: agricultural decline, immense wealth, urban poverty, criminality and insanity, political and corporate corruption, commercialism, moral decay, and working-class unrest. While their goals and strategies varied, self-described progressive reformers shared a collection of attitudes. They had faith in the ability of humans to ameliorate social problems by means of collective action, economic and political efficiency, and scientific expertise. They were especially dedicated to the principle of engineering optimal environmental conditions to produce constructive outcomes. During the 1890s, this environmentalism emerged as a governing view in the United States, challenging the pessimistic inferences of hereditary determinism. Historians of the Progressive Era have explored disparate sites of reform, from playgrounds to factories and farmers' markets to urban slums. The settlement movement was a conspicuous reform program aimed at reducing class antagonisms by establishing communes of middle-class reformers in poor city districts. Meyer was a regular visitor to Chicago's well-known Hull House, where he developed intellectual alliances with its founders, Jane Addams and Julia Lathrop, and internalized the American reform impulse. Uniting progressive reform efforts in the United States—including Meyer's in psychiatry—was an abiding confidence in the interconnectedness and malleability of the individual and society, and, importantly, a belief that deliberate action guided by qualified experts led inevitably to progress.[26]

During the Progressive Era, a transformation within American medicine that had begun earlier in the century reached its fullest expression in the modern university hospital. The early-nineteenth-century hospital was a social welfare institution, an extension of the almshouse that had provided care for the destitute and dying. As the decades passed, hospitals increasingly became associated with scientific expertise and active treatment. Medical research, elite careers, and professional influence gradually concentrated in the university hospital. When the Johns Hopkins Hospital and School of Medicine opened in Baltimore in the early 1890s, it conspicuously embodied the new scientific medicine.[27]

The appearance of Johns Hopkins on the American medical landscape grew out of institutional, technical, and cultural changes in European medicine during the previous century. Some broad contours of those gradual and heterogeneous developments are particularly important to the story of Meyerian psychiatry. For centuries, medical practices had been based principally on the humoral theory, in which illness was specific to the sufferer's unique constitution and environment. Using this interpretive framework, healers diagnosed and treated patients, without having access to events taking place inside the body. In eighteenth-century Europe, studies by skilled anatomists

resonated with new moral and philosophical perspectives of the Enlightenment. They performed their work on executed criminals and the unclaimed
bodies of paupers and occasionally conducted autopsies for legal or teaching
purposes. A convergence of diagnostics, surgery, and dissection in medical
training at the start of the nineteenth century fostered what historians
have described as a solidist and localist ethos in Western medicine—an
emerging view of the sick body as an assemblage of distinct and diseased
organs.[28]

Unique circumstances in postrevolutionary Paris engendered a fundamental reconceptualization of sickness. The secularization of the city's
massive Catholic hospitals provided elite physicians with unrestricted access to patients' bodies before and after death. Systematically, physicians
examined living patients and, when these patients died, the physicians performed autopsies on the same bodies they had examined earlier. No longer
a Christian hospice, the crowded ward was now a source of research material and was renamed *la clinique* after the Greek word for sickbed. In a dedicated dossier for each patient, these new clinicians meticulously documented
clinical signs observed at the bedside and pathological lesions discovered in
organic tissues at autopsy. Using comparative and statistical methods to
study thousands of recorded cases, they were able to discern previously un-
perceived patterns of pathology. This novel clinical-pathological method
was empirical and analytical compared to the interpretive and scholastic
character of humoral medicine. A single patient or diseased organ showed
nothing, the Paris clinicians argued. Only an analysis of a large number of
cases and specimens revealed objective identifiers of distinct disease, in the
living and in cadavers. Tissues were newly conceived as the building blocks
of the body and the physical sites of illness. Although humoral medicine
continued to thrive throughout Europe, the published researches of the Paris
Clinic found an ever-expanding readership. These publications described
and diagrammed the progression of disease in dissected organs, verifiable
by readers who could conduct their own autopsies. This ontological view of
disease—as an entity that exists apart from the body it devastates—was a
fundamental shift in conceptions of health and sickness. It made the hospital a critical space for medical training and discovery.[29]

The disease model became the paradigm of medical science in the course
of the nineteenth century. As Paris medicine elevated the morgue and scalpel to sacrosanct status, anatomical pathology emerged as an autonomous
science focused on diseased tissues. Pathologists soon were searching for

disease inside the body on multiple fronts, aided by technical improvements to the microscope that brought new sites and objects of inquiry to the fore after midcentury and fostered laboratory disciplines such as histology and bacteriology. Once again, institutional conditions were integral to change. Rulers of German principalities began to view academic science in terms of political power and industrial progress, and Germanic universities established specialist institutes that competed with each other for funds and prestige. Eventually, dynamic physiological models of disease outpaced those based on static anatomical pathology, and experimentalism overshadowed dissection as a mode of inquiry. Within scientific medicine, laboratories took pride of place, producing pioneering techniques and research. Trends in German philosophy encouraged dualism and materialism in medical thinking and popularized the view that all phenomena were reducible to physical or chemical principles. Experimentalists across Europe made the 1880s the decade of discovery, affirming germ theory and creating vaccines for half a dozen killer diseases. At the end of the nineteenth century, many medical practitioners and investigators considered the quest to locate and eradicate disease—whether at the bedside, through the microscope, or in the morgue—medicine's primary task.[30]

Relative to other scientific institutes, the clinics and professorships in psychiatry came late to German universities. They emerged in the 1870s with unharmonious institutional roots in both traditional asylums and new university hospital wards for neurological diseases. Psychiatry in this period was associated with the study of abnormal mental states, such as psychosis, dementia, and delirium. Neurology was viewed as the science of the nervous system, including its anatomical structures, normal functioning, and neurological problems like paralysis, stroke, and chorea. In new psychiatric clinics devoted to clinical-pathological research, the enterprises of psychiatry and neurology tended to merge. Neuropathologists developed dissection and microscopy techniques to examine nerve tissues and fibers. In general terms, they conceptualized the nervous system as a mechanical apparatus made up of sensory and motor nerves that conveyed nervous impulses between the brain and periphery the way wires conveyed electricity. The leaders of this new academic discipline were confident that a unique histological basis would be identified for distinct forms of psychosis and neurological disorders. As Meyer entered medical school in Zurich in the 1880s, investigators in Vienna were confirming correlations between lesions in localized areas of the brain and symptoms of aphasia, a neurological impairment involving loss of language abilities. Based on these concrete findings, diverse theories of similar cerebral localizations soon proliferated in the attempt to explain various other psychopathological states.[31]

When the young neurologist and pathologist arrived in the United States in 1892, however, no lesion had materialized that could account for the symptoms exhibited by the patients in mental asylums. Theoretical and methodological discord characterized the young specialties of psychiatry and neurology. New research in physiology and biology, as well as the advent of the neuron theory, challenged assumptions about the anatomical basis of insanity and nervousness. Clinical psychiatrists throughout Europe discussed other possible etiologies, such as auto-intoxication caused by malfunctioning glands or organs, and psychological causes (psychogenesis) such as emotional trauma. In the United States, Meyer found no academic discipline devoted to psychiatry. Most of the American physicians calling themselves neurologists were private practitioners, not university researchers—and they were in high demand, due to an epidemic of nervous breakdowns associated with the diagnosis neurasthenia. In general, he found the American scientific ethos dominated by the materialist and dualist assumptions that had characterized his German medical training. Americans embraced experimentation, he observed, and showed a profound reverence for laboratory equipment and techniques, but, from his perspective, most of his superiors and fellow specialists appeared comparatively indifferent to the scientific data that laboratories were meant to collect and test.[32]

He also discovered that American mental hospitals had not participated in the transformation that had made the modern general hospital an essential part of medical training and research. Over the course of the nineteenth century, Americans built public and privately funded institutions to provide protection and medical treatment to persons deemed "lunatics" or "insane." To confine a man or woman in an "asylum," as such institutions were called, typically required a legal declaration of insanity based on testimony from kinship networks as to lack of resources, financial or human, to care for an actively symptomatic individual with mental illness whose behavior placed the sufferer or community in jeopardy. (While terms such as *lunatic*, *insane* and *asylum* carry negative, even repellent, connotations today, this was not the case one hundred years ago. My extensive use of such terminology is strictly historical and in no way discounts the seriousness of people's experiences with mental illness and psychiatric institutions, past or present.) Superintendents of mental asylums, called alienists in this period, represented the country's oldest professionalized medical specialty. In 1875, there were approximately sixty mental asylums with 30,000 patients. By the 1890s, this hospital network had expanded, and the institutions themselves had become massive structures to accommodate a national patient population that exceeded 90,000. The medical objectives of alienists still reflected the original social function of asylums: first, to provide a restorative physical and psychological environment for patients who could potentially recover— so-called moral therapy; second, to garrison those who did not recover and lacked the resources to be cared for elsewhere. The larger asylums became, the more their superintendents were preoccupied with administrative rather than moral governance. Reports of overcrowding, physical restraint, and false confinement tainted public perceptions of asylums. They operated more like prisons than hospitals; in most cases, a legal declaration of insanity was a prerequisite of admission. Additionally, the beliefs that insanity was incurable and was the result of poor genetic inheritance were pervasive. Two menacing ideas, then, shaped cultural notions of mental illness at the beginning of the twentieth century: the brutality and finality of the asylum and the indelible hereditary stigma of insanity in the family tree. For most Americans, the lunatic asylum was a terrifying place of last resort.[33]

Asylum superintendents had faced criticism from fellow physicians for decades. In 1874, the neuropathologist and alienist Edward Spitzka remarked that these busy institutional managers had become "experts at everything except the diagnosis, pathology, and treatment of insanity."[34] Twenty years later, the prominent Philadelphia neurologist Weir Mitchell lambasted alienists for isolating themselves from the rest of medicine and, despite their professed expertise, contributing little to the scientific understanding of mental diseases. By the 1890s, most alienists recognized the need for their specialty and their institutions to modernize. In 1892, the Association of Medical Superintendents of American Institutions for the Insane, founded in 1844, changed its name to the American Medico-Psychological Association (in 1921, it was renamed the American Psychiatric Association). There were scattered efforts to integrate asylum practices into existing structures of medical science. The specialty was in transition—but conflict, not unity, was the rule. As historian Gerald Grob emphasizes, American psychiatry emerged as a byproduct of these institutional circumstances.[35]

This, then, was the situation in psychiatry when, in 1893, Adolf Meyer became the pathologist at a large regional asylum in Illinois. He found that the superintendent was eager to support his brain research. It also became clear that his new chief thought that pathology dealt exclusively with autopsies, chemical analyses of bodily fluids, and curious perversions of nature preserved in glass jars. When Meyer performed autopsies, he had no clinical observations of patients with which to correlate and confirm his findings. Indeed, no individual medical records were kept on the patients, and the small staff of physicians perceived no connection between their medical duties and Meyer's work in the morgue. He soon discovered that within North America his institution was typical. While many superintendents expressed interest in establishing pathological laboratories in their hospitals, the process of doing so was not straightforward. When Meyer toured several northeastern asylums in 1895, he saw expensive laboratory equipment unused due to lack of training or interest. Moreover, he gradually intuited an unspoken belief among Americans, including many of his colleagues, that the alienist's moral obligation to safeguard insane individuals included protecting them from scientists seeking experimental subjects. Again and again, Meyer encountered superintendents who were eager to associate their institutions, but not their patients, with medical science. He confronted, therefore, another strange reality in the New World: those who would pioneer scientific psychiatry in the United States in the twentieth century planned to abandon half of the clinical-pathological method. His perplexity about this state of affairs launched him on a career path for which he had not planned.

Chapter Overview

The first two chapters of this book explain its title. The focus of Chapter 1 is Meyer the pathologist and his efforts to transform asylum medicine into a clinical science of psychiatry. He believed that, in order for the seeds of psychiatric science to bear fruit, the specialty must first undertake the demanding and often monotonous work of clearing and preparing the soil. Repeatedly, he managed to convince his superiors to implement time-consuming practices that were foreign to asylums, such as maintaining individualized medical records and systematically examining patients. Support for these changes was bolstered by significant new research by the German clinician Emil Kraepelin. In several studies examining the historical trajectories of

American mental hospitals, Gerald Grob concludes that Meyer's influence led to systemic and irreversible changes to asylum procedures throughout the United States. This chapter examines the contingencies of that change from the perspective of Meyer's own opportunities and objectives.[36]

Chapter 2 is devoted to Adolf Meyer's biological theory of mind. It is a comprehensive explanation of psychobiology, the conceptual framework he devised to operationalize psychiatry as a clinical science. It reoriented psychiatric theory and practice away from the diseased brain and toward the disordered personality. Psychobiology embodied an idea that already had traction at the end of the nineteenth century: that the mind had evolved by means of natural selection and that mental activity was thus a function, not a byproduct, of a biological organism. It was in this sense that Meyer spoke of a biological mind and of *psycho-biology*. The only explicit analyses of Meyer's ideas are those of Ruth Leys, who consistently identifies theoretical instabilities in his formulations of psychobiology. Evaluated solely as the product of a thinker, psychobiology might reasonably appear incoherent and unconvincing. Trying to explore the concept from Meyer's perspective as a brain researcher, pathologist, clinician, professor, and physician, I conclude that—for him—its primary function was clinical, medical, practical. He conceived psychobiology as an interpretive framework for studying, diagnosing, teaching, and treating mental disorders as biological dysfunction. This chapter examines three aspects of Meyer's psychobiology: as a laboratory science rooted in comparative neurology and neuron theory, as a philosophy of science based on the philosophies of pragmatism and common sense, and as a unifying interpretive framework for clinical practice.[37]

The fundamental premise of psychobiology was that an action of the body's nervous apparatus (at once anatomical and physiological) *and its expression as mental activity and behavior* constituted a single adaptive response of the human organism. Meyer called this type of adjustment, as biologists termed an adaptive maneuver, a psychobiological reaction and made it a critical unit of analysis. Therefore, he conceptualized mental disorder not as an ontological disease but as a failure to adapt, resulting from multiple and interdependent anatomical, physiological, and mental variables. Using clinical methods, he believed, many of these variables could be identified, ranked for causal significance, and, in some cases, modified for experimental test or therapeutic ends. The core of Meyerian psychiatry was this conflation of the anatomical, neural, mental, and behavioral as a single, irreducible sensory-motor response. A failure to grasp this point and, especially, the tendency to equate the descriptor *biological* with *physical, bodily,* or *somatic,* can render anything Meyer said or did unintelligible.

The remaining chapters explore how Meyer attempted to combine pathology as method and mind as biology at Johns Hopkins between 1913 and 1917. The focus of Chapter 3 is the primary instrument of his influence, the Henry Phipps Psychiatric Clinic. The Phipps Clinic was an eighty-eight-bed teaching and research hospital designed by the architect Grosvenor Atterbury according to Meyer's directives. Like its parent institution, it served nonpaying, subsidized, and private patients in wards segregated by gender. I explore the various purposes of the institution from the perspectives of Johns Hopkins authorities, members of the American public and medical profession, and Meyer's requirements for a clinical science of psychiatry based on psychobiology.[38]

Chapter 4 is an analysis of the investigational and diagnostic practices Meyer mandated at the Phipps Clinic. Adapted from the methods of Emil Kraepelin, his clinical system represented his attempt to harmonize the methods and expectations of scientific medicine with a biological view of mind. His pragmatic interpretation of experience as a site of pathological processes deprived him of a physical specimen, such as a tissue or cell, traditionally studied by pathologists. His solution was to place at the center of his approach the case history, a clinical tool that was already central to other medical disciplines. The case history was a dossier comprising all the data relevant to the development of the patient's illness—life experiences, clinical observations, laboratory results, and insights derived from psychotherapy. Meyer employed it as a technical and conceptual device to convert ephemeral experience into a permanent object of study on which to base collective clinical practices. The case history, I argue in this chapter, was to Meyerian psychiatry what the corpse was to the Paris *clinique.* Similarly, the French intellectual Michel Foucault theorizes that psychiatrists in the nineteenth century employed formalized techniques of examination and history taking in order to establish the patient's life as "a tissue of pathological symptoms." Foucault suggests, whereas I do not, that this allowed the psychiatrist— whose medical dominion included neither a body nor a cure—to circumvent subjective symptoms and act objectively upon disease.[39] By contrast, Meyer's reading of experience as instrumental meant that he considered the intersubjectivity of patient and psychiatrist essential to his medical inquiry and therapeutics. He used the case history to transform subjective experience into data that could be shared and agreed upon by trained personnel, which constituted his definition of an objective fact. Ruth Leys situates another of Meyer's techniques, the Life Chart, within the Progressive Era quest for a science of the individual—a discipline to reconcile the tension between the socially determined individual (who was subject to scientific study) and the free individual (who acted spontaneously).[40] This chapter analyzes the technical devices of modern psychiatry in their Meyerian form and in terms of their practical function in Meyer's quest to establish a clinical science of psychiatry.

Chapter 5 analyzes Meyerian psychiatry in action, operating in a perpetual cycle of observing, recording, analyzing, and applying the data of psychobiology. I call this innovative approach Meyer's "therapeutic experiment." In a reversal of the institutional circumstances that had so surprised him in the American asylum twenty years earlier, at the Phipps Clinic Meyer united medical inquiry and therapeutics into a single clinical enterprise. This chapter shows how a daily routine of bed rest, recreation, good nutrition, and social interaction—conventional remedies traceable to antiquity—were embedded in, and their significance transformed by, the machinations of the therapeutic experiment. By framing medical treatment as a kind of social rehabilitation and suggesting that the environment of the Phipps Clinic was itself a therapeutic milieu, Meyer implemented a model already widely in use at tuberculosis sanatoriums and diabetes hospitals, in which patients learned how to live with their diseases in the outside world. Experiences of patients and staff are woven throughout the chapter, and Meyer's use of occupational therapy, hydrotherapy, sedation, and restraint are examined in some detail.

Meyerian psychotherapy, the focus of Chapter 6, was based on instrumentalist principles of American pragmatism. Meyer framed the psychotherapeutic encounter as a dynamic interaction between an organism (the psychiatric patient) and its environment (of which the therapist was a part). As such, its objectives revolved around the patient's life experiences and adaptive habits. He deemed psychotherapy to be a collaboration of psychiatrist and patient. His objective was to reconstruct causal events that led to each unique form of psychopathology and, ideally, to modify subsequent developments in the direction of recovery. In a period that was culturally conducive to talk therapy, Meyer was a central figure in the advent of medicalized psychotherapy in the United States. The popular New Thought, Christian Science, and Emmanuel movements all advertised the healing power of talk. The Progressive Era also gave rise to America's love affair with self-improvement and self-fulfillment. It was in this context that most Americans, including physicians, heard about yet another talking cure called psychoanalysis. Meyer employed psychoanalytic techniques in the Phipps Clinic, even though his rendering of psychotherapy as a collaborative endeavor involving psychiatrist and patient was incompatible with Sigmund Freud's theoretical formulations. Historians of psychoanalysis cite Meyer as a principal importer of Freudian ideas to Anglo-American psychiatry, and an important teacher of the first generation of American psychoanalysts. Yet, by the 1930s he was an outspoken critic of Freudian doctrine. In the course of elucidating Meyer's particular brand of psychotherapy, this chapter contextualizes his early support for and later disapproval of psychoanalysis.[41]

After sixteen years working in large American mental asylums, Adolf Meyer had earned a reputation as a skilled neuropathologist, productive investigator, experienced clinician, and effective teacher. When Johns Hopkins suddenly needed a chief of psychiatry in 1908, these credentials marked him as a worthy candidate. His determination to establish psychiatry as a clinical science and branch of biology was, in turn, validated by the disciplinary power and institutional resources he wielded as one of the Hopkins chiefs on par with William Osler and William Welch. At Johns Hopkins, he secured the opportunity to mold the burgeoning specialty around his vision for American psychiatry: pathology as method and mind as biology.

Throughout his forty years as a clinician and professor of psychiatry, Meyer's objectives remained consistent: to discern pathological processes, to understand the multifaceted causal relationships that account for them, and to actively apply those findings to research, teaching, and therapeutics. He suggested that each patient's unique form of psychopathology was best understood as an experiment performed by Nature. Mental disorder, he reasoned, was the end result of multiple variables that created chains of cause and effect. Causal chains could be observed in a planned experiment in the laboratory, he explained, or in a series of tissue specimens that showed the development of organic disease. He insisted that life experiences and adaptive behaviors, too, developed from discernible chains of cause and effect. His constant refrain to trainees and colleagues was that, by using established clinical methods, it was possible to differentiate patterns of pathology, not only at the level of tissues or cells, but also at the level of the person. "For every step there are adequate causes, usually causes which would not have upset you or me, but which upset the patient," he proposed to colleagues, one month before he was offered the most powerful position within American psychiatry, and he posed the question "Now, what makes the *difference* between her and you and me?"[42] Discerning the answer to this question, and acting upon it, was the raison d'être of this pathologist of the mind.

CHAPTER 1

Pathology as Method

Adolf Meyer's Vision for a Clinical Science of Psychiatry

Once he had finished reading his paper on brain pathology to the members of the American Medico-Psychological Association in Denver in 1895, it must have been with a mix of relief and satisfaction that Adolf Meyer carefully packed up the specimens that had accompanied it. After all, this was not just his inaugural scientific presentation to the historic organization. It was the first time he had had the money or time to attend since arriving in America two years earlier as an inexperienced neurologist from Switzerland and taking a job as an asylum pathologist. There was good reason to feel a sense of accomplishment. Nevertheless, when the president of the association, Edward Cowles, approached him afterward, Meyer hardly could have anticipated that it was to offer him a job.[1]

Cowles was the superintendent, or alienist, of the McLean Hospital for the Insane near Boston, a large private institution founded in the 1820s to provide for the mentally ill of Massachusetts. Meyer was familiar with Cowles' pioneering laboratory research on physiological reflexes during states of exhaustion. Cowles explained to Meyer that state authorities in Massachusetts wanted a similar research program established at a state asylum in Worcester and that he had been asked to survey potential candidates at the meeting. Soon thereafter, Meyer received an official job offer from the superintendent of the Worcester State Lunatic Asylum. Eager to advance his own research objectives and to connect himself with prominent alienists like Cowles—as well as influential neurologists and psychologists at nearby Clark University and Harvard University—Meyer accepted and ventured to Worcester.[2]

Fourteen years later, in 1909, Meyer was invited to Clark University to receive one of several honorary degrees conferred during celebrations of the institution's twentieth anniversary (a gathering noteworthy today as the occasion of Sigmund Freud's lone visit to the United States). By then, he was no longer directing the scientific work at nearby Worcester asylum, however, and had recently resigned as director of the New York Pathological Institute, having held both positions for seven years. During that time, he had earned a commanding reputation as a modernizer of mental asylums. When Meyer returned to Worcester in September 1909, he did so as the newly appointed first psychiatrist-in-chief of Johns Hopkins Hospital and director of its recently endowed psychiatric clinic. As he joined his fellow honorees at Clark University one week before his forty-third birthday, it hardly mattered that he had yet to see a patient or give a lecture at the nation's foremost medical research university. His new rank as a Hopkins chief singled him out as America's preeminent psychiatrist. Less than sixteen years after impulsively emigrating to a country in which he had few personal contacts and no obvious employment prospects, Adolf Meyer occupied the most powerful institutional position within the fledgling discipline of American psychiatry.

This dramatic professional ascent was contoured by his attempts to transform asylum medicine in the United States, under fire for its outmoded practices, into a clinical science of psychiatry. Historians of medicine have long credited Meyer as the principal importer of scientific methods into American mental asylums, introducing and implementing them in key institutions between 1893 and 1908.[3] But, how did an inexperienced foreigner manage to convince established alienists to adopt new practices? And, what were his own reasons and capacity for becoming a crusader for reform? During Meyer's employment in three large American institutions before his move to Johns Hopkins—at Kankakee in Illinois, Worcester in Massachusetts, and in the New York State asylum system—he worked persistently to implement new clinical methods based on the work of the German clinical psychiatrist Emil Kraepelin. These efforts garnered him repute as a scientific reformer, an accomplished neuropathologist and clinician, and a leader within the emerging field of psychiatry. His motivations were twofold and interrelated. First, he experienced a reorientation of his own scientific interests away from brain research (what he called comparative neurology) and toward work with mentally disordered patients (something he distinguished as psychiatry). Meyer's reform impulse was fueled by a passionate conviction that if the problems of asylum patients were essentially medical (upon which alienists and their critics agreed), psychiatrists must approach those problems using the investigational methods of pathology, the only basis upon which medicine could claim to be scientific. Concurrently, his objectives were influenced by his developing biological conception of mind.

Meyer's elite training—acquired in the medical universities and scientific cultures of Zurich, Paris, London, Edinburgh, Berlin, and Vienna—predisposed him to a model of medical practice and research based on pathology, the study and diagnosis of disease. Pathology had important origins in early-nineteenth-century France. The Paris school, as it is often called, relocated medical training from lecture halls to the hospital, or clinic, and emphasized hands-on learning rather than rote memorization from authoritative texts. Empirical observation and firsthand experience were guiding principles for the Paris school. The approach was characterized by novel methods of medical practice and investigation born of unique institutional circumstances. The secularization of traditionally religious hospitals in postrevolutionary France provided Parisian physicians unprecedented access to the bodies of thousands of patients (before and after death) in the city's massive hospitals. These conditions allowed them to examine and interview patients systematically, and they documented their observations in a dossier created for each patient. Autopsies also became systematic, to identify diseased tissue, called a lesion. The Paris clinicians added detailed descriptions of these pathological findings to the dossier.

These new clinical methods became the basis for Western medical training and research and, eventually, for Meyer's vision of twentieth-century psychiatry. Using the data collected in thousands of individual records, the Paris clinicians used statistical methods to ascertain correlations between their clinical observations of living patients and the pathological lesions discovered postmortem. The clinical-pathological method elucidated for the first time the relatively fixed and predictable courses of several common diseases, from early signs to death. Patients whose individual symptoms appeared unalike were often found to share the same disease based on new clinical criteria. This new knowledge—generated by a methodical comparison of data from all available cases—was readily applied in the clinic to the bodies of individual patients. In some cases, physicians could now envision the pathological processes at work in the body based on the clinical signs they observed while examining patients. This knowledge thus began to inform decisions about diagnosis, prognosis, and therapeutics and to guide research questions. Moreover, systematic and ongoing documentation of clinical and pathological data became an instrument for assessing the accuracy and efficacy of those decisions regarding individual bodies. The value that the Paris clinicians placed on observation and experience also inspired innovative techniques of physical diagnosis (for example, auscultation with the newly developed stethoscope and new modes of palpation) that generated additional clinical data to be correlated with pathological findings in the future. Novel modes for generating knowledge, then, relied upon the reciprocity of clinical observations and postmortem pathology findings, and the continued correlation of the data of individual cases and those generated by comparative and statistical analyses of all cases. The clinical-pathological method created opportunities to detect previously indiscernible patterns of disease. It was this clinical model that Meyer was determined to apply to psychiatric disorders.[4]

Throughout the nineteenth century, medical training in European universities increasingly occurred in the clinic and autopsy room, where would-be doctors learned to differentiate the normal from the pathological in living patients and in organs and tissues at autopsy. Within this context, the new medical discipline of pathology emerged. In the second half of the century, theories and methods promulgated by Rudolf Virchow opened up promising new avenues for differentiating pathological processes—not according to symptoms and lesions (the end results of disease) but by their causes. Virchow, widely regarded as the father of modern pathology, was influential in German medicine. His foundational research and teaching on the cellular mechanisms of disease placed the microscope at the center of the new discipline, and innovative laboratory technologies followed that enabled pathologists to examine and experiment upon the structures and behavior of cells. The microtome, for example, sliced tissue so finely that it became translucent under the microscope, and stains and fixatives applied to the specimen exposed various elements of intracellular organization. Autopsy-based anatomical pathology and microscope-based cellular pathology developed as autonomous disciplines with specialized knowledge distinct from that of the clinician. Virchow insisted, however, that laboratory pursuits remain contingent upon ongoing correlations of clinical, postmortem, and experimental data—carefully observed, meticulously recorded, and methodically collected. He termed this triad "general pathology" and he considered it the only path to understanding disease. The hope of a general pathology of insanity—of thousands of records filled with observations from the clinic and laboratory to be compared and correlated—energized Meyer's efforts in American asylums.[5]

Supported by the work of general and specialized pathology, university medicine in Europe became an increasingly collective enterprise. Case records were maintained and utilized by personnel in multiple departments; diagnostic reasoning was discussed in case presentations to one's peers; clinical procedures were demonstrated to students using living patients; and laboratory work was conducted by groups of researchers. This infrastructure was necessary for the collection and correlation of clinical and pathological data, and collective scrutiny was increasingly considered an additional measure of scientific rigor. Young Americans with sufficient means crossed the Atlantic to receive advanced medical training in European clinics and laboratories. In the 1880s, small centers of academic medicine began to develop in the United States and, by the time Meyer arrived in 1892, the clinical and laboratory methods that were fundamental to his own elite training were being utilized in some American hospitals.

In the final quarter of the century, several prominent discoveries demonstrated the causal role of microorganisms in disease; this placed further emphasis on the enigmatic pathological processes taking place in the body. Virchow denounced bacteriology as a return to the medical speculations of German romanticism. Similarly, prominent physicians in the United States who were loyal to the empirical ethos of Parisian medicine disparaged any so-called rational therapies that emerged from experimental sciences (of which bacteriology was only the most recent). Nevertheless, the recognition of the cellular basis of disease as well as the causal role of microorganisms—both contingent on the microscope and experimental techniques—fortified the notion that medicine's role was to identify localized sites of disease in the body using laboratory methods. In German medical universities, which dominated European medicine during the second half of the century, this trend was reinforced by the rise of materialism and positivism in German science.[6]

The focus on bodily pathological processes taking place in tissues and cells contoured German psychiatry. Although the wholesale medicalization of madness intensified over the course of the nineteenth century, the investigation of insanity using the clinical-pathological method was problematic. Throughout Europe and North America, people with mental illnesses were congregated in asylums, planned as sanctuaries, isolating them from urban centers of academic medicine. One exception was Paris, where alienists had unimpeded access to patients and cadavers in massive urban institutions built to house large numbers of indigent persons deemed insane. German physicians in regional asylums did document and compare large numbers of cases over extended periods, but in the 1870s academic physicians derided this "alienist science" and convinced state authorities that psychiatric research and patients belonged under the purview of medical universities. Berlin professor Wilhelm Griesinger was an instrumental reformer determined to base German psychiatry on the premise that every disease had a single specified cause. He proposed that each form of insanity could be traced to a localized lesion somewhere in the tissues of the nervous system. The theory of cerebral localization led logically to the merger of neurology and neuropathology as a single science. A controversial figure who was interested equally in pathological research and the rehabilitation of patients, Griesinger inspired a whole generation of researchers to adopt his dictum: "mental diseases are brain diseases." His students placed the structural mechanics of the nervous system and the search for anatomical lesions at the core of German psychiatry in the closing decades of the nineteenth century. They became the first directors of a new type of elite research institution in Europe, the psychiatric clinic, and the teachers of clinicians such as August Forel and Emil Kraepelin, who will soon enter this story.[7]

Griesinger's student Theodor Meynert directed the psychiatric clinic at the University of Vienna. A master anatomist, Meynert conceptualized a model of neuropathology based on the idea of a fixed and closed network of nerves that connected various "centers" of nervous functions in the brain. A student of his, Karl Wernicke, applied his mentor's conclusions to studies of aphasia (a disorder of language and speech) and confirmed two localized regions in the brain responsible for specific dysfunction. Based on these results, Wernicke devised a theoretical model that provided plausible explanations for a variety of mental disturbances. The theories and textbooks of Meynert and Wernicke became hegemonic in German neurology and psychiatry. By the 1890s, many physicians and alienists accepted their hypotheses as proven fact when they discussed and diagnosed mental disorders. Meynert and his students pioneered techniques of brain research eventually mastered by young neuropathologists like Meyer. Their speculative claims, however, eventually came under attack and were labeled "brain mythology." August Forel, who became Meyer's mentor in Zurich, spoke about the disappointment of working in Meynert's lab, where he confronted the painful reality that many alleged discoveries of nerve tracts connected to specific brain centers were in fact products of the great anatomist's imagination.[8]

The circumstances that Meyer encountered upon arrival in the United States in the 1890s were different from those in Europe, although they reflected many of the same developments. In general, he found the American scientific ethos suffused with the experimentalism and materialism of German science. He also discovered that the alienists superintending mental asylums constituted the nation's oldest professionalized medical specialty, which had evolved steadily through two waves of asylum building, first in the 1820s and again in the 1860s. By the 1880s, however, the specialty had come under criticism as being out of touch, secretive, and unscientific. Its defining practices reflected the asylum's original purpose as a humane shelter for the indigent insane, along with midcentury theories of insanity according to which the institution's combined architectural and moral environment was itself therapeutic. Alienists faced pressure from state governments to increase efficiency and accountability and simultaneous demands from other medical specialists to align asylum practices with scientific medicine. Most superintendents recognized the need to modernize but were unequipped to effect change. Public confidence in their practices had waned and their overcrowded institutions resembled prisons more than hospitals. A handful of alienists were committed to redefining the specialty in accordance with scientific principles. Edward Cowles led by example, establishing at McLean the first laboratory in the United States to combine experimental research in neurology and psychology. It was because Cowles thought Meyer able to do the same at another Massachusetts asylum that he recruited him in 1895.[9]

Medical Training

Adolf Meyer was born in Switzerland in 1866, in the parish house of a farming village called Niederweningen near Zurich. His father, Rudolf Meyer, was a Zwingli minister in the established Protestant church, well read and intellectually curious. His mother, Anna Walder Meyer, was sensible and approachable, and she encouraged the interests of her children Adolf, the oldest, Anna, and Hermann. One uncle was a medical doctor, and the children sometimes accompanied him when he called on patients. As a young man, Adolf was attracted to matters that touched upon the role of mental experiences within the natural world. Conversations and experiences with his father and uncle regarding the wellbeing of souls and bodies frequently appeared at odds with theoretical principles ubiquitous in his German education such as vitalism, dualism, and materialism. Meyer later expressed the belief that his childhood environment fostered in him an intuitive conviction that what some revered as the soul and others dismissed as metaphysics was, in reality, a natural phenomenon. At nineteen, Adolf pondered two obvious career choices, the clergy and medicine. He decided on medicine, writing in his diary at the time, "I am glad that I have decided to study the whole of man." Throughout his life Meyer would paint this precocious remark as prophetic and point to it as an early sign of his anti-dualist and anti-materialist outlook. In truth, the youngster had only a vague intuition that the mental and physical realms were intimately connected. If he expected his university studies to bring clarity to the matter, he would soon be disappointed.[10]

Most of his medical training at the University of Zurich was in the Germanic model and, indeed, many of his professors were German. His courses in anatomy, pathology, zoology, obstetrics, chemistry, pharmacology, and surgery combined traditional lectures on medical theory with bedside teaching and laboratory training. Planning to join his uncle in general practice, Meyer paid close attention to obstetrics and surgery. He discovered he had a talent and passion for dissection and histology. He studied advanced pathological anatomy with Edwin Klebs, discoverer of the diphtheria bacillus and former assistant to Rudolf Virchow, and he completed Hermann Eichhorst's multi-year courses called Pathology and Therapy (a coupling that constituted Meyer's definition of scientific medicine). He also took Eichhorst's course on diseases of the nervous system (and later regretted not having paid closer attention in class).[11]

He remembered more robustly his studies in neurology with the pathologist Constantin von Monakow, especially a series of small graduate seminars in neuropathology. He recalled that Monakow's demonstrations of various structures of the nervous system were fascinating but rushed and not adequately explained. For this reason, the attendance rate often slumped. Indeed, Monakow cancelled one course midsemester when Meyer was the only student who remained. Although no one knew it at the time, another student in Monakow's lab, the American Henry Donaldson, would play a pivotal role in inaugurating Meyer's career in America.

Meyer's affinity for the intricate handwork of pathology increased. He became especially adept at the delicate task of using a microtome to slice paper-thin sections of tissue, then staining and pressing each specimen between two glass slides for microscopic examination. His professors, however, provided no satisfactory answers to his questions about the relationship between mind and body. One teacher simply avoided talking about the brain, he wrote a few years later, and another explained that it was nothing more than "a dish of macaroni." A third proposed "a mysterious mixture of spinal mathematics and a thalamic soul-center." When he earned his medical degree in 1890, Meyer was a capable pathologist and histologist but had not yet gleaned any scientific insight into the role assigned to mental experience in the natural order. He concluded that this lack need not affect his professional plans: "I was bent on general medicine."[12]

After his initial medical training, Meyer embarked on a year of postgraduate studies that led him to reconsider becoming a country doctor. In France and Britain he was exposed to approaches in neurology that focused on the dynamic functioning of the nervous system. This struck him as very different from the emphasis on static anatomical structures that was at the heart of German neuropathology. In Paris he attended the lectures of famed neurologist Jean-Martin Charcot, whose ability to demonstrate the relationship between a patient's symptoms and life history Meyer admired but whose showmanship and dogmatism he disliked. Meyer also frequented the laboratory of the married researchers Joseph Jules Dejerine and Augusta Marie Klumpke, who were undertaking clinical research on the localization of particular nervous functions in the brain. Dejerine issued Meyer a permit granting the novice free access to the wards of the famous Bicêtre mental asylum, and instructed him to observe closely the wide variety of abnormal thoughts and behaviors exhibited by the inmates there. In the afternoons, Meyer attended teaching clinics at various Paris hospitals or went instead to the morgues to observe autopsies over the shoulders of experienced pathologists. "I am uncommonly glad to be able to know French life and especially French medicine," he wrote to his mother. He lamented that his training thus far had been shaped solely by the "German influence" of anatomical pathology compared to the "geniality and elegance" of French neurology and physiology. "I shall go away from here most reluctantly," he admitted to her. Determined to learn English, he extended his study period and traveled to Britain.[13]

The focus in British evolutionary biology on how the nervous system functioned physiologically rather than how it was constructed anatomically, as well as its emphasis on empirical observation rather than theoretical dogma, resonated profoundly with Meyer. In London he joined other students shadowing the epilepsy researcher John Hughlings Jackson on his clinical rounds at Queen's Square Hospital. Meyer was captivated by Jackson's biological view of the nervous system based on the evolutionary principles of Herbert Spencer. He collected Jackson's writings and studied them eagerly, in addition to those of Spencer and the Darwinist Thomas Henry Huxley. In another London dispensary for epileptics, he found clinical neurologist William Gowers taking notes in shorthand as he listened to patients' medical histories, so that no detail would go unrecorded. This made an impression on Meyer, as did Gowers's textbook on the interrelations of nervous structures and their functions in the spinal cord. Meyer became intoxicated by the possibilities of studying a dynamic nervous system, rather than its static anatomical landscape. In 1899 he recalled of his British teachers, "I liked them better than the Prussians who stood over my medical cradle." Throughout his career, Meyer would remember his contact with British medicine and science as the element of his postgraduate explorations most significant to his intellectual development.[14]

Evolutionary biology became foundational to Meyer's concept of psychobiology, as did British philosophies of science. Thomas Henry Huxley's definition of science as "organized common sense" so appealed to Meyer that he resolved thereafter to define the scientific endeavor for himself in the same way. Science, he was convinced, was not defined by the principles of physics or chemistry, nor by experimental techniques, but by the application of scientific rigor to observing, documenting, comparing, and ordering data. As Charles Darwin and other naturalists had demonstrated, scientific integrity did not depend upon mathematical formulae or controlled experiments. Though he did not know it yet, Meyer himself would soon embark on a sea voyage that would deliver him to his own Galapagos—the American mental asylum.

At the end of his study period, he felt inundated with new ideas about the relationship of mental and bodily phenomena. German pathology and neurology had left his desire to study "the whole of man" unsatisfied. His time in France and Britain convinced him that concrete methods did exist for doing just that. He was filled with youthful enthusiasm and naïve optimism. As Meyer boarded a train homeward, he began to imagine his own grand contribution to the scientific understanding of man. What Gowers had accomplished for the spinal cord he would produce for the brain: a comprehensive text that elucidated the interrelations of all cerebral structures and their functions. Somewhere between London and Zurich, he decided to devote the rest of his life to studying the brain.[15]

With a newfound interest in neurological research, he returned to the University of Zurich to complete his doctoral thesis under the supervision of August Forel. Forel's combined expertise in laboratory research and psychotherapy also made a significant impact on Meyer's intellectual orientation. Forel had completed his own training in neuroanatomy under Theodor Meynert in Vienna and afterwards became the assistant to another pioneering researcher, Bernhard von Gudden in Munich. In 1879, Forel was appointed professor of psychiatry at Zurich and, concurrently, director of the Burghölzli mental hospital there. By the 1880s, he had also been recognized for experimental research that helped confirm the cellular basis of the nervous system. Forel took a strong, practical interest in making his patients well. He had traveled to the University of Nancy in France, where several members of its medical faculty were investigating the effects of hypnosis and suggestion. Unlike Charcot, who hypnotized patients to study hysteria, and the Nancy experimenters, who investigated how hypnosis worked, Forel mastered the technique in order to arrest symptoms in his patients. When Meyer entered the University of Zurich in the late 1880s, Forel was a leading specialist in what was termed the Nancy school of psychotherapy. Forel had instituted reforms at the large university hospital, including the successful use of hypnosis. Added to his neuroanatomical discoveries, these practices made both his name and that of the Burghölzli renowned. Under the influence of Forel and his successor Eugen Bleuler, the integration of psychotherapeutics and laboratory research typified what became known as the Zurich school of psychiatry.[16]

Meyer proposed to Forel an ambitious project to study the structures and functions of the human brain. A decade later, he described his supervisor's response: "There they come," Forel roared, students who "want something to get through within four or six weeks and spoil that material and I must write the bulk of the thesis myself in order to make it fit to pass!" Meyer endured the unexpected tirade and, when "the storm was over," Forel agreed to supervise the project. Forel's laboratory housed a large collection of serial sections of animal brains—glass microscope slides housing sequential layers of the brain that were stained to reveal fine anatomical structures. Meyer was disappointed that there was no human brain material for his thesis research. Instead, Forel chose the brain of a chameleon and assigned his new trainee the problem of identifying the nucleus of the motor nerves of its multidirectional eye. At first, Meyer worked diligently in Forel's lab at the Burghölzli. Eventually, however, he became irritated by his chief's condemnation of his "blasphemous" sectioning and staining techniques (somewhat surprising, perhaps, since Forel's own groundbreaking research emerged from trying unorthodox staining methods). Convinced of their value, Meyer refused to discard his methods. He moved his research into the family kitchen. It was there that he took on his first lab assistant: his mother. During the experiment, the time came for his mandatory military service, so he left his specimens in her charge. His correspondence contains a letter in which his mother asks how long she should continue to refresh the liquid preserving his reptile brains. Their work paid off. He was proud of his completed doctoral research, confiding to his brother that it was "the first result of a comparative anatomy of the lower animals—in brain anatomy no one comes close to me." In the end, Forel also was pleased.[17]

Civil and academic duties fulfilled, Meyer traveled to Austria, where he attended the final lectures of the aging Meynert, the patriarch of German neurology and his teacher's teacher. He returned to Zurich to discover that, by working with Forel, he had vexed another professor, who reneged on the promise of a position in his laboratory. Forel offered an assistantship, but Meyer refused it. "I had already made up my mind to migrate or to settle in practice, probably the latter only in order not to scare my mother." Nevertheless, after talking with a friend who had a position in the United States, he decided hastily to emigrate, in order to pursue a career in neurological research. It was, Meyer later acknowledged, "a lasting and fatal attack of emigration-fever."[18]

His now-widowed mother was indeed upset. His married sister had succumbed to tuberculosis a year earlier, and his younger brother was also leaving home, to begin business training. Meyer was resolute, however, and began making travel plans. Regrettably, the day before his departure, the family attended the funeral of his maternal grandmother, which took place only hours after the death of his great aunt. The latter was dear to his mother, who had nursed the invalid relation. As he departed, Meyer was cognizant that his mother's world had been utterly depopulated—first her husband and daughter, and now her mother, convalescent charge, and both sons gone. He hesitated, but his passage was paid—unfortunately, as it would turn out, with money borrowed against his mother's already limited assets. They said farewell.[19]

The twenty-six-year-old Swiss doctor and neurologist bound for the New World was, naturally, a product of all these experiences and environments: a liberal upbringing in a loving and learned home and democratic republic; a German medical education emphasizing laboratory pathology, supplemented by an apprenticeship at the Burghölzli (where patients mattered as much as their diseases); his confidence in the methods of French and British clinical medicine; and, finally, the distressing familial circumstances of his departure.

Emigration

En route to America, he again stopped in France and Britain. He stayed a month with the Dejerines, where he showed Madame how to operate her new microtome and earned from Monsieur a letter of recommendation in addition to the ones he carried from Forel and Monakow. Meyer's scientific aspirations for himself were coming into focus. He envisioned founding a research institute devoted to the biological development of the brain—a museum of specimens from diverse animal species to be used for training and research, instead of the "unintelligible descriptions" found in textbooks. It was a lofty goal buoyed equally by the boundless aspirations of youth and practical aims. "It was my conviction," Meyer recalled in 1899, "that the study of neurology was made extremely difficult because each worker worked with his own methods." The idiosyncratic inconsistency of technique and terminology of Meynert, Eichhorst, Forel, Monakow, and Gowers struck Meyer as a barrier to progress. As the director of a brain institute, he would "accumulate serial material which would enable me to control data furnished by others," with the goal of standardizing comparative neurology. He pondered the handful of suitable parent institutions in the United States for such an endeavor, most of which were in their infancy: there was Johns Hopkins in Baltimore and Clark University in Worcester (both roughly a decade old); the newly incorporated Columbia University in New York City; and, there was Harvard in Boston. He decided he would elicit advice as he headed westward.[20]

A young American staying in Paris showed him a news item about the exploits of William Rainey Harper, president of the University of Chicago. The new research university was funded by John D. Rockefeller, and Harper had surreptitiously pillaged the faculty of Clark University in Worcester by offering higher salaries and intellectual autonomy. When Meyer read that the neurologist Henry Donaldson had been lured away from Clark's biology department (along with most of its other members), he immediately wrote to the president of Clark University, Stanley Hall, to apply for a teaching position. Hall informed him, however, that Donaldson would also continue to teach neurology courses at Clark. Meyer was disappointed and, evidently unaware that Chicago and Worcester were 1,200 miles apart, he had little reason to doubt Hall.[21]

As luck would have it, Meyer learned that Henry Donaldson was in London attending the Second International Congress of Experimental Psychology, so he decided to attend the conference himself. He hoped that an introduction to Donaldson might result in some advice about where to settle in America and, perhaps, even a job prospect. Though the congress was Meyer's first exposure to experimental psychology, the names of the men to whom he was introduced were familiar: Francis Galton, James Mark Baldwin, William James, Edward Titchener, Pierre Janet, Hippolyte Bernheim, Hugo Münsterberg, Cesare Lombroso. He also attended the British Medical Association meeting, where he "had some bad luck" in his introduction to William Osler, being presented to the chief of medicine at Johns Hopkins by someone Osler disliked intensely. Osler was "very chilly towards me," Meyer recalled, "so that settled Baltimore." A second introduction to Osler a few years later was met with a warm and enthusiastic response, prompting Meyer to reflect that "it certainly amuses me now to think how such questions are settled." Victor Horsley, the well-connected University College professor of surgery and pathology, advised the inexperienced Meyer to go to Chicago. The East was too thickly occupied by neurologists, Horsley warned, and a private practice would falter.[22]

Meyer's meeting with Donaldson yielded two revelations. First, much to his surprise, he recognized Donaldson from Monakow's laboratory, even though he had "never betrayed with a word who or what he was." Donaldson also disabused him of the notion that he would continue to teach courses at Clark University. Meyer realized that the reply to his employment query had been "one of those regrettable constitutional prevarications" for which Hall was known. Donaldson promised him introductions in Chicago, though nothing more. Meyer continued to deliberate on his final destination. Clark University, he concluded, was undesirable because of Hall's dishonesty. Feeling that he had at least one firm and trustworthy contact in Donaldson, the chair of neurology at a new, wealthy research institution, Meyer settled on Chicago. As he sailed for America, however, he rejected the advice of Forel and Dejerine to go into psychiatry and clinical research with patients. "I felt it required much more ability than I had for verbal expression," he recounted years later, "rather than opportunity for concrete demonstration through action." That opportunity, he believed at the time, resided in comparative neurology, not psychiatry.[23]

Meyer arrived in Chicago in autumn of 1892, amid the chaos of labor strikes, the muck of urban poverty, and severe economic depression. Yet, the city was preparing a colossal display of cultural and industrial wealth for its upcoming World's Columbian Exposition. Chicago emerged in this period as a center of progressive reform efforts to mitigate the excesses of capitalism and develop social programs. In 1890, Jane Addams had founded Hull House, the most famous of several Chicago settlements established in this period to address urban poverty. Meyer became a regular visitor to Hull House. He formed lasting ties with Addams and her associate, Julia Lathrop, and internalized the principles of the settlement movement. He also joined the Swiss Socialist Society, the Swiss Athletic Club, and the local Pathological Society.[24]

Henry Donaldson arranged an unpaid fellowship for Meyer at the University of Chicago that provided access to laboratories and, importantly, a university title (docent) that would enrich a private medical practice. Donaldson's labs were not equipped for comparative neurology, and Meyer rarely made the long trolley trip to campus. He did prepare a prospectus for his brain institute that Donaldson presented to the president of the university. Harper appeared to give it real consideration. Although Donaldson was interested in recruiting the young neurologist to his department, he did not champion the proposal. Meyer was so confident in the venture, however, that he announced to members of the American Medical Association in 1894, "[S]uch a museum will, I hope, be one day the pride of Chicago and furnish the best ways of instruction." He also pitched his plan to the president of Rush Medical College but received no reply.[25]

Immediate financial needs superseded efforts regarding brain research. His medical practice consisted of only a few patients until he was discovered by a representative of a workmen's sick fund, which increased business. In time, he was hired to oversee a small neurological dispensary, where he gave a weekly teaching clinic. He also sold microscopes and medical equipment to supplement his income. Throughout, he attempted to continue his brain research in his rented rooms. Jars containing preserved brains and spinal cords crowded a makeshift laboratory that doubled as his consultation room for patients. One visitor described it as a "small and neat neurological institute" above a shoe store. "I am lodging at present an opossum," he informed his mother in one letter. A week later, he reported that he had chloroformed the creature and dissected its brain, adding wistfully: "I should like to come into the kitchen and show you." Eventually, Donaldson offered him an assistant professorship in histology for the upcoming academic year, with a $1,000 salary. Meyer was dissuaded from accepting by the neurologist C. L. Herrick, who assured him that paternalist politics at the University of Chicago inhibited its faculty and their research programs. Instead, he accepted a position as the pathologist at a new state mental asylum sixty miles south of Chicago. There he anticipated ample time, space, and freedom for his scientific pursuits.[26]

Kankakee, Illinois

The forceful campaigns of reformer Dorothea Dix on behalf of the indigent insane in the 1850s had convinced many states to provide for the safe-keeping of such individuals. The legacy of her activism was lasting popular support for custodial protection of those afflicted with mental disorders. During the latter half of the nineteenth century, this social mandate led to widespread expansion of mental institutions, in both number and size, to the extent that the efficient management of their populations overshadowed therapeutic aims. The Illinois Eastern Hospital for the Insane at Kankakee operated like an independent village. In addition to its 2,100 patients (women and men housed separately), it was populated by roughly 300 nurses and attendants, kitchen and laundry staff, stonecutters, clerks, engineers, farmhands, and livestock—plus its superintendent and five assistant physicians—all situated on forty acres. Its patient population represented a mix of chronic and acute conditions, with patients grouped in pavilions according to the nature or severity of symptoms. Meyer was struck by the liberal financial support granted mental asylums in the United States. By now, his own mother was hospitalized in the Burghölzli; and, comparatively, he thought the provisions for patients' nutrition, safety, and recreation were superior at Kankakee.[27]

The hospital's superintendent was Clark Gapen. Gapen's expectations of his new pathologist were typical: to perform autopsies required for legal purposes; to analyze urine, blood, and sputum for evidence of noteworthy microorganisms; and, to investigate curious postmortem findings. Gapen was enthusiastic about Meyer's brain research. The senior assistant, on the other hand, immediately informed him haughtily that he had no use for pathology. Initially, this did not bother Meyer. At the Burghölzli, physicians were both responsible for a ward and expected to undertake experimental research. Only five assistants—none of whom had laboratory training—were responsible for Kankakee's 2,100 patients. He was eager to take advantage of the gulf that separated the clinic and its patients from the morgue and laboratory. As the sole pathologist, Meyer anticipated continued access to human brains with plenty of time to dissect, slice, serialize, and scrutinize.[28] He wrote to his brother excitedly about the prospects for specimens: "I am in the clover now!"[29]

He soon realized, however, that it was impossible to perform his pathology duties because of traditional divisions between patient care and scientific inquiry. The asylum's original social mandate, to safeguard the insane, included protection from scientists who might torment them in the name of experimentation. Many Americans considered the responsibilities of alienists and motives of scientists to be mutually exclusive. His first autopsy at Kankakee was observed by the entire staff with "critical curiosity." He detected traces of syphilitic lesions on the skull, but none of the onlookers could confirm the patient's symptoms or behavior before death. When he asked to see the clinical observations gathered during the patient's hospitalization, he was met with blank stares. His fellow physicians informed him that each patient was examined once, upon admission, usually while he or she was washed in the bathtub. Some patients were institutionalized for years or decades without further attention from a physician. The senior assistant bragged to Meyer that he could complete this exam in less than five minutes. As in other asylums, information about patients was kept in a daily log book. When Meyer consulted these registers, he found that they contained only accounts of the trouble the patients caused for the attendants. "The worst and fatal defect," he would complain on the eve of his departure from Kankakee in 1895, "was that I was expected to examine brains of people who never had been submitted to an examination." In the logs, he found descriptors such as "untidy" (with feces) and "unable to go about," yet no one could tell him if a paralysis of the sphincter or of the legs was associated with these behaviors. At Kankakee, Meyer confronted the disturbing reality that American psychiatry had abandoned half of the clinical-pathological method.[30]

Meyer explained to Gapen that autopsies were useless without a corresponding medical history and ongoing observations. His boss was skeptical. Meyer pestered Gapen to authorize the use of a mimeographed template for recording clinical data, and eventually Gapen relented. Getting the assistants to use this "blank"—by then commonplace in general hospitals—was a challenge. The senior assistant resented the extra work and refused, but the others were ordered to complete the extra paperwork requested by the new pathologist. Even though it required a few hours to complete, Meyer noted that the younger physicians appeared eager for "real medical work." He also asked Gapen for permission to collect clinical observations himself, much to the astonishment of the staff, since "a pathologist interested in patients was a rare thing at Kankakee." Gapen consented, and the attendants promised to summon Meyer when a patient's death seemed imminent. This earned him a nickname among the amused patients and staff: the ominous crow. "I got so disgusted over this role," he wrote a few years later, "that I refused playing the crow any longer." Determined to abolish the notion that pathology dealt solely with autopsies and the contents of bodily fluids, Meyer and Kankakee's new infirmary physician began visiting all the patients in a single ward, regardless of condition or prognosis, discussing possible causes and diagnoses in the presence of other patients and staff. These improvised clinical rounds emulated the practices of the well-known clinicians under whom Meyer had trained. Gapen, however, begrudged the pathologist's challenge to his authority over the living bodies under his care. Meyer did not appear to care. "The great Chief has no inkling," he complained to his brother, "and is always meddling." It was the same combination of naïve self-confidence and stubbornness that had driven him to move his doctoral research into the family kitchen.[31]

Gapen had his own ideas about what it meant for his institution to be scientific. He decided to exploit his pathologist's European training to enhance his hospital's reputation as a progressive center of science. He ordered Meyer to teach a series of lectures and advertised it in national medical journals. The *Cincinnati Lancet-Clinic* proclaimed, "the opening of a summer school in neurology and psychiatry in the Kankakee Hospital for the Insane is a rift in the clouds of darkness which have hung over our great caravansaries known as insane asylums." The medical journal condemned Ohio authorities for banning clinical teaching in mental institutions, claiming that "their autocratic closure to such purposes" was "wholly unworthy of the last decade of the nineteenth century." In the end, no one except Kankakee physicians registered for the course. For Gapen, this mattered less than the public recognition that his institution was, as the Ohio writer implied, a symbol of scientific enlightenment. Several American asylums in this period were building and equipping laboratories, but, as Meyer discovered on a tour of several northeastern institutions, they often sat idle due to lack of interest or knowledge on the part of the asylum's physicians. To Meyer, Gapen's view of science was characteristic of American alienists; it was based on an empty veneration of laboratory techniques without any interest in the clinical-pathological correlations such methods were intended to produce.[32]

Gapen also asked Meyer to teach a course on brain anatomy for the staff, a task the latter took up enthusiastically, even though he had reservations about his ability to communicate clearly. He decided to leave out all hypothetical concepts in his lectures. While he maintained deep respect for Meynert and Wernicke and valued their substantiated findings (which were numerous and significant), he abhorred the fact that their speculative interpretations were accepted as conclusive. He was determined to teach without relying on brain mythology. He created a comprehensive catalogue of German, Spanish, Swiss, French, and English research on the nervous system, a resource he would use for the next twenty years. Teaching Gapen's assistant physicians the way he had been trained, however, was impossible. With approximately 500 patients each, it was unrealistic to expect them to master the technical intricacies of neuropathology. As an alternative, he prepared in advance serial sections of various anatomical structures using autopsy material from the morgue. The physicians were interested in the specimens but were befuddled when they examined the stained brain tissue under the microscope. Their American medical training had been devoid of brain anatomy. Meyer realized that they did not comprehend the relationship between the specimen and the brain from which it derived. He was sympathetic and began to fashion replicas of the anatomical structures under discussion from modeling clay, a pedagogical device he used for the rest of his career.[33]

It infuriated Meyer that his students readily rehearsed the theoretical mechanisms described by Meynert to explain their patients' symptoms yet had never dissected a brain or examined nervous tissue microscopically for themselves. This experience reinforced his distrust of theory and his commitment to empiricism. In 1895, he declared, "[R]ecords which are not written to dictation while everything is seen by several people, are not worth the paper and should not be recognized as a basis for further work." He became increasingly preoccupied with basing psychiatric research on what he described as "sharable" or "actually accessible facts of observation," as an antidote to brain mythology.[34]

Despite institutional inertia, Meyer attempted to introduce collaborative clinical practices at Kankakee. He implored Gapen to abandon the log books for individual medical records. He pestered him to equip special rooms for the routine examination of patients and to schedule staff conferences. The superintendent, however, continued to frustrate efforts to transform his American asylum into Meyer's European clinic. Gapen bragged to Meyer that in 1875, he had supervised 300 patients without help and kept good notes. Meyer insisted that this was not scientific medicine, which he defined as the systematic application of the clinical-pathological method. Defiantly, he fetched a new patient, a woman, and escorted her to the staff residence where his colleagues were occupied with leisure activities. He began a comprehensive neurological examination. He continued the practice with a different patient every evening thereafter. Most of the staff became committed to participating in the daily ritual. Some commenced similar examinations on the wards using the diagnostic techniques Meyer had demonstrated. According to Meyer, the effect was "simply splendid" and he felt encouraged. "The confidence of the patients was gained and many points overlooked in the ward were found out." Adhering to Virchow's principles of general pathology, he once again outlawed brain mythology during these improvised clinical rounds. "They cannot use any academic expressions," he wrote to Forel, "the description must always be of something clinically observed." These collective clinical practices—the maintenance of individual medical records by nurses and physicians, the presentation of patients for teaching and collaborative evaluation, and case conferences—were not common practices in American asylums.[35]

Meyer's scientific vision of a great brain institute was rapidly being revised in the face of his increasing preoccupation with clinical examinations and teaching. Contrary to his initial worries about his communication skills, he found that he possessed a natural ability to teach and to develop a rapport with a patient. He read widely to supplement his new interests: Wilhelm Wundt and his American students, Stanley Hall and William James, along with Pierre Janet, Theodor Ziehen, Edward Spitzka, and the work of the Nancy school. He found a textbook written by Emil Kraepelin particularly helpful in his encounters with patients.[36] "The more I can throw myself into the clinical field, the better it will be," Meyer wrote to his brother, only a year after boasting about his prospects for brain specimens; "that is what offers the most possibilities."[37] He felt sufficiently confident in his new role to define the ideal specialist in the institution's 1894 report to the Illinois governor: "[T]he alienist must have a broad general education, a good knowledge of the life and social condition of the patients, and very thorough clinical training." Without clinical accuracy, he concluded in an article for the *American Journal of Insanity* in 1895, "pathological anatomy remains a science of

the dead." Meyer continued his comparative brain research and never suggested that autopsy or laboratory findings were unimportant. Quite the op-
posite, he became a fierce advocate of Virchow's general pathology, insisting
that clinical, experimental, and postmortem data were interdependent and
necessary to the advancement of psychiatric knowledge.[38]

Meyer's reorientation toward living patients was influenced by another
set of meaningful experiences during this period. Soon after his arrival in
America, he received news that his mother had fallen into a delusional depression. Despite all reassurances to the contrary, she believed that her el-
dest son, Adolf, was dead. She became deranged and was admitted to the
Burghölzli hospital under the care of her son's famous teacher, Forel, who
informed Meyer confidently that she was incurable. Meyer was certain that
the mental pressures of the financial debts incurred to finance his emigration, combined with her tremendous personal losses, had caused the nervous breakdown. He was guilt-ridden and worried. His unease grew as he
encountered many patients in similar states at Kankakee. Significantly, his
mother did recover, as did some of the similarly afflicted patients at Kankakee. Three decades later, he remarked that this experience made psychiatry
"real" to him. It also incited an interest in the circumstances in which patients became unwell and then recovered, as well as an enduring distrust of
absolute and fatalistic diagnoses. It reinforced his belief that mental disorders could not be investigated solely by examining brain tissues extracted
from cadavers. It now appeared obvious that any thoughtful naturalist could
deduce the events that had led to his mother's psychopathology.[39]

Meyer's unplanned foray into the reformation of the American asylum
was still in its first year when the neurologist Weir Mitchell lambasted alienists during fiftieth anniversary celebrations of the Medico-Psychological
Association. Mitchell accused them of medical negligence due to their collective apathy toward scientific methods. According to Meyer, his boss returned
from the meeting enraged because his own scientific reforms at Kankakee
had gone unnoticed. Meyer did not attend the meeting, but when he addressed the same association thirty-four years later as its president, he admitted that, at the time, the recommendations that Mitchell made in conjunction
with his accusations had struck him as rather aimless and unconnected to
the realities of clinical-pathological work. Nevertheless, shortly after the
meeting, he forwarded Mitchell's address to the governor of Illinois, an action
he claimed led to the first competitive examinations for internships in the
state's mental hospitals.[40]

Perhaps Mitchell did not hear about the scientific work under way in eastern Illinois, but others did. At the same meeting the following year, in 1895, Edward Cowles of the McLean Asylum in Boston approached Meyer about implementing a program of scientific research at the Worcester State Lunatic Asylum in Massachusetts. "The experience at Kankakee," Meyer wrote not long after, "had shown conclusively that pathology begun on the post-mortem table failed to make its point along almost the whole line."[41] He was now convinced that in order to study and treat insanity scientifically, the work of the pathologist must begin long before the autopsy. Worcester provided an opportunity to prove it.

Worcester, Massachusetts

Meyer had good reason to be optimistic that the piecemeal practices he had attempted to introduce at Kankakee could be implemented as an organized clinical system in Worcester. On the one hand, the state of Massachusetts was well known for its pioneering efforts on behalf of the insane. The Worcester asylum, built in 1833, marked the beginning of what became a comprehensive, though overcrowded, state hospital system. In the 1890s, humanitarian reformers began calling attention to inadequate conditions, and the Board of Lunacy and Charity was formed to oversee improvement. It was this body that requested a pathologist for the Worcester asylum. In addition, Worcester's superintendent, Hosea Quinby, promised Meyer funds and autonomy to conduct scientific research at his hospital. "There is abundant material and exceptional facilities here for good work," Quinby assured Meyer in a letter that accompanied the job offer, "we only lack the man to organize it." In addition to a program that combined pathological and psychological research, Quinby wanted to establish a connection with nearby Clark University to train biologists and psychologists interested in nervous diseases. The plan struck Meyer as progressive, even as it contained echoes of the traditional rift between the clinic and laboratory. Quinby proposed that, while the new pathologist would supervise the daily work of the assistants (medical examinations and autopsies), he would be "relieved of all purely routine duties" to focus on laboratory research. Meyer accepted the appointment after Quinby guaranteed him unrestricted access to patients in conjunction with his supervisory duties and research programs.[42]

Meyer could now envision a phoenix rising from the old alienist's administrative specialty—a clinical science of psychiatry grounded in general pathology, supported by a medical university, and oriented toward the living patient. He immediately set out to standardize procedures for medical examinations, history taking, and ongoing clinical observations, to encourage collaboration, and to correlate clinical and pathological data. He recognized, nevertheless, that he lacked the specialized knowledge necessary to stoke the fire of change. His postgraduate training was in anatomical pathology, comparative neurology, and evolutionary biology. What little he had learned about generating data from living patients he had gleaned by observation, albeit by trailing talented clinicians such as Forel, Jackson, Dejerine, Meynert, and Charcot. There was one clinician in particular, however, who had already demonstrated that scientific research on insanity was possible, indeed fruitful, by studying patients themselves: Emil Kraepelin.

The following summer, in 1896, Meyer was granted leave from Worcester to travel to Heidelberg, where Kraepelin was undertaking a pioneering method of psychiatric research. Kraepelin meticulously documented the medical history of every patient admitted to his clinic. His research agenda was to develop reliable diagnostic techniques and stable nosological categories based on the courses of diseases. To facilitate this goal, Kraepelin transformed his clinic into a transit station through which the greatest possible number of patients were admitted, observed, and diagnosed. Afterwards, they were distributed to other mental institutions. Because his university clinic and the regional asylums belonged to a single German state system, he was able to track the progress or decline of his cases for years, even decades.

By comparing thousands of medical histories, Kraepelin discerned distinct symptomatic patterns (not symptoms themselves) and grouped them into nosological categories. Particularly noteworthy was his delineation of the common psychoses into two types. According to Kraepelin, the disease course of what he termed dementia praecox was typified by psychotic episodes in early adulthood with continued mental deterioration and no recovery. Other cases showed a different pattern of psychosis, in which the patient experienced limited periods of delusional despair or excitement that interrupted often lengthy periods of health—a disease course he classified as manic-depressive insanity. When the patient was in the hospital, he emphasized, the symptoms of each were often indistinguishable. The differences in periodicity were objective markers of distinct diseases. Kraepelin's conclusion was immediately useful to physicians and alienists, because it provided a diagnostic measure of psychotic behavior (a mainstay of the state asylum in any country) based on the patient's medical history. His quantitative method of distinguishing new forms of psychopathology is widely considered the first application of scientific principles to the study of psychiatric disorders. The practical and methodological innovations Kraepelin introduced did much to increase the professional authority of psychiatry at the beginning of the twentieth century.[43]

Kraepelin's methodology—collecting and comparing observations of all cases to reveal otherwise indiscernible pathological patterns—appealed to the pathologist in Meyer. It was this clinical model that he attempted to institute at Worcester. Meyer would later reflect, "At a time when others thought they could get at the nature of disease by arduous section-cutting and microscopy, Kraepelin realized that the great need was a careful sifting of the symptoms according to their importance [in] determining the nature of the disorder." When Meyer later implemented this approach at Johns Hopkins, he would refer to it as "rag-sorting." In Heidelberg, he was most impressed with the scientific rigor of the clinicians. At the same time, he found Kraepelin's system limited; it prioritized diagnosis and psychological experimentation but marginalized what he considered the pillars of scientific medicine: pathology and therapy. Nevertheless, he agreed that the insights published by Kraepelin in 1896 represented a singular achievement that eclipsed any extant laboratory discovery about psychopathology. "Physicians should be exceedingly careful to make a somatic and psychiatric correlation in the study of all mental cases," he reminded members of the Philadelphia Neurological Society a few years later, since "no striking progress has been made except that obtained from well-observed patients." His survey of other European clinics also stimulated and solidified his thinking. In 1898 he reflected, "Seeing so many standpoints convinced me more than ever of the correctness of my principle . . . that scientific medicine does not begin on the post-mortem table." In terms of both individual patients and programmatic research, he believed that general pathology—the correlation of clinical, laboratory, and autopsy findings—was the path to advancing psychiatric knowledge.[44]

During this trip, Meyer was also reunited with his mother, who had recovered from her mental breakdown, despite Forel's grim prognosis. It appeared that her time in the Burghölzli hospital and an improvement of her financial and family situation had contributed to her recovery. Once again, Meyer reflected on the relationship between the changing circumstances of her life, the content of her delusions, and the reappearance of her warm and even-tempered personality. His European sabbatical brought renewed clarity, focused by his experiences of Kraepelin's clinic and his mother's rehabilitation. When he boarded the ocean liner bound for America this time, his professional identity was clear to him: he was a clinical psychiatrist.[45]

He returned to Worcester determined to emulate the rigorous observation and history taking conducted at Heidelberg. In contrast to Kraepelin's focus on diagnosis and epidemiology, Meyer explained, his aim was "to go as far as possible at the study of the facts in each case, and to put the emphasis on the living patient and the problems of the determination of his condition, the causes, and remedial measures."[46] He intended to analyze large numbers of case histories to discern distinctive causal patterns in the development of mental disorders. As was the case in other medical specialties, he expected these comparative analyses to inform diagnostic reasoning, direct future research, and, especially, to indicate possible medical interventions. August Hoch, a fellow Swiss and a pathologist at nearby McLean Hospital, also had trained with Kraepelin. Friends and intellectual allies, Meyer and Hoch campaigned successfully to convince their superiors of the value of the Kraepelinian methodology.[47]

Implementing Kraepelin's methods at a large asylum rather than a small clinic was not a straightforward process. Indeed, Kraepelin had warned Meyer it would be impracticable. "The wards of hospitals," he reiterated in 1902, "must answer for a clinical institute until a special psychopathic hospital is erected."[48] He hoped that such a clinic would eventually be established for him at Worcester. Until then, existing institutional infrastructures and attitudes posed challenges. As at Kankakee, individual medical records did not exist in the Worcester asylum. First, then, Meyer demanded that an adequate medical history be created for all patients currently in the institution. In addition, he wanted each new dossier abstracted, indexed, and cross-referenced according to symptoms, causes, and treatment results (a research tool that had been recommended to him by William Osler before he worked with a similar system in Heidelberg). When he arrived at Worcester, four physicians were responsible for the institution's 1,200 patients, so he requested additional staff to generate the new records.[49]

Despite Quinby's assurances of autonomy and resources, however, Meyer met with difficulty. Like Gapen at Kankakee, Quinby appeared eager to invest in the asylum's image as a scientific institution but indisposed to providing for the everyday clinical work. Meyer discovered that Quinby was happy to approve funds for laboratory equipment and scientific journals, but he viewed requests for payroll increases with decided skepticism. "New instruments could easily be shown," Meyer grumbled about the chief's concern with public attitudes, "while the work of men and women is more difficult of demonstration." There were other parallels to Kankakee. Quinby, too, wanted Meyer to publish reports of unusual cases and pathological oddities. Meyer tried to explain why all cases must be analyzed systematically. His goals were also impeded by what he deemed erroneous beliefs about scientific methods. In his 1898 report to the Worcester trustees, he complained that "the larger principles of general pathology seem superfluous" to many alienists and that pathological research was not "a few technical methods largely relating to the microscopic examination of dead tissues and excreta."[50] He remained frustrated that laboratory techniques were mistaken for scientific methods.

His persistent requests eventually garnered him four new junior physicians, resulting in one doctor for every 150 patients. This unusually low ratio baffled visiting physicians and officials, one asking with some skepticism, "How do you keep the boys busy?" What Meyer called his Worcester Plan, however, required manpower: "We must make a study of groups of our cases and compare them with the classical pictures of the literature; find out the differences and the problems suggested by them; the methods of treatment, of psychological and medical observation should always be arranged so that they could be confronted with the present standing of the knowledge in that field and assimilated, or where they show something new we should become conscious of it and recognize the relative value of new observations."[51] He taught the staff how to perform physical, neurological, and mental examinations systematically on every patient and to record their findings using standardized terminology reflecting only their own observations (once again he banned allusions to brain mythology). They gathered at bedsides to interview patients and took notes in shorthand. Finally, every morning he convened the newly enlarged staff—eight assistants, himself, and Quinby—for a staff conference to discuss new cases and ongoing developments on the wards. As had been the case at Kankakee, he observed that most of the physicians were interested and engaged in this work.[52]

Less than three years after he had been declined such a position, Meyer took up teaching duties at Clark University, which maintained strong graduate programs in biology and psychology. Its president, Stanley Hall, was himself a psychologist eager to apply laboratory methods to the problems of

psychopathology. Hall had urged Cowles to do so at McLean. Meyer's lecture courses between 1896 and 1901 reflected his interest in the intersections
of psychiatry, biology, and psychology. In previous years, Hall had brought
his doctoral students to the Worcester asylum to observe "illustrative demonstrations," which Meyer described as "curiosities as morbid exhibits."
Meyer replaced this practice with a weekly teaching clinic during which he
presented a broad range of neurological and psychiatric problems by examining and interviewing patients. Eventually, the graduate students of the
Harvard psychologist William James also began attending these clinical
demonstrations. Meyer's intellectual development and professional advancement benefited from his association with these academic psychologists.[53]

Meyer's time in Worcester was personally and professionally transformative. Convening case conferences and conducting bedside rounds, as well as
his teaching clinics and university lectures at Clark University, provided opportunities to refine his clinical intuition and improve his communication
skills. Since returning from Heidelberg, he had emulated Kraepelin's method
of creating a personal catalogue of abstracted case histories, handwritten by
him on five-inch-square index cards. By 1901, he boasted an impressive specimen collection unlike any in the United States. "As I look over my psychiatric career," Meyer recalled at its zenith in 1921, "I cannot help considering
my Worcester period as one of the soundest, and in a way the most solidly
useful phase of my work—since I see in it the period during which I collected
the most lastingly valuable and most substantial material for work."[54] Nevertheless, Meyer was often frustrated, miserable, and lonely in Worcester.
Quinby's partiality for administrative order rather than scientific goals
infuriated the idealistic reformer. For example, Meyer's request to hire a
druggist was denied because there was no available place at the dining table
assigned to staff members of a druggist's social rank. In addition, his status
as a bachelor relegated him to noisy quarters shared with younger physicians
"whose choicest moments," he remarked wryly at the time, "are associated
with the banjo and a hot time in old Town." There was no transportation
between the country asylum and the town of Worcester, no exercise in the
winter, and no "intellectual relief from the humdrum existence." Were it
not for his financial obligations to his mother, he reflected privately, "my
psychiatry would have come to an end." By 1900, however, he had repaid the
family's debts and his mother appeared healthy. His social prospects brightened in the fall of 1901 when he was introduced to Mary Potter Brooks, a
poised and well-educated schoolteacher. In a jubilant report of their court-

ship and engagement, he explained to his mother that she was an American who had been educated in Germany. "One of those rare daughters of this land," he waxed, with "the wide grasp of life which is an absolute necessity to us sons of the old world."[55]

Events unfolding in New York during his time at Worcester also buoyed Meyer's confidence in his work and future prospects. In 1895, authorities in New York State had responded to public pressure to improve asylums, in part by establishing a scientific institute for investigating phenomena related to nervous diseases and insanity. The first of its kind in the United States, the New York Pathological Institute was established to provide training in brain anatomy and methods of diagnosis to the state's alienists. Ira Van Gieson, a reputable and well-connected neurologist, was appointed director. However, Van Gieson was preoccupied with autopsy work and appeared uninterested in asylums or the physicians and patients in them. For years, he clashed bitterly with alienists. In 1899, Van Gieson lunched with Meyer and August Forel, who was visiting from Switzerland. "We had long discussions about the aims which he and we represent," Meyer wrote to his brother, allying himself with Forel's Zurich school. He advised Van Gieson then that any plan of research without benefits to asylums was impractical. The state had apparently come to a similar conclusion. With New York authorities threatening to close the state-of-the-art institute, an evaluation committee was appointed in 1900 that included Edward Cowles. The committee summoned August Hoch and Meyer to provide recommendations. Meyer outlined his "ultimate plan in Massachusetts"—a longitudinal study of all patients at Worcester modeled on Kraepelin's clinical research. Afterward, he described Cowles as somewhat astonished. "He grasped the aim only halfway," he told his brother a few days later, "nevertheless, the seed was sown." Another committee member, the Harvard pathologist William Councilman, was so impressed with Hoch and Meyer that he impulsively suggested they establish a psychopathic institute at Harvard; he promised to make the recommendation to the university's president immediately. George Alder Blumer, the committee's third member, subsequently wrote to Meyer: "Your viewpoint was of the greatest value for the committee. You will surely be contented with our opinion that the Institute shall continue but be reorganized as a teaching institute, and that the Director must be first of all a clinical psychiatrist." At Meyer's urging, the committee also agreed that a course in clinical psychiatry be a precondition for a physician's entrance into the state asylum service. He was elated, although he confessed to his brother that he felt sorry for Van Gieson, whom he respected as a researcher. Meyer's views had been taken seriously by powerful players in the specialty. Proudly, he wrote home, "I believe this week to have done more for the reform of psychiatric enterprises in America than ever before." Now, he confided optimistically, he was sure to get a "little clinic" for himself at Worcester so he could apply Kraepelin's research model effectively over the next several decades.[56]

Meyer instituted significant change at the Worcester asylum. Despite the challenges, by 1901 he was satisfied that the new clinical system, however imperfectly implemented, had produced results. Clinical case histories increased both in quality and quantity—approximately 2,500 "careful records" had been created. He enumerated the overall improvements in his annual report to the trustees: accuracy of record keeping, systematic testing of blood and urine, and greater uniformity in autopsy work and archiving specimens. The most important outcome of the physicians' use of Kraepelinian methods, he emphasized, was the expansion of their clinical knowledge "concerning the interpretation of many phenomena." Systematic and methodical work, in other words, had taught them to recognize important clinical signs in their interactions with patients. In his study of a hundred years of history at the Worcester mental hospital, historian Gerald Grob concludes that Meyer's reforms were so fundamental that a reversion to traditional methods became practically impossible. Another important historical development emerged from Meyer's time in Worcester. A substantial number of young specialists trained by Meyer remained committed to a clinical approach and went on to make substantial contributions in psychiatry, psychology, and neurology.[57] They included Isador Coriat, Allen Diefendorf, George Kirby, and Albert Barrett. Henry Cotton, who was later accused of performing unnecessary operations on patients at the New Jersey State Lunatic Asylum in Trenton, also began his career at Worcester. Meyer supported Cotton publicly throughout the controversy.[58]

Meyer had gone to Kankakee as a young pathologist to begin a great collection of brain material; at Worcester he became a collector of clinical case histories. Whereas at Kankakee his efforts proceeded by trial and error, at Worcester he implemented a clinical system. Next, he would attempt to bring it all together and make his system operational on a wide scale.

New York State

In the end, it was Meyer who was named the new director of the New York Pathological Institute in 1901. He had hoped for a small clinic that worked in conjunction with a custodial asylum, so that, like Kraepelin, he could follow patients for extended periods and produce longitudinal case histories. He deliberated carefully about accepting responsibility for the clinical and laboratory work, including training, at thirteen public institutions. Eager to play a role in effecting widespread reforms in psychiatry, Meyer moved with his new wife to New York. His task was to establish and coordinate research and training programs for all the mental hospitals in New York State from the central laboratory in Manhattan equipped for Van Gieson. He had achieved his dream of overseeing a great institute devoted to the study of the brain—only now he saw little need for it. Autopsy results were invaluable, he emphasized, when correlated with clinical observations. "The success of neuro-pathology depends largely on accurate and critical clinical observations," he had declared a few years earlier. "Pathological anatomy is working blindly without it, and can not hope to elucidate the apparent contradictions between physiological experiment and anatomical lesions unless the clinical data are more carefully collected." This redefinition of the ideal research model for psychiatry—a program based on Virchow's general pathology and Kraepelin's clinical methods—shaped Meyer's objectives in New York. His first decision was to relocate the laboratories to an abandoned bakery building on the grounds of the Manhattan State Hospital. Pathology work and the retraining of New York's alienists would not take place in an isolated Manhattan laboratory but at the nation's largest mental asylum (4,400 patients), located on Ward's Island in the East River.[59] It was absolutely necessary, he explained to puzzled onlookers, that laboratory research be proximate to clinical work with patients.

His priority was to establish a standardized routine of clinical and laboratory work at each of the state's thirteen asylums. First, he had to explain and, perhaps more importantly, sell his clinical vision to seasoned alienists with decades of experience managing mental hospitals and patients. He could only assume that, like Gapen and Quinby, many of them perceived scientific research as largely unrelated to caring for patients. The superintendents were summoned to Ward's Island on 1 December 1902 for a week of clinical training under Meyer. Some of them likely anticipated that the new thirty-four-year-old director would echo Weir Mitchell's condemnation of their outdated practices and hospitals. Of course, their bitter disputes with Van Gieson were still fresh. Meyer deliberately introduced his clinical reforms using concrete examples from his experiences at Kankakee and Worcester. The purpose of systematic clinical and laboratory work, he told them, was to drive research *and* to inform daily medical decisions with patients. During the week on Ward's Island, the alienists were trained in the Kraepelinian methods he expected them to mandate at their institutions.[60]

Next, he toured their hospitals, spending a week at every state asylum in New York. He led case conferences, teaching clinics, ward rounds, and demonstrated examination and history-taking procedures. In order to emphasize the shortcomings of studying only anomalies and curiosities, he reviewed fifty sequential cases at each hospital. He noted that some of the assistants resented the extra work that occupied hours previously devoted to "tennis and drinking" but that others displayed enthusiasm. In his retirement years, he recalled how one superintendent told him: "You have done more with the staff in one week than I have been able to do in ten years. Now they talk about their cases while eating lunch!" Eventually, Meyer inaugurated inter-hospital conferences that took place six times a year and rotated among the hospitals.[61] D. K. Henderson, one of Meyer's assistants at the Pathological Institute (and later recruited to Johns Hopkins), remembered that these meetings created a spirit of competition that raised personal and collective standards of practice. "It was done by kindness, by example," he recollected of his longtime mentor and friend, "not by preaching." He credited Meyer's leadership abilities with inspiring interest in the new clinical procedures rather than antagonism.[62]

Meyer's operational overhaul of the New York state hospitals and research institute irked some alienists, even as it attracted talented young physicians to the field. The well-known psychoanalyst Abraham Brill began his career at Ward's Island, where he was encouraged by Meyer to explore the work of Carl Jung and Sigmund Freud. In 1912, Brill stated that Meyer's tenure had launched a new epoch that revitalized the specialty. "Despite the grumbling of old timers," Brill commented, "the old way of writing a one-line note about the patient's mental and physical condition every three or six months had to stop." On the other hand, in 1904, Alexander MacDonald—a self-described patriarch of American asylum medicine—publicly disparaged the reforms taking place in New York and encouraged fellow alienists to resist them.[63] While Meyer certainly faced opposition to change, he did not view it as the greatest challenge to reform.

In part because of the growing prominence of Kraepelinian ideas in American psychiatry, there had developed a critical mass of established alienists and neurologists, as well as young physicians, who were interested in clinical psychiatry. One year after their initial summons to Ward's Island, the superintendents were reconvened by Meyer at the end of 1903. His address to them makes clear that his main obstacle was not institutional inertia or generational resistance—it was satisfying their demands for more clinical training for their physicians. They requested that he spend more time at their hospitals, demonstrating and supervising the new procedures, or send another qualified clinician. His dismay was patent. It was impossible to distribute himself any further, he protested. And, there was no other clinician "independent enough" to go to a different hospital each week, examine fifty patients, digest the clinical and pathological data, and then immediately use that material for teaching. It seemed there was not enough of Meyer to go around. The only solution, he told them, was "to share the inconvenience of these impossibilities." The point of highlighting this exchange is not to reinforce the traditional image of Meyer as a heroic reformer. Rather, it is to underscore that there was an existing demand for advanced clinical training in New York that coincided with his own training, propensities, and objectives. The right reformer was in the right place at the right time to effect widespread and lasting change.[64]

Meyer was contented and productive in New York, both professionally and personally. In 1904, he was appointed professor of psychiatry at Cornell Medical College. He and his wife enjoyed a satisfying social life and a respected civic presence in New York City. Journalists sought out Meyer for his comment on social issues and developing news stories. Indeed, in news coverage of a sensational legal saga, Harry Thaw's trials for the murder of Stanford White in 1906, his testimonial insights ran on the front page, along with those of other so-called insanity experts who were testifying. After seven years at the helm of the New York Pathological Institute, he affirmed in his final annual report in 1909 that his basic mandate to install general pathology as the guiding principle in the state's asylums had been fulfilled. He had instituted systematic neurological and mental examinations, medical histories, and laboratory tests for all patients; had standardized individual records; and had created a culture of ongoing clinical observation, routine autopsies, and staff conferences. "In all the hospitals," he reported, "these are matters in which a safe and profitable routine is established." In the most extensive state hospital system in the country, Meyer had established the basic infrastructure of American psychiatry in the twentieth century.[65]

The American asylum changed Meyer's mind about the appropriate methodological trajectory for psychiatry, and, in the course of struggling with its limitations, he had earned a reputation for himself in the New World. The institute on Ward's Island became an important center of psychiatric research and training in the United States. Meyer's trainees, moreover, began transporting Kraepelinian and Meyerian ideas to other key institutions in North America, Britain, and Asia. Since emigrating, Meyer had published more than fifty scientific articles that reflected his diverse interests and work in comparative neurology, clinical psychiatry, social reform, medical epistemology, diagnostics, hospital organization, and therapeutics. His profession-building efforts extended his ideas beyond New York. He became a founding member of the American Psychopathological Association and the National Committee for Mental Hygiene, and within older organizations he was recognized as an emergent leader. When Meyer returned to Worcester in September of 1909 to receive an honorary degree from Clark University, his efforts as a scientific reformer, his experience as clinician and teacher, his leadership role within psychiatry, and his reputation as a serious researcher had all been validated.

Between 1893 and 1908, Meyer was simultaneously developing his own conceptualization of the structure and function of the nervous system—what he would eventually call psychobiology. Convincing his fellow specialists to adopt his concept of a biological mind, however, proved far more difficult than winning them over on methodological issues.

CHAPTER 2

Mind as Biology

Adolf Meyer's Concept
of Psychobiology

Lady Macbeth: Here's the smell of the blood still:
all the perfumes of Arabia will not sweeten this little hand.
Oh! oh! oh!

Doctor: This disease is beyond my practice: . . .
More needs she the divine than the physician.

William Shakespeare, Macbeth, Act V, scene 1

In the spring of 1908, Adolf Meyer conjured this famous Shakespearean scene for his peers at the annual meeting of the American Medico-Psychological Association. As director of the New York Pathological Institute and professor of psychiatry at Cornell Medical College, he occupied a leadership role within a specialty increasingly known by the term *psychiatry*. He concluded his paper with the sleepwalking of Lady Macbeth to drive home his explicit message that psychological factors, in addition to bodily ones, played a contributing role in some mental disturbances. "The greatest master of Anglo-Saxon thought has given us in Lady Macbeth's dream-states a marvelous picture of a psychosis," Meyer remarked, "in this case the living over of troubling episodes." As he observes her bizarre behavior, Lady Macbeth's physician deduces its connection to the king's murder: ". . . unnatural deeds Do breed unnatural troubles" (V.i). For Meyer, it was evident—as evident as it was to Shakespeare's physician and audiences—that the experience of participating in murderous treason played a decisive role in Lady Macbeth's frantic midnight ablutions.[1]

The dramatic medical consultation also reiterated the other, more provocative, of Meyer's contentions. Lady Macbeth's doctor contemplates the implications of his observations: "My mind she has mated and amazed my sight—I think but dare not speak" (V.i). Assuming that his patient's troubles are spiritual rather than bodily, Shakespeare's practitioner excuses himself and refers the case to a priest. "He sees the plain facts," Meyer declared incredulously to his own audience, "and he thinks but dares not speak." This was a dilemma faced by any physician constrained by dogma, he proclaimed, and Meyer accused his peers of approaching mental disorders no more scientifically and no less superstitiously. He contended that various forms of dualism adopted by scientists in the previous century (paradoxically, in an effort to purge their disciplines of superstition and metaphysics) had engendered a blind faith in the anatomical basis of all disease that was crippling the advance of psychiatric knowledge. Valuable clinical data, Meyer pleaded, were overlooked because of "misleading dogmatic ideas about mind" that dominated medicine in the United States. Medical practitioners and researchers must think medically about mental disorders, he preached, "yea, teach that it is our duty to think, and to act." Explaining the mind abstractly was to be left to philosophy—medicine faced the urgent task of using its own proven scientific methods to generate practical knowledge of various forms of psychopathology.[2]

Between 1893 and 1908, Meyer worked to develop a biological view of mental activity that corresponded with that of the biological body already under medicine's purview. In 1897, he insisted that mind and body did not—indeed, could not—occupy separate domains if one accepted the Darwinian explanations of natural selection and heredity. Following the lead of prominent thinkers, Meyer viewed mental activity as a biological function that had evolved in concert with the body's anatomical arrangements and physiological processes, according to its use as an instrument of adaptation. In humans, he concluded, the result of this biological integration was the indivisible "person" in constant adaptation to its environment. "From the fertilized ovum of a mother," he explained, "the development of the mind goes hand in hand with the anatomical and physiological development, not merely in parallelism, but as a oneness with several aspects." Around 1908, Meyer adopted the term *psychobiology* to express this "oneness with several aspects"—the conceptual core of his vision for a clinical science of psychiatry. He would spend the next forty years explaining, promoting, utilizing, teaching, and defending it.[3]

Psychobiology was Meyer's solution to what he considered obstacles preventing psychiatry from operating as a clinical science and medical practice. He defined medicine as the merger of pathology and therapy. He believed that the established methods of pathology disciplined medicine as a science.

Medical practice, concurrently, was constituted for him by the physician's attempt to understand, prevent, and ameliorate sickness using insights derived from pathology. To cleave one from the other, he maintained, debased the medical endeavor. Psychobiology medicalized the biological mind and cast mental disorders not as distinct diseases, as a great majority of Meyer's peers speculated in this era, but as failed adaptation. Accordingly, a patient's reaction to environmental demands was of paramount importance in Meyerian psychiatry—a natural phenomenon subject to clinical investigation and a promising site of medical intervention. With a focus on factors shaping the development of mental disorder, rather than etiological mechanisms or classification, Meyer deemed psychobiology to be a suitable interpretive framework for pursuing what he considered the three pillars of pathology: the identification of abnormalities, the investigation of how they developed, and the modification of contributing factors. "Psychiatry *has* problems essentially medical," he assured his peers in 1897. "We can not afford to disregard any side of the biological unit in the patient, but must use psychological as well as physiological and anatomical methods under the guidance of general pathology." He declared confidently in 1908, the year he was appointed to Johns Hopkins, that psychobiology would liberate psychiatry from the "useless puzzles" and "narrowing straight-jacket of traditional assumptions" that had convinced many of his medical peers that psychiatry could not operate as a clinical science.[4]

My examination of psychobiology is divided into three sections, a separation of which Meyer would certainly disapprove. His aspiration for psychobiology was that it would synthesize methodological and ideological viewpoints. My parsing it, then, into laboratory science, philosophy of science, and clinical science may seem contradictory. Like his strategy for studying the indivisible person without losing sight of its constituent components, my goal is to illuminate key facets of psychobiology in order to understand its complexity as a whole. As a laboratory science, psychobiology combined an evolutionary view of the nervous system, the emerging theory of neurons, and Meyer's own insights derived from experimental and comparative neurology. He hypothesized that a sensory-motor action of the nervous system and its functional expression as thought and behavior constituted a single psychobiological reaction. Psychobiology's pragmatist-inspired "common sense" epistemology, Meyer asserted, revealed adaptive thinking and behavior to be natural phenomena. This philosophy of science, he contended, made mental disorder subject to the methodologies of natural history and, especially, to the clinical-pathological method of general pathology. As a clinical science, he proposed, psychobiology acted as a unifying interpretive framework for investigating the relative significance of multiple causal factors in mental disorders. The last section previews how Meyer conceptualized psychobiology as a vehicle for a clinical science of psychiatry, as explored in the remaining chapters.

What follows is a composite exposition of psychobiology as Meyer employed it at Johns Hopkins Hospital in the prewar era, one drawn from his scientific publications, public addresses, and private notes from 1893 to 1918. It is neither a critical analysis of psychobiology nor an intellectual history that traces the evolution and originality of Meyer's ideas. Psychobiology was composed of many moving parts—some conceptual, some epistemological, some methodological—and Meyer's unsuccessful attempts to explain the relationship among them in print led to widespread misconceptions about its meaning and his intentions. His dreadful writing style merely compounded more significant barriers to a clear formulation of psychobiology. First, psychobiology was rooted in neurological research, pragmatic philosophy, and clinical experience with patients—it was difficult to grasp without a solid grounding in all three. Second, psychobiology was a deliberately open-ended interpretive framework, not a theory Meyer sought to prove. "I have no dogma in my teaching," he declared proudly in 1904, but he remained doctrinaire about his distrust of any closed theoretical system that limited scientific inquiry. Definitive resolutions and guidelines for practice, like those Sigmund Freud provided for his adherents, were simply not on the Meyerian agenda. Significantly, however, the ideas and claims contained in Meyer's published and private statements about psychobiology during the quarter-century period 1893–1918 remained utterly consistent. The conscientious reader finds him neither equivocal nor inexact, and it is this conceptual continuity that this chapter is aimed at elucidating.[5]

Psychobiology in Context

In 1898, Meyer recorded privately that his ideas, while they were becoming more intelligible to others, seemed rather to "move away from, than towards the doctrines of the current schools."[6] Within medicine, especially the disciplines of neurology and pathology, Meyer did challenge orthodoxy: first, the ontological view that disease was a tangible entity that could be located somewhere in the body, and second, that the nervous system was a fixed and passive apparatus for conveying nervous impulses. Yet, psychobiology reflected trends in American society and medicine of the Progressive Era, especially the growing conviction that environmental factors impinged on health and disease. An overview of five conceptual landscapes that are far more complex and heterogeneous than can be described here will contextualize how Meyer saw psychobiology as a solution for psychiatry; they are: the association of ideas, the reflex-arc, mechanistic science, cerebral localization, and evolutionary biology. All of these were contoured by philosophical discourses in ways that also shaped Meyer's views.

The association of ideas was a prevailing explanation of mental activity throughout the nineteenth century. According to this theory, complex ideas were compounded from simple and fixed associations between sensory impressions (which were assumed to result from interactions between objects in the environment and the five sensory organs that were governed by the laws of physics) and motor movements in the body. Around the middle of the nineteenth century, physiologists revised association theory to reflect the new insight that the nervous system was divided into *sensory* nerves, which transported sensations, and *motor* nerves, which conveyed movement (these were also termed *afferent* and *efferent* nerves, respectively). Several experimentalists, notably Marshall Hall, demonstrated that sensory and motor operations were mediated at each corresponding segment of the spinal cord, forming a reflexive sensory-motor unit called a reflex-arc. Inspired by its ability to explain physical activity, philosophers and psychologists adopted the physiological reflex-arc as a promising theoretical explanation for complex mental activity, too.[7]

In the mid-nineteenth century, German physicists, including Hermann von Helmholtz and Emil Du Bois-Reymond aspired to demonstrate that all living things operated according to causal mechanistic principles of physics and chemistry. Integral to this so-called mechanical science was the notion that all forms of energy, such as motion, electricity, and nervous impulses, were equivalent. From this perspective, the body was imagined as a machine. Rudolf Virchow, the de facto leader of German pathology at this time, asserted that the nervous system operated on electrical processes similar to those inside telegraph wires. Like associationism and the reflex-arc principles, mechanical science emphasized the reductive elements of natural phenomena (for example, atoms and cells) because they could be measured and quantified. By focusing on the structural mechanics of the human body rather than its essential nature, these self-described organic physicists successfully distanced their agenda from the radical materialism associated with the failed European revolutions of 1848. By the close of the century, however, many German-speaking intellectuals and scientists acknowledged that, while mind and body were undoubtedly physical systems, this did not justify their reduction to passive mechanical principles. The most conspicuous of these challenges were various forms of holism to which Meyer was exposed during his university and medical training.[8]

When Meyer entered medical school in the 1880s, neurological research in the German-speaking world was dominated by mechanistic theories developed by Theodor Meynert. Meynert, then director of the psychiatric clinic at the University of Vienna, was a master anatomist and conducted pioneering studies on the structures of sensory and motor nerves. He deduced that psychopathology in humans was the result of anatomical lesions that interrupted dedicated tracts serving localized centers of functioning in the brain. His student Karl Wernicke confirmed two such centers by correlating clinical observations of patients with language dysfunction (aphasia) and anatomical lesions located at autopsy. The concept of cerebral localization allowed medical students and investigators to visualize the machinations of the brain and provided a plausible mechanical explanation for other forms of psychopathology. As physical analogues for mental activity, the reflex-arc and cerebral localization acquired enormous significance in neurology, physiology, and psychology in the final decades of the nineteenth century. These concepts were integral to advancing research programs in these disciplines and became the focus of scientific papers and textbooks. Subsequently, many physicians and alienists routinely assumed that the concepts had been verified experimentally, which was rarely the case.[9]

Concurrently, conceptual models based on biological evolution became increasingly important to scientific thinking in Europe and the United States, calling into question a passive mechanical view of the body. The works of Charles Darwin brought the idea of the progressive evolution of living phenomena to the fore in the latter half of the nineteenth century and forced scientists to acknowledge it as a theory that could be addressed or disproven using scientific methods. Discourses about its applications and implications were fueled by various, sometimes contradictory, interpretations of Darwin's ideas, especially those forwarded by Thomas Henry Huxley, Herbert Spencer, and Ernst Haeckel. The concept of developmental mechanics, elaborated by the German embryologist Wilhelm Roux and developed by his students, also revised scientific discourses on organic processes. Their experiments on embryonic cell masses demonstrated that environmental factors—internal and external—influenced cellular growth and differentiation. The cellular development of an embryo, Roux's work suggested, was by no means predetermined but was influenced by interrelated strings of causal chains shaped by specific environmental conditions—a concept that would become integral to Meyerian psychobiology. Increasingly, the human was studied as an organism and animal subject to the laws of nature, and robust discourses emerged on the primacy of heredity or environment in adaptation.[10]

The contours of these conceptual landscapes were shaped by tensions among philosophical traditions. Within the German intellectualism that guided Meyer's formal education in Zurich in the 1870s and '80s, this was especially true of positivism and Kantian-inspired idealism. In general terms, positivists insisted that valid knowledge was derived only from direct experience acquired via the sensory organs and that the quantifiable (and, thus, objective) findings of physics and chemistry represented ultimate knowledge of the world. Idealists rebelled against this empiricism, suggesting that nothing was knowable independently from a mind's perception of it. Helmholtz propounded a positivist view of neurophysiology. He argued that if subjective experience emerged from the lawful interactions between the sensory organs of the bodily machine and objects in the environment, then experience itself was reducible to physiological mechanics. Some argued that scientific methods were applicable only to what physically exists (physicalism), while others insisted that all natural phenomena were reducible to matter or energy (materialism). All of these positions faced serious challenges from diverse nonreductionist and holistic approaches. In neurology and psychology, the emergence of psycho-physical parallelism—the notion that bodily and psychological processes occurred simultaneously and were unrelated—reflected attempts to acknowledge the existence of nonmaterial mental activity without corrupting objective data with metaphysics. Western practitioners and investigators, ultimately, were concerned with diagnosing and identifying disease in bodies and bodily processes. For some, this somaticism was an ideological choice, whereas others simply took it for granted. In Britain and America, strong empirical traditions tempered reductionist tendencies. Deterministic views about heredity also exercised considerable influence in these countries, entrenching a prevalent belief that insanity resulted from a family's defective genetic "stock."[11]

In the 1890s, ideological and methodological discord characterized the natural and human sciences. Medical explanations embedded in associationism, reflex physiology, anatomical pathology, and hereditary determinism were dismantled and reinterpreted. Critics questioned cerebral localization, for example, since aphasic patients regularly regained lost functions. Some neurologists and psychiatrists dismissed the anatomical models of Meynert as "brain mythology" when his alleged discoveries could not be verified.[12] The confirmation of distinctive nerve cells, given the name neurons, suggested that the architecture of the nervous system was not a closed mechanical network but a nonlinear landscape of cells that were not physically connected. The experimental physiologist Claude Bernard proposed that living phenomena were not reducible solely to the laws of physics and chemistry, and he showed how higher organisms created their own internal environments regulated physiologically according to a multiplicity of ever-changing variables, such as temperature, oxygen supply, and sugar or chemical levels in blood. He argued that a stable interior environment (later termed *homeostasis*) was critical to health and independence and that it granted higher organisms a degree of autonomy within the determinism imposed by natural law.[13] Biologists and philosophers explored the interaction between an organism and its external environment as a dynamic exchange and focused on the adaptive capacities of animal and human nervous systems. Environmentalism, more generally, emerged as a guiding principle for American intellectuals and reformers during the Progressive Era.[14]

Psychobiology was a product of all these diverse developments in science, philosophy, and medicine at the end of the nineteenth century. It emerged from experimental findings that Meyer believed disproved the notion of a mechanical nervous system, from his insistence that mental functioning was a natural phenomenon and his intolerance for all forms of reductionism and dualism, and from his conviction that, for purposes of scientific inquiry and medical practice, the critical conceptual frame for understanding psychopathology was not philosophy, anatomy, physiology or psychology. It was biological adaptation.

Psychobiology as Laboratory Science

Meyer's gradual reorientation away from comparative neurology and toward clinical work with patients began soon after his emigration from Zurich in the early 1890s. Within a few years, he had promised himself privately to develop a clinical science of psychiatry in the United States based on the concept of a biological mind. In 1898, he published a detailed explanation of the "biological working hypothesis" that would become the basis of psycho-biology. It merged two well-known models of the nervous system, that of the British neurologist John Hughlings Jackson and the emerging theory of neurons. To this merger Meyer added several key insights derived from experimental research. The result was his own reconceptualization of the nervous system which he called the segmental-suprasegmental model.

"We must necessarily take some position in the general method from which to approach neurology," Meyer wrote in the majestic plural customary in scientific papers of the time, "and we choose the one of evolution."[15] This was his declaration of dynamic biology's triumph over a scholastic, static, and descriptive anatomy. It set the tone for the 1898 manifesto on which he based psychobiology. Like several of his closest contemporaries—Sigmund Freud, Eugen Bleuler, and later Pierre Janet—Meyer gravitated toward the evolutionary model of the nervous system elaborated by John Hughlings Jackson.[16] In the 1880s, Jackson proposed that the human nervous system was more than a topographical landscape of anatomical structures. He concluded that it was organized hierarchically according to evolutionary development—from oldest to newest structures and their specific functions. When Meyer taught brain anatomy, he likened Jackson's idea to the concentric circles in the trunk of a tree. The oldest layer was a core of simple mechanisms capable of simple actions that the human shared with its earliest ancestors. Engendered by biological adaptation, newer structures appeared progressively around this primitive core. The most recent acquisition, the cerebral cortex (Latin for "brain bark") emerged at the surface of the brain's principal structures like the bark on a tree.[17] Jackson believed that his clinical findings supported an influential premise that had been popularized by the Darwinist Herbert Spencer: that progressively more specialized (and structurally complicated) nervous mechanisms had evolved from simpler ones.[18]

Extrapolating from his clinical-pathological studies of epilepsy and aphasia, and from the results of published animal experiments, Jackson established three distinct "levels of evolution" in the nervous system. These were conceptual divisions based on correlations between specific nervous dysfunctions observed in patients and anatomical lesions discovered in those patients postmortem. The lowest level represented primitive mechanisms responsible for involuntary reflexes (for example, the kicking motion produced when the doctor raps your patella just below the kneecap). The middle level represented an area called the motor cortex responsible for voluntary movements, especially those activated by instinctual impulses. The frontal lobes of the cerebral cortex—the most recent and most complex structures—Jackson proposed, corresponded to the highest level associated with psychological states.

From this organization, Jackson surmised an explanation of psychopathology called dissolution that Meyer would adopt for psychobiology. Jackson theorized that more evolved and complex mechanisms in the brain belonging to higher evolutionary levels exerted an inhibiting influence on the reflexive and impulsive activity of lower, primitive ones. During the course of evolution, this process facilitated the development of ever more complex nervous structures and functions. Jackson concluded that the symptoms displayed by his epileptic and aphasic patients indicated a dismantling of these evolutionary upgrades. In labeling this pathological process dissolution, he borrowed a term from Herbert Spencer. Jackson speculated that dissolution occurred when a higher mechanism was destroyed or impaired by an anatomical lesion, causing a double loss of function: the miscarriage of normal functions performed by the damaged mechanism as well as its ability to inhibit lower nervous activity. For example, he observed that an impairment of complicated nervous functions such as muscle control, coordinated movement, or language typically coincided with the appearance of primal nervous responses, such as incontinence or nonverbal emotional vocalizations.[19]

Jackson's model was applied broadly. In the United States, Meyer joined a handful of American advocates, including his patron Stewart Paton, a Baltimore neurologist, as well as Boston's E. E. Southard, who regularly disagreed with Meyer on just such matters.[20] As a young neurologist visiting Jackson's Queen Square clinic, Meyer had been astounded by the accuracy and utility of these correlations. Unlike mechanical models, Jackson's schema offered a uniform view of the dynamic functioning the nervous system. When Meyer returned to London forty years later to deliver the Maudsley Lecture, he described the English clinician as "the functional pathologist *par excellence*."[21] From the beginning, however, Meyer found aspects of the model untenable. Jackson considered the nervous system a sensory-motor machine powered by reflex physiology, a view that was consistent with cerebral localization. To prevent metaphysics from distorting his scientific data, Jackson adopted the parallelistic position that mental states emerged during (but not because of) motor movements in the brain.[22]

Meyer believed that new experimental findings demonstrating the cellular basis of the nervous system at once confirmed the evolutionary hierarchy of Jackson's nervous system and disproved the dualist safeguards meant to protect it. The neuron doctrine, as it came to be known, emerged from the laboratories of several European researchers in the final decades of the nineteenth century. Multiple investigators established the existence of cells with features particular to the nervous system: a cell body with a profusion of fine nerve fibers. The advent of the optical microscope at midcentury had offered the first detailed views of the tangles of delicate fibers in nerve tissue, but the relationship between these fibers and cells had remained inconclusive. The introduction of a novel technique for staining tissue by the Italian pathologist Camillo Golgi revealed finer structural details of nerve cells and fibers, because it selectively blackened some and not others. Using Golgi's staining method, August Forel (Meyer's advisor at the University of Zurich) observed in 1886 that all nerve fibers emanated from nerve cells and, significantly, never connected to other cells or their fibers. In the same year, the Swiss histologist Wilhelm His arrived at similar conclusions while studying the development of nerve cells in embryos. Soon afterward, the Spanish researcher Santiago Ramón y Cajal confirmed the observations of Forel and His experimentally.[23]

The neuron doctrine challenged firmly established views of the nervous system that were based on the concept of a network of nerve fibers. Throughout the nineteenth century, philosophers and experimentalists had compared the human nervous system to a railway system with a central station (brain), main line (spinal cord), and regional routes (peripheral nerves). This system conveyed sensations from the physical world to the brain (via sensory nerves), where they were translated into responses that were sent to muscles (via motor nerves). By the 1870s, textbook chapters by authorities such as Meynert affirmed the notion that nerve fibers formed a network akin to a fisherman's net. The assertion by the neuron theorists that nerve cells were not physically connected sparked debate among Europe's most exceptional investigators about how sensations and nervous impulses were conveyed between brain and periphery. Some of the most emphatic opposition came from Camillo Golgi, whose staining technique had led Forel, His, and Cajal to their conclusions. As more researchers employed the Golgi stain to prove and disprove the autonomy of neurons (the term included the nerve cell and its fibrous appendages, dendrites), the overlapping disciplines of pathological anatomy, neurology, histology, and psychiatry teemed with neologisms, conflicting theories, and forceful speculation. This contentious and confusing discourse fueled the generational revolt against "brain mythology" and widened the gulf between these disciplines and the rest of medicine.[24]

Meyer, having trained with August Forel, took autonomous neurons as a conceptual starting point when he chose to specialize in comparative neurology. Under Forel's tutelage, he became skilled in neuroanatomical and histological methods and well informed about the controversies in the field. As a young émigré in 1893, he showed his allegiance in his first English publication. With attitudinal swagger not unheard of in the young and idealistic, he juxtaposed the conclusions of His and Forel with the "wild speculation" of Meynert's "dry and often unsatisfactory humiliating study of descriptive anatomy." His own generation, he crowed, was concerned with "problems of physiology and of evolution."[25] Five years later, in 1898, he was more thoughtful and thorough. He published a one hundred–page article on neuron research in which he reviewed the results of British, French, German, Swiss, Italian, and Spanish researchers who were working to prove and disprove the theory. By presenting all viewpoints, Meyer hoped to show American medical readers that the neuron theory was equivalent to "modern neurology." The article was characteristically dense and repetitive, but those who made it to the last twenty pages were introduced to his own novel contribution to the field.

Meyer's segmental-suprasegmental model of the nervous system entwined evolutionary levels and neurons with some critical insights. He accepted the premise that the nervous system evolved by way of increasing specialization and inhibition, and he took for granted that this development was biological, the result of cellular growth and regulation. When these two views are harmonized, Meyer concluded, "the obscurity of nervous diseases seems to give way."[26] Two critical conceptual modifications, explained below, occurred to Meyer when he generalized principles from experimental results. First, he hypothesized that the evolutionary levels of the nervous system were structurally and functionally integrated. Second, he asserted that this integration was achieved by the selective interaction of neurons rather than by their being connected as a network. These insights did not stand out in the theoretical chaos that consumed neurology and psychiatry in the 1890s, but Meyer considered his reconfiguration of the nervous system innovative and valid. After its formulation and publication in 1898, it became the neurobiological premise of psychobiology, and it guided

Meyer's clinical practices, teaching, and discipline-building efforts for the next five decades.[27]

In 1895, Meyer surmised that there must be a mechanism that integrated vertically all the evolutionary levels of the nervous system. Jackson had proposed that evolutionarily higher mechanisms in the cerebral cortex exerted an inhibiting force on lower ones and that dissolution released this horizontal control. "According to my own view," Meyer explained, "the highest level would consist in an intra-cortical and inter-cortical mechanism."[28] He believed that morphological differences between newer structures of the brain and the primitive core of the nervous system indicated the emergence of such an integrative mechanism. As was well known, the lowliest reflexive mechanisms developed from segmentation. A segment is a biological building block used by a developing embryo to construct a mature body. While some organisms remain segmented throughout development (a rain worm, for example), segmented patterns are modified in most species to achieve specialization (this is why hands and feet are similar, but different). Meyer thought it significant that the principal components of the human brain—the hindbrain (cerebellum), midbrain, and forebrain (cerebrum, including the cerebral cortex)—did not develop from segmentation but were "anatomically lifted out" of the segmented neural stem. Because they developed "above" the segmented core in this way, Meyer labeled them "supra" segmental. He concluded that the suprasegmental brain was the intracortical and intercortical mechanism he had hypothesized a few years earlier. It not only inhibited lower mechanisms horizontally, as per Jackson's hypothesis, but also integrated the hierarchical evolutionary levels vertically.[29]

Meyer posited a link between the gradual loss of neural segmentation and the centralization of nervous control in the brain. At Worcester between 1896 and 1900, he dissected a variety of species along the evolutionary spectrum—opossums, sheep, bats, cows, pigeons, lizards, and rats—to represent the evolutionary development of the human brain. As experimental psychologist and historian Cheryl Logan explains, Meyer hypothesized that during the course of evolution a shift occurred from decentralized control across a wholly segmental system—in a rain worm, for example—to centralized control by suprasegmental mechanisms. Before this transition, specialized functions were constrained within segmented structures that acted autonomously. After the appearance of the nonsegmented brain, many of those mechanisms no longer functioned independently of its control. In simplistic terms, this explained why a rain worm cut in half could develop into two worms and why less-evolved slaughtered chickens frequently run around without heads, but headless humans on the move are the stuff of ghost stories. Centralized control had spurred specialization in higher animals, Meyer reasoned, culminating in human traits such as consciousness, memory, imagination, and language. He concluded that the evolutionary purpose of the suprasegmental brain was to integrate and regulate all but the lowest reflexive mechanisms of the nervous system.[30]

Meyer took for granted that this integration was achieved by characteristic neurons found in various structures of the brain and the peripheral nerves. By extrapolating and synthesizing experimental results from neuropathology, embryology, and neurophysiology, Meyer inferred that neurons did not merely convey nervous impulses to nearby neurons; they selectively interacted with different types of neurons throughout the nervous system. Constantin von Monakow, his pathology professor in Zurich, performed degeneration experiments on nerve cells by removing a small portion of the brain of anesthetized animals. Monakow observed that the cells and their fibers in the removed portion were destroyed, but so too were cells in intact areas that sent their fibers into the portion that was cut out. He described the latter as "dependent elements," cells that originated elsewhere but whose fibers terminated in the removed part. Meyer generalized this principle and suggested that the fibers of sensory and motor neurons extended upwardly, downwardly, and laterally to integrate segmented and suprasegmental mechanisms and functions. For anatomists, he suggested, this vertical structural integration was "the most stimulating principle" for future research. He later commented that his conception of psychobiological functioning as "mentally integrated" and the English physiologist Charles Sherrington's concept of "integrative function" coincided and that the two theories were "particularly congenial." Drawing on studies by experimentalists Wilhelm Roux, Claude Bernard, and Victor Horsley, Meyer theorized that a nerve cell did not receive its action from another cell. Rather, depending upon its characteristics, including state of excitation and of nutrition, each nerve cell reacted to certain stimuli while other influences left it passive. He also deduced that nerve cells participated in multiple actions. Neurons, Meyer suggested, were choosy and versatile. In studying neurophysiology, he deemed the differentiating capacity of neurons more prominent than their connectivity.[31]

Selective interaction of neurons challenged traditional views. The nervous system did not, Meyer proposed, comprise centers of functioning served by linear networks transporting sensory-motor reflexes. He dismissed the possibility that a complex nervous function such as language was housed in a single location in the brain. The old "center," he editorialized was "a sort of homunculus, a mysterious pigmy who acts his part as a little man would, accumulates images and energies and discharges them, sends them along the wires of fiber-paths to his superiors and inferiors who discharge them again on others." He argued that correlations between language dysfunction and damage to a particular area of the brain did not indicate a center but merely the involvement of one of Monakow's dependent elements. As for the idea that the reflex-arc was the elemental unit from which all nervous operations were compounded, Meyer was equally irreverent. No longer could nervous activity be compared to a piece of meat, he scoffed, deposited into one end of a sausage machine and emerging at the other as a sausage. Nor could it be likened to an electrical current seeking the path of least resistance, despite the evident electrical properties of the nervous system.[32]

"It is more correct to say," he asserted, "that a chain or *complex* of nerve-elements gets into a state of coordinated activity" and "the cooperation of a *whole* chain represents the neural activity in any reaction." He described this coordinated neural activity as a "wave of agitation" that passed through segmented and suprasegmental terrain via neurons and their lengthy fibers. Moreover, once a neuron participated in a particular neural action, he assumed, it was predisposed to do so again. Frequent repetition of coordinated neural activities, he added, accounted for "the formation of systematic acts," the neural substrata of mental and behavioral actions performed without conscious awareness. With other researchers, Meyer suspected that neurons possessed both nutritive and functional components bound together by metabolic processes and, morphologically, by the nucleus. He suggested, provocatively, that instead of organizing the nervous system around topographical divisions or cerebral localization, it might one day be possible to organize it functionally according to the interrelated activity of all neurons involved in a specialized function—that is, the sensory neurons, motor neurons, *and the intermediate neural chain that connected them.*[33] At the time of his death in 1950, several neurophysiologists commented on the originality of Meyer's conclusions fifty years earlier. By suggesting that hierarchical nervous mechanisms were integrated vertically, and by proposing to reorganize the nervous system according to the differentiated activities of neurons, he had departed radically from contemporary views in neurology at the end of the nineteenth century.[34]

Meyer emphasized to his readership, however, that mapping such neural function was presently unachievable. In 1898, it was impossible to observe the activity of neurons. It was rare to examine a complete neuron on a single visual plane because the length of nerve fibers (also called dendrites) ranged from millimeters to meters. Experienced researchers, he explained, built up composite views of neurons by microscopically studying multiple sectional planes of the same brain structure or region. Nevertheless, physicians and investigators regularly talked of the links between neural activity and specific psychiatric symptoms. Meyer condemned this as no better than brain mythology: "To try and explain a hysterical fit or a delusion system out of hypothetical cell alterations which we cannot reach or prove is at the present stage of histo-physiology a gratuitous performance." Until neural activity could be observed directly—which he did not anticipate in his lifetime, if ever—he suggested that the only scientific way to proceed was to study its expression as function.[35]

A coordinated neural activity and its functional expression, he argued, formed a single biological reaction of the organism. This was a fundamental and motivating truth for Meyer—the oneness that he would eventually term psychobiology—and the most widely misunderstood or overlooked principle in Meyerian psychiatry. "What we know clinically as function," he explained, "is always the expression of the activity of a whole *mechanism*."[36] Citing the Darwinian principle that "all biological function is an adaptation," every sensory-motor response and its functional expression as behavior was, by definition, adaptive.[37] Meyer described the brain as the "mechanism of sensory-motor plasticity," a responsive and dynamic apparatus constituted by anatomical structures and physiological processes.[38] The activity of this apparatus was expressed functionally as the mind, which he identified as the "mechanism of adaptation and behavior."[39] Here, then, was the oneness with several aspects that he called psychobiology. "Mental reactions are taken as complete phases of adaptation," he explained, that included both the "physical" and the "mental" (his quotation marks) as "reactions of adjustment of the person as a whole."[40] This conflation of the anatomical, neural, mental, and behavioral as a single adaptive response became the enduring basis of Meyerian psychiatry.

A person constantly adjusted and readjusted to environmental demands by means of "the most complex of biological regulations—those of mentation." Mentation was the mind in operation, the functional expression of the centralized brain's activity. It regulated a complex suite of adaptive resources accumulated during human evolution: morphological characteristics, physiological systems, physical abilities, instinctual drives, learning by association, emotions, memory, perception of reality, and abstract reasoning.[41] Mentation variously coordinated, combined, and inhibited these resources "so as to form concrete conduct and behavior," a process that Meyer described as psychobiological. His interest in a science of human behavior emerged at the end of the nineteenth century. After reading the work of British writers William McDougal and Charles Mercier, he began using the terms *conduct* and *behavior* to describe observable expressions of a person's adaptive performance. Casting adaptive behavior as a window onto mental and neural activity, he believed, paved the way for a general pathology of mental disorders.[42]

Meyer acknowledged that his segmental-suprasegmental nervous system was speculative. "I do not hesitate to call it so," he conceded, adding smugly that mechanistic explanations based on the reflex-arc were also speculative and "evidently less hampered by facts."[43] He regarded the flexibility of psychobiology as its greatest virtue. It made conceptual room for both functional regeneration—when a patient regained a lost function after damage to the brain—as well as cellular regeneration. And it provided a shared biological framework for the problems of anatomical and cellular pathologists, experimental psychologists, psychiatrists, and general physicians.[44] Why he buried the presentation of his biological working hypothesis, as he called it in 1898, in the final pages of a lengthy literature review is unclear. He may have been apprehensive about challenging conventional views. In the opening remarks of his article, Meyer claimed to present "certain facts not usually considered" that were intended to replace "Meynert's time-honored plan of the brain by one more in harmony with modern views." For an unknown neuropathologist working in Massachusetts, it was a bold challenge to the discipline's patriarch.[45]

Psychobiology as Philosophy of Science

Psychobiology was a deliberate departure from orthodoxy. It discarded elemental sensations and the association of ideas, questioned cerebral localization, and relegated the reflex-arc to involuntary responses. It was biological, not mechanical—based on a reconceptualized "apparatus of biological plasticity, the nervous system" that powered a living organism, not an organic machine. Meyer envisioned psychobiology as a branch of biology that revolved around the dynamic relationship between organism and environment. Pragmatism contoured these views significantly, and he believed that endowing psychobiology with a pragmatic epistemology rendered it operational as a clinical science. The ideas of pragmatist philosophers, especially those of William James and John Dewey, helped Meyer to mediate a biological conception of mind with the goals and expectations of clinical medicine. Whereas medical researchers such as Meynert and Jackson had separated physical from mental phenomena for the sake of clarity, Meyer (often at the expense of clarity) considered their interfunctionality to be the legitimate subject matter of what he called the "new psychiatry."[46]

By asserting that both evolutionary biology and comparative neurology demonstrated the fallacy of a separate mind and body, Meyer was drawn into philosophical discussions. In an analysis of psychobiology's theoretical foundations, historian Ruth Leys rightly describes Meyer's pragmatism as neither rigorous nor entirely coherent. Indeed, Meyer was no logician. The purpose of psychobiology was to study mental disorders in living patients using the methods of pathology. From his perspective, achieving this goal entailed overthrowing the old German reductionist dogma in medicine and installing in its place an Anglo-American empiricism that supported "the only logical true method" of pathology. What he required from a philosophical system—and found in American pragmatism—was a working epistemology that rendered human behavior and experience legitimate scientific data. Meyer's contemporaries often remarked on his intellectual breadth, and his Old World elite education and passion for etymology made him well versed in the *Geisteswissenschaften*. Nevertheless, when he talked philosophy in his scientific papers, it was campaign-style rhetoric intended to unseat orthodox views—always emphatic, often patronizing. He painted each side with broad strokes, freely interchanging the terms *scientific, biological, modern, functional, dynamic, genetic,* and *pluralistic* in triumphant contrast to the similarly transposed *mystical, mechanical, dogmatic, anatomical, atomistic, deterministic,* and *reductionist*. He was no philosopher, to be sure. He was a pathologist on a mission.[47]

The strong orientation toward material reductionism in European and American science posed the greatest challenge to Meyer's campaign to invalidate reductionistic and dualistic frameworks. In 1899, Meyer lamented that medicine remained on a "pre-biological, materialistic standpoint," something he associated with the somatic view. A decade later, he compared his medical peers to Lady Macbeth's irrational doctor. Many physicians who observed psychological phenomena in their patients, he complained, "rapidly acquire[d] one of the traditional exclusive standpoints dangerously near certain mystical concepts." He cited anatomical explanations of hypnotism and hysteria as good examples, as well as schematic diagrams of the brain's various "centers" that far exceeded the limited evidence for cerebral localization. He discounted other somatic theories in circulation in the 1890s: irritated or depleted nerves, hereditary degeneration, and auto-intoxication (metabolic poisoning).[48] "Little has been achieved," Meyer declared in 1902 about these lines of research, and "the little is difficult to understand." Only small groups of skilled investigators could substantiate each other's results. It inflamed Meyer when unverified theories of causation were published in journals and textbooks to become the brain mythology recited by unwitting medical students and practitioners.[49]

For Meyer, the most destructive development wrought by "the materialistic one-sidedness of medicine" was the misconception that the term *pathology* referred to a material entity such as a microbe, chemical reaction, or diseased tissue. He dismissed this ontological view of disease as a fetish rooted in the futile pursuit of the Kantian *Ding an sich* (the so-called thing-in-itself that Kant termed the "noumenon," as opposed to the thing as it is perceived by the observer). "The whole movement of modern thought is one of distrust of the noumena back of things," he persisted, "but medical discussion finds it difficult to outgrow the old habit." He charged that "the experimental neatness" of bacteriology made investigators too fastidious or too lazy "to develop standards and methods of other lines of pathology." Consequently, he observed, most practitioners and investigators assumed that there was "no pathology of insanity as yet." Psychiatry, he urged, possessed valuable beginnings of a general pathology of insanity and a specialized discipline of "mental pathology." To convince his peers that the scientific integrity of data collected in the clinic matched that of the autopsy table and laboratory, he lobbied for a pragmatist philosophy of science that he dubbed "common sense."[50]

At the end of the nineteenth century, pragmatism emerged in the United States as a challenge to positivism and holism. Pragmatist thinkers developed epistemologies based on biological adaptation as alternatives to mechanical explanations of human behavior premised on associationism and the reflex-arc. Pragmatists foregrounded the interaction of organism and environment, a choice that reflected an enthusiasm among American intellectuals in this period for dynamic rather than deterministic interpretations of adaptation. Another common goal of pragmatist projects, including psychobiology, was to pursue academic research that yielded practical applications for the individual and society.

Meyer's engagement with pragmatism began immediately after his arrival in the United States and intensified thereafter. He became acquainted with the well-known Harvard philosopher William James between 1896 and 1900. James had trained as a physician and completed doctoral studies with the experimental psychologist Wilhelm Wundt. He taught biology and physiology at Harvard before he was appointed the university's (and nation's) first professor of psychology. His textbook, *Principles in Psychology*, was published in 1890. Its evolutionary and empiricist orientation appealed broadly to Americans, popularizing James's ideas and bringing him wide recognition. Meyer's position at nearby Clark University as well as their attendance at professional meetings in Boston brought them into regular contact. Eventually, James began sending his graduate students to the Worcester asylum to attend the young pathologist's clinical lectures.[51] The two men came to regard each other highly. Shortly before his death in 1910, James declared of Meyer to a mutual friend, "[T]hat man has the levelest mind on psychological matters that I know!" In 1902, Meyer confided in his wife that he was at home in the psychology of James and, in this connection, could envision a "thoroughly Meyerian" and American psychiatry.[52]

Jamesian philosophy structured Meyer's conclusion that the phenomenon commonly referred to as mind was a biological function. As a medical student, Meyer became convinced that the mind could "make things happen" when he observed August Forel's successful use of therapeutic hypnosis at the University of Zurich.[53] In 1879, James challenged T. H. Huxley's assertion that all animals were automata, organic machinery expelling a mental byproduct. Nicknamed Darwin's Bulldog, Huxley defended biological determinism as the result of natural selection. James argued, by contrast, that the mechanism of natural selection proved that consciousness was not epiphenomenal, since the only explanation for its emergence was its utility in biological adaptation. In other words, if consciousness was selected, it had a function and, therefore, causal agency. Mind, James insisted, played an active role in adaptation. This premise was a persuasive challenge to Huxley's determinism.[54]

James theorized that consciousness evolved to perform two interrelated functions: to discriminate between worthwhile versus unessential environmental stimuli in relation to a specific goal, and to give comparative weight to the most promising of all possible responses to the selected stimuli. A goal might relate to instinct, emotion, curiosity, strategy, or might arise spontaneously—securing food or a mate, fighting or escaping, problem solving or socializing. Importantly, James reasoned that when multiple responses to a single stimulus were possible, consciousness gauged which response might prove most advantageous by comparing potentialities against previous experience. Conscious attention was focused on a demand or goal when no other nervous response answered automatically (for example, a reflexive, instinctual, or habitual reaction) and, typically, when multiple automatic responses conflicted. James also argued that the value of any sensory-motor action depended upon its usefulness to the situation in which it emerged and was applied. An action was considered useful or "true" if it efficiently attained the individual's desired end. Therefore, knowledge and truth were not universal; they were always dependent upon a specific interpretive context. Since the evolutionary purpose of consciousness was to produce the most effective sensory-motor response, James maintained, conscious thought was incomplete until discharged as an act. Human consciousness, he declared, exerted a constant pressure in the direction of survival, an adaptive advantage that he compared to gambling with loaded dice.[55]

James argued that human experience was progressive and ever-expanding. The possibility of multiple nervous responses to a single stimulus nullified a strictly mechanical and associating nervous system. James suggested that voluntary sensory-motor responses were shaped not only by the association of ideas but also by their dissociation. He described dissociation in this regard as the capacity to differentiate a specific characteristic from its concomitants in a situated object (for example, shape, color, gender, temperament) and to exploit, subsequently, the idea of it in new and unrelated situations. Mental activity, James proposed, was both synthetic (associating) and analytic (dissociating), which allowed the individual to readjust his or her knowledge of the objective world.[56] Meyer believed that the interactivity and selectivity of neurons supported this hypothesis. "In the neurological cant, we are accustomed to speak of connections of neurons for the purpose of association." It was more appropriate, he proposed, to "think of interrelations of neurons for the purpose of *dissociation* and *readjustment.*" The reality of multiple nervous responses to a single stimulus was important to James's objective to establish the existence of free will. It became the basis of Meyer's insistence that common forms of psychopathology, previously thought irreversible and incurable, might be modified by manipulating environmental conditions for the purposes of scientific testing and, ideally, therapeutics.[57]

Meyer appreciated the middle ground forged by James between positivistic empiricism and holistic idealism. James suggested that neither philosophy acknowledged the multiplicity and utility of lived experience. Diverse realms of reality mutually interpenetrated, he proposed, and required different analytic approaches. Meyer regarded this pluralism, as he called it, to be essential to psychiatry's success as a clinical discipline, because multiple streams of data, derived from fundamentally different phenomena— anatomical, physiological, and experiential—were all essential to "an objective conception of the nature and work of the nervous mechanisms." He advertised psychobiology as an integrating framework that maintained the specificity of disparate methods and data, such as those associated with autopsy, neurology, bacteriology, serology, endocrinology, psychology, anthropology, or psychotherapy. Rather than manipulating dissimilar data to make them fit a single theoretical system, Meyer contended that theory and methods must change according to the nature of the data.[58] His persistent efforts to integrate related disciplines, despite their incongruities, were inaccurately perceived as eclecticism. Leo Kanner, trained by Meyer and then appointed by him to establish a program in child psychiatry at Johns Hopkins in the 1930s, disputed this perception: "He gratefully acknowledged contributions from all sources so long as they stemmed from concrete, factual observations." The idea of Meyer nibbling here and there, selecting only tidbits that were to his liking, Kanner scoffed, was "pathetic" and "entirely wrong." Unlike Kraepelin or Freud, Meyer refused to systematize psychobiology, and according to Kanner, "therein lay his strength." The pluralism of psychobiology, he insisted, was neither arbitrary nor uncritical.[59]

According to James, the common sense philosophy of Scottish Enlightenment thinkers reflected this instrumental character of all knowledge. When Meyer emigrated in the early 1890s, his intellectual position reflected a standard view of this philosophy, what he described in 1898 as "the naïve realism of common sense." Over and above the ordinary meaning of *common sense* (a cluster of beliefs considered by most people to be true), Scottish common sense was an epistemological stance: "I exist; I am the same person today that I was yesterday; the material world has an existence independent of my mind; the general laws of nature will continue, in future, to operate uniformly as in time past." James argued, however, that (like truth) common sense was not universal, since it evolved according to its use by individuals with subjective goals and beliefs. He redefined it as effective judgment consistent with the ultimate arbiter, individual experience. Thus, all forms of experience (not just that accumulated via the five tactile senses) produced valid subjective knowledge with sharable qualities, he insisted. By 1908, Meyer had adopted James's radical empiricism and instrumentalist common sense as epistemological ballasts for psychobiology: the test of any thought or act was its workability in effecting the person's desired end satisfactorily.[60]

James underscored the importance of habit formation in the evolution of human consciousness. In 1839, when Darwin concluded in his account of the *Beagle* voyage that Nature made habit omnipotent and its effects hereditary, he was adding to an existing discourse on habits. By the end of the nineteenth century, habit formation was an especially resonant idea in discussions of human development and social progress, especially in the United States, and the discussions revolved around Darwinian evolution. In 1884, Darwin's collaborator, George Romanes, mobilized an argument similar to that of James's to contend that mental habits played a role in adaptation. Like morphological habits, he suggested, mental habits developed from the repetition of "consciously intelligent adjustments," and they were performed with such precision and force that they were indistinguishable from reflexes or instincts. Romanes cited the practice of wearing clothes in many human societies. Clearly not an innate trait, such modesty was rather a "taught habit of mind," executed without conscious consideration.[61]

James went further and suggested that the complexity of such habits attested to mind's causal agency. Protracted conscious effort was required to master sensory-motor responses such as walking, riding a bicycle, or using language. Almost all human abilities—physical, intellectual, social—were the result of painful test, failure, and practice before they became habitual. "If practice did not make perfect, nor habits economize," James concluded, the developing individual would be in a sorry plight. Meyer utilized this rationale to bolster his claim that mind was a biological function. "If we admit that practically every sensory-motor reaction, even the most complicated ones, can be unconscious-automatic," he wrote in 1898, "we conclude that a further mechanism of differentiation and association must enter into activity to allow the quality 'psychical' to come in as an additional biological phenomenon." During the course of evolution, the complexity of unconscious-automatic acts, or habits, and the development of analytic (dissociative) thinking had been mutually reinforcing. James also observed that habitual and instinctual responses were not fixed but were often modified by the individual to suit the exigencies of new situations.[62]

The high valuation of lived experience in Jamesian philosophy resonated with Meyer yet James regarded mind as an object in relation to other objects, a view that did not articulate with psychobiological integration. Other pragmatist philosophers profoundly influenced by the ideas of James proposed that thought and action were a single stream of activity. This principle guided the work of philosopher John Dewey and an interdisciplinary group of researchers at the University of Chicago in the 1890s. These researchers made action the essential object of study in their exploration of the function of mind in adaptive strategies, an approach that became known as functionalism. According to the functionalist view, how the human organism interpreted an event depended upon the activity it pursued and, subsequently, the outcome of that activity transformed the event. In other words, stimuli and responses were endlessly reflexive—like the maneuvers of two tennis players volleying a ball—always producing new knowledge that was used, in turn, to respond more advantageously in the future. Perception and action were a continuum of experience that constituted reality. "Things," Dewey decreed, "are what they are experienced as." This shifted the focus of pragmatic inquiry to the dynamic interaction between an individual and his or her uniquely constituted environment—what Meyer called a psychobiological reaction.[63]

Meyer met John Dewey at the University of Chicago in 1893. Dewey was synthesizing current laboratory research on the nervous system and consulted physiologists and neurologists at the university, including Jacques Loeb, Henry Donaldson, C. L. Herrick, and Meyer. Dewey drew insights from their laboratory studies to stimulate and bolster his pragmatic epistemology.[64] The work and interests of Meyer and Dewey were complementary, especially regarding the issue of an organism's dynamic and spontaneous responsiveness versus the deterministic operations of a sensory-motor machine. In 1895, Meyer added an additional mechanism to Jackson's model that integrated and regulated the entire nervous system (what he eventually called the suprasegmental brain and its function, mentation). In a series of lectures the following year, Dewey also emphasized that the specialized activities of the nervous system were arbitrated by "an interrelating or coordinating structure." In 1896, Meyer and Dewey both suggested the same nonlinear revision of the traditional reflex-arc: a unit comprising not only sensory and motor responses but also the intermediary connections between them. Meyer framed his revision in terms of the selectivity of autonomous neurons and what he called the biological plasticity of the person. Dewey's critique of the reflex-arc posed a serious and lasting challenge to mechanical explanations of human behavior in American psychology.[65]

Following James, Dewey maintained that consciousness emerged in the absence of a reflexive, instinctual, or habitual response. "Thought arises," Dewey proposed, "through the need of striking a balance, that is, of discovering the course of action which will reconcile a number of conflicting minor activities." He suggested that emotional responses, such as grief and sorrow, altered a person's perception of the world, diminishing his or her capacity to complete action and thereby achieve goals. Conscious effort, Dewey reasoned, was employed instinctively to restore the unity of perception and action. Similarly, the happiness and satisfaction that accompanied completed goals were indicative of this restoration. Meyer wedded his biological conception of mind to this view of perception and action as a single continuum of experience, and he promoted psychobiological balance as an important concept for psychiatric research and practice.[66]

Meyer's contact with Dewey was renewed when Dewey joined the faculty of Columbia University. Between 1904 and 1908, a lasting friendship and intellectual alliance developed between them. During this period, Dewey emerged as a leader in a group of like-minded progressive social reformers that included the director of the New York Pathological Institute. Every Sunday, Meyer and his wife dined with Dewey and the pragmatist historian James Harvey Robinson and their spouses. After Meyer moved to Baltimore in 1909, he typically traveled to New York by train every Friday to consult patients, conduct business, and lunch with Dewey. Dewey's project was to apply a pragmatic approach to pedagogy. Meyer used it to frame psychobiology as the scientific study of failed adaptation and a medical approach to mental disorders.[67]

In thoroughly pragmatic fashion, Meyer selectively appropriated pragmatist concepts that achieved his goal of making psychiatry a clinical science. He formulated a working epistemology for psychobiology that he dubbed "common sense." By the time the Phipps Clinic opened in 1913, he was using the concept of common sense consistently in three related ways: as a sensory modality subservient to mentation, as a measure of normal psychobiological reactions, and as a tool of scientific inquiry. When he retired in 1941, "common sense psychiatry" had become a popular and often misunderstood shorthand for the Meyerian approach.

Broadly, Meyer considered common sense a sensory modality that served mentation, including those mental tasks described by James and Dewey. Implicit in this phrase, which he used ubiquitously, were synthetic and analytic thinking, as well as the constant interpretation and reinterpretation of sensory-motor adjustments. In addition to the five tactile senses, common sense supplied sensory knowledge of environmental conditions that required evaluation according to experience. Following James, Meyer believed that the ability to dissociate gave human mentation its distinct character, in contrast to the largely associating capacities of other animals. Reiterating one of James's well-known examples, Meyer proposed that a person might imagine the idea of fear without becoming fearful and, alternatively, the idea of terror could evoke a response that might accompany genuine fear (such as an increased heart rate). Consequently, unlike a physiological reflex, which remained contingent on a proximate stimulus (spatially and temporally), a psychobiological reaction involved a "more extensive scope of potential links and interrelations" among stimuli and potential responses that spanned past, present, and prospective experiences.[68]

The environment to which a person adapted, then, was constituted simultaneously by an external reality filled with situated objects (including other persons) and an internal mental life populated with associated and dissociated ideas. Both environmental spheres were dominated by stimuli in the form of symbols. The use of symbols, Meyer suggested (as had others), facilitated the use of ideas as social tools by representing their commonly agreed upon meanings. "Language is the objective expression of what is common property and common reaction and of what common sense and haphazard human experience have turned into a currency of stable symbols." Language was one of countless symbolic forms, including gestures, writing systems, currency, clothing, class constructs, and social expectations, that represented shared meanings attributed to ideas. "We usually and inevitably act as members of a herd or like it, but uncritically and without much discrimination," he explained in 1913, "and the number of spontaneous and independent and truly original actions in a lifetime is not very great in the majority of people." Meyer took for granted that persons instinctively utilized this collective but tacit knowledge for individual advantage in day-to-day adaptive adjustments. This intuition about social relationships and expectations—what Meyer deemed common sense—evolved as a complementary sensory modality to the human's sight, hearing, smell, touch, and taste. Like James and Dewey, he regarded symbols as transactional and provisional, since subjective goals and experience shaped and reshaped their shared meanings. For example, selecting the most efficient response to finding a banknote lying in the street entailed a psychobiological calculus in which the inserted values were symbolic and the formula was common sense. Spend, surrender, or ignore the unclaimed money? The most advantageous response was contingent upon a person's goals and experiences—a millionaire and a tramp might agree on the monetary value of the banknote yet interpret and then utilize this shared knowledge to serve different individualized ends.[69]

"Conduct and behavior," Meyer declared in 1911, "constitute the life of the individual." He believed that the pragmatic modification of common sense as an instrument of adaptation rendered conduct and behavior—in other words, the patient's life—legitimate and productive sources of clinical material for psychiatry. Common sense was a tool of social interaction and evaluation. "We are social beings and members of a family and of a community," he proclaimed in 1914, "and act as a rule as agents of a common sense consensus." Like individual experience, the "common sense consensus" functioned as a gauge by which a person selected self-interested goals and determined which responses would most likely achieve them. Through trial, error, and repetition, conscious psychobiological adjustments often became psychobiological habits. This transition from a conscious to an automatic act occurred not only for repetitive actions (walking or using a fork, for example) but also for actions determined by tacit social practices (such as wearing clothes or delaying urination until socially acceptable). "When it comes to the mechanism of behavior itself, to what we call our mind," Meyer explained in 1912, "we find that behavior is regulated by feelings, by fears and desires, by knowledge and wisdom, by personal desire or social custom." The common sense consensus therefore circumscribed deeply internalized cultural norms that influenced the workability of myriad adjustments in conjunction with innate abilities, instincts, emotions, and experience. If a person aimed to pay for something in a shop with rocks instead of money, for instance, the prospective workability of the goal was minimal, since the relative values of money and rocks were regulated by the common sense consensus. Meyer's interests revolved around determining how and why his patients were unable to trade in this currency of symbols and social expectations.[70]

If Meyer was determined to orient psychiatry methodologically towards pathology, he would require an essential conceptual device, a standard of normal. The common sense consensus served this function in Meyerian psychiatry. A traditional pathologist recognized disease by deviations from healthy tissues or functioning. Only by this comparison was it possible to discern the pathological. Meyer suggested that the psychopathologist could learn to differentiate efficient from inefficient psychobiological reactions using the common sense consensus. Healthy mental functioning facilitated problem solving, he explained, and led naturally to feelings of satisfaction. "Our comparative measure of the various disabilities," he wrote in 1908, "is the normal complete reaction or adjustment." Failure to achieve satisfaction was common in humans who adjusted to disappointment with strategies such as distraction, avoidance, daydreaming, forgetting, or finding fault with others. A healthy coping mechanism, however, could develop into "a miscarriage of the remedial work of life" and an inability to profit from the common sense consensus. It was this functional impairment that Meyer deemed pathological.[71]

He did not consider it abnormal to repudiate the majority view or subvert cultural norms, nor did he advocate severe social homogeneity. Rather, the common sense consensus was a measure of individual social function. Consider again, for instance, other potential responses to encountering the unclaimed banknote: burn it, mail it to the governor, bury it in the backyard, or use it as toilet paper—none of which, though abnormal, would earn someone an admission to the Phipps Clinic. But what of the person who insisted that it was not money? When Meyer visited his mother in Zurich in 1912 (just a few months before her death and the opening of the clinic), that was precisely what she told him when he placed coins in her hand. Convinced that she was destitute and bound to remain so, she insisted that the money he gave her was counterfeit. In a contemporaneous letter, Meyer promised his mother that their family was financially secure. "Believe me," he assured her, "even if your illness gives you other ideas." His mother was unable to adapt to environmental demands dictated by the common sense consensus. Meyer deemed this impaired social functioning a pathological reaction and a medical condition. "The normal and efficient reaction of the individual" was inherent in this evaluation, as was Meyer's belief that each person was an *agent* of the common sense consensus. As a comparative standard of normal behavior, the common sense consensus played a key role in operationalizing psychobiological psychiatry as a clinical discipline.[72]

Meyer also argued that clinicians could be trained to employ common sense critically, an idea he attributed to T. H. Huxley. As a young brain researcher, Meyer had enthusiastically adopted Huxley's definition of scientific inquiry as "trained and organized common sense." In an essay expounding the scientific integrity of natural history, Huxley suggested that a scientist "uses with scrupulous exactness the methods which we all, habitually and at every moment, use carelessly." A scientist's common sense, Huxley proposed, differed from that of the average man only as the trained swordsman's skill compared to that of a savage wielding a club.[73] Huxley's common sense, however, reflected the scientific naturalism that characterized Victorian science. In such a system of thought, the scientist was trained to divest personal experience of meaning and value in order to study phenomena "objectively." Meyer's definition of "critical common sense" retained Huxley's assertion that scientific training disciplined common sense, and it integrated the pragmatic tenet that scientific knowledge cannot be isolated from experience or interpretive contexts. Echoing Huxley but intoning James, Meyer claimed that critical observations of "our *own* mental activity *and* that of others in practical life" generated objective facts admissible to the domain of science and its "task of *unbiased systematization of experience.*" A mental phenomenon, he insisted, "is as observable and as objective as any other fact of natural history."[74]

Too often, physicians surrendered their "common sense attitude" and translated psychological phenomena "into a jargon of wholly uncontrollable brain-mythology," Meyer complained, "with the conviction that this is the only admissible and scientific way." He characterized materialism in 1897 as an "undue exaggeration of the importance of our tactile senses" and, in 1907, he deemed idealistic monism, materialism, and solipsism to be embarrassments to scientific psychology and psychiatry. "I should quit being a physician and a teacher," he declared to members of the American Medical Association in 1915, "if I felt compelled to doubt the possibility of my studying and knowing your minds and those of my patients well enough to draw practical conclusions." He reminded his peers that it was the cohering of consensus around observed results of (otherwise unobservable) mechanisms that conferred on them the status of fact. "By just as searching evidence as physics demands for sound waves or electro-magnetism, we certainly do all we can to pin down the evidence of its *existence* and *nature* and to prove it *before others* as well as before ourselves." He contended that, as the functional result of unobservable psychobiological mechanisms, maladaptive behavior constituted legitimate material for a general and specialized pathology of mental disorders.[75]

Thirty years after its initial formulation, Meyer explicitly acknowledged psychobiology as a pragmatist project. "Fortunately there is a positive constructive philosophy," he remarked in 1928, "that of a John Dewey, that of Progressive education, that of common sense psychiatry."[76] Supported by a pragmatic epistemology, he was confident that his psychobiological framework allowed clinical psychiatrists to interpret mental disorders without falsely dividing mind and body. "By dropping some unnecessary shells and traditions," he implored those physicians he had compared to Lady Macbeth's doctor, it was possible to reorganize psychiatry around a biological unit divided "into adjustments of the person as a whole and adjustments of individual organs." Confident in its scientific and philosophical moorings, Meyer asserted that psychobiology rendered all forms of psychopathology subject to the methods, objectives, and collective practices of clinical medicine.[77]

Psychobiology as Clinical Science

"Internal medicine cultivates the pathology and therapy of special organs," Meyer suggested in 1913, reporting on the opening of the Phipps Clinic at the International Congress of Medicine in London. "Psychiatry has to add to this the study of the broader integrations, not only of individual organs, but also of the person as a whole in a system of social adaptation."[78] By the time he was installed at Johns Hopkins, Meyer had endowed the personality with a metaphorical organicity and placed it at the center of a psychiatry newly conceived as a clinical science. Modifying his Huxleyan definition of common sense, he had announced in 1902, "[P]athology to me means organized practical medicine."[79] Casting observable behavior as a series of psychobiological reactions provided an interpretive framework for identifying, investigating, and intervening upon abnormalities—Meyer's trifecta of pathology. In his discussions of clinical practice, he mobilized the concept of psychobiological equilibrium as a signifier of mental health, on the one hand, and the notion of dissolution to conceptualize psychopathology on the other. Together, these broad conceptual frames of mental health and illness provided a basis for clinical reasoning, research questions, and treatment efforts—without reference to unknown essential causes. Confident in the integrative and plastic nature of the segmental-suprasegmental nervous system, Meyer claimed that every form of psychopathology represented a constellation of causal factors with varying degrees of significance. If this were not true, he proposed, all chronic alcoholics would develop paranoid delirium, and every case of syphilis would progress to mental derangement and paralysis, neither of which was true.[80] The task of clinical psychiatry, he argued, was to scrutinize all available data and use the methods of pathology to differentiate contributory from incidental phenomena. Like the other clinical departments at Johns Hopkins Hospital, the purpose of Meyer's work at Phipps was to produce insights about the treatment and prevention of medical conditions and to study the results of therapeutic efforts. He harbored no expectations of curing cases of insanity at the Phipps Clinic using psychobiology; at the same time, he was adamant that scientific medicine was constituted by the union of pathology and therapy.

Meyer often compared the healthy personality to a strong heart that returned to its regular rhythm naturally after strenuous exercise, deep sleep, or imminent danger.[81] Physiologist Claude Bernard and his students had established self-regulated equilibrium as the efficient working state of the body. Similarly, Meyer defined mental health as a state of psychobiological balance that occurred naturally. Mentation regulated what he called the psychobiological household—the integrated suite of physical, physiological, instinctual, emotional, intellectual, and social resources possessed by a human organism. A person adapted to extremes of experience by instinctively selecting stimuli and responses that restored psychobiological balance. Social interaction and common sense were integral to the maintenance of mental equilibrium. "No one is fit to be absolutely independent," he explained. "[W]hy should we not be able to harmonize with our dignity the conviction that there are times in every life when we had best accept the consensus of common sense rather than our own temporary feelings?" Meyer considered the profitable integration of individualism and social solidarity to be the royal road to mental health.[82]

He defined psychopathology as a "disharmony of those regulations which shape a well-balanced economy" in the personality.[83] Normal and abnormal psychobiological reactions resided on a common spectrum. "Behavior may be sufficient or insufficient," he explained. "[I]f it is insufficient, it gets [a person] into trouble, and he will have to use the ways of nature in trying to get out of the trouble." He was confident that all mental states had evolved for an adaptive purpose. For example, every person occasionally experienced the desire to "crawl into a hole," and Meyer surmised that the listlessness that often accompanied grief served to incapacitate a person so as to forestall reckless impulses. Excessive anger, fear, dread, and love were also normal extremes that a healthy personality was able to utilize for individual gain or withstand temporarily.[84] The ability to adapt to extremes of emotion and instinct efficiently was supported by "the apparatus of biological plasticity, the nervous system." There was a wide range of overlapping and nebulous symptoms that might indicate that a person's capacity in this regard was deficient or compromised.[85]

Within the psychobiological framework, no mental state was inherently pathological. If the self-regulating function of mentation was impaired and unable to restore balance (for whatever reason), that situation signaled dissolution. "Whenever the functions and the organ of behavior become morbid or sick," Meyer explained, "it is because the one or the other of the adaptive functions becomes unruly, unable to balance, over-assertive or too little assertive." Dissolution brought about the double loss of function noted by John Hughlings Jackson. It interfered with the regulating function of mentation, compromising the person's capacity to integrate past, present, and prospective experiences (and, thus, to interpret symbolic stimuli and choose an appropriate response); additionally, those primitive sensory-motor responses that were normally inhibited by mentation reared up as insufficient proxies unsuited to navigating the common sense consensus.[86] For instance, a thirty-year-old artist from New York City who suffered from debilitating phobias was admitted to the Phipps Clinic. "What is it you want to retreat to?" Meyer asked her, to which she replied, "some kind of safety, where I can at least be alone." He explained to her, "the impulse to hide is like that of a wounded animal," and the woman agreed that she was, indeed, just such a beast. Yet she was not a wounded animal, he assured her, but merely a bruised person. "A human being finds reliance in social confidence," he informed her; "you must find a way to get out of your primitive reaction." Despite the absence of any imminent danger, the artist could not be reassured that her fear was unwarranted. For Meyer, this signaled her inability to utilize common sense to her advantage. It explained why she reacted by hiding rather than adjusting according to the reality of the common sense consensus (that she was safe). His prescription was that she engage both physically and socially with the external environment she shared with others. By doing so, he told her, "you open a way to a give-and-take that is thoroughly human, and a very different thing from the primitive instinct to get away." Within the explanatory frame of dissolution, all abnormal psychobiological reactions were unregulated adaptive maneuvers, regardless of severity, etiology, symptomatology, or typicality.[87]

For Meyer, it was critical that psychiatry adopt a conceptual framework that did not demand causal reductionism as a prerequisite for medical practice or research. He thought that psychobiology provided that framework. He demanded a system that allowed the disparate data of specialized pathology—whether anatomical, cellular, or mental—to coexist and form a general pathology of psychiatry. "The knowledge of lesions is but *one* of the resources of the formula of real pathology," was his persistent message. His formula remained consistent. First, distinguish what aspect of the psychobiological reaction was under study: Was it expressed anatomically as a brain tumor or physiologically as a toxic disturbance originating outside the body (alcohol or opium) or inside the body (infection or poisoning by a malfunctioning organ)? Or, was it expressed functionally, as an impairment of emotion, instincts, or intellect? Second, identify the conditions that shaped the reaction's development. Third, determine to what extent those conditions were modifiable and why.[88] "The ideal," he wrote in 1908, was the "creation of comparative standards with the same denominators." The goal was not to reduce or fuse those anatomical, physiological, and mental phenomena that constituted a psychobiological reaction. Unlike the study of an individual organ such as the heart or liver, Meyer declared, echoing William James, in order to study the workings of the personality the "wholesome pluralism of practical life" had to be maintained. "I have too much respect for the spheres of histology and study of behavior, with their respective laws," he explained, "to encourage the hybridization." He stated that only trained neurologists and histologists could contribute useable data about the neural basis of psychopathology. Physicians and psychologists meanwhile must collect the data of mental life and avoid empty talk of "psychological histology." It was the task of clinicians to bring it all together under a general pathology of psychiatry.[89]

Meyer lobbied so persistently to redress the disparity between bodily and mental phenomena in psychopathology that he was obliged to clarify his position on causality. He rejected somatic and psychological reductionism, both of which were incompatible with psychobiological principles. He took the example of abnormal mental states that accompanied infections or brain lesions: "In *these*," he emphasized, "the *mental* facts are the *incidental* facts of the experimental chain." His intention was to underscore that every form of psychopathology represented at once an organic and a functional pathological process. "I should consider it preposterously absurd to try to explain an alcoholic delirium merely on fears and psychogenic factors," he remarked in reference to a case similar to that of Lady Macbeth, "and I consider it equally absurd to disregard [her] experience." His faith in the pluralistic composition of mental activity made subjective experience a compelling source of clinical material. In promoting his belief that miscarriages of the personality were no less biological than those of bodily organs, Meyer was also motivated by practical concerns of methodology. More often than not, he urged, the dysfunctional result—for instance, Lady Macbeth's strange behavior at her empty washing bowl—was the only clinical material available for study and testing. He hoped that, by obviating the choice between bodily and mental phenomena, psychobiology would reorient clinical psychiatry toward the differentiation of causal constellations of specific mental disorders.[90]

In some forms of psychopathology, mental factors indeed emerged as leading causes, Meyer proposed, and he participated in the robust medical discourses on psychogenesis under way during this period. Prominent figures such as Pierre Janet, Morton Prince, Eugen Bleuler, and Sigmund Freud forwarded various psychogenic theories. Meyer offered his own psychobiological spin on psychogenesis: that some mental disorders originated with the misinterpretation of experience. In 1895, he described an experience from his own childhood that illustrated his meaning: "I remember how one day I heard my father read of the murder of a whole family by burglars." The idea of such a massacre lingered in the young Adolf's mind. He began to dream regularly of the murder of his own family. "If this cause of the disturbance of my sleep had been discovered, if the suggestion of the frightful story had been corrected," he reasoned, "perhaps many unwholesome tendencies of my growing disposition might have been modified." If his faulty interpretation had been "discovered" and "corrected," Meyer proposed—if his father had reassured him that their family would not meet a similarly gruesome end—he would have been spared the (unspecified) harm that resulted. During childhood, of course, such misinterpretations were commonplace and usually temporary, he acknowledged. When a faulty interpretation remained misaligned with common sense, it could become a recurring criterion for subsequent adjustments to new situations.[91]

In pragmatic terms, what Meyer described as a "pathological experience" was a defective adaptive tool employed unwittingly by the person.[92] In some individuals, such a tool became the basis of a progressively pathological and habitual mode of reacting. The idea was analogous to a mason who is building a massive brick wall and does not realize that his level (a tool used to ensure that each brick is plumb) is slightly miscalibrated. The effect of the miscalibration is at first imperceptible, but with each new brick, the defect becomes more obvious, difficult to correct, and hazardous. Meyer suggested that psychogenic disorders were rooted in such pathological habits. "Many individuals cannot afford to count on unlimited elasticity in the habitual use of certain habits of adjustment," he explained, and "instincts will be undermined by persistent misapplication." For these forms of psychopathology without an evident organic basis—commonly called functional disorders in this period—he advocated his concept of habit-disorganization as a working hypothesis that acknowledged the pluralistic nature of psychobiological reactions. Meyer discerned what he concluded was a common mode of reaction in psychogenic disorders. The patient's adjustments were habitually incomplete (did not bring satisfaction), unsuited to the situation at hand, and ill controlled. There was "a tendency away from the contact with reality and self-correction, a scattering of the personality," and "through it all a stultifying of the instincts which are essential for balancing in the complex demands of life." The concept of pathological habits, one also explored by Pierre Janet, emerged as fundamental to Meyer's clinical practices and therapeutic strategies at the Phipps Clinic. (His approach to psychogenic disorders is the focus of Chapter 6.)[93]

Meyer regularly used a weaving metaphor to emphasize the extent to which such "unconscious-automatic" acts, as he called habits in 1898, became a function of the personality, performed without conscious awareness. When a twenty-nine-year-old theology student was admitted to the Phipps Clinic, she exhibited extreme religiosity and excited sermonizing. She was particularly distressed by what she perceived as the harmful physical and moral consequences of her longstanding practice of masturbation (a common belief in this period). Meyer noted that she was "driven by it into great activity—especially along religious lines, to the exclusion of social contacts." Faced with the belief that her health and soul were being irrevocably jeopardized by "self-abuse," the patient attempted to adjust by seeking redemption (an adaptive maneuver surely undertaken by countless contemporaries who did not develop mental disorder). Because this pursuit of redemption brought temporary satisfaction, she reemployed this strategy in new situations. She made a valiant effort while at the clinic, Meyer reported, "to unravel her fantastic fabric, but seemed unable to completely free herself." The patient's debilitating religious fantasies were assumed to be woven into the fabric of her personality. Psychobiology, he promised, could scrutinize "what habits we find interwoven and with what effect." He was confident that, in many cases, causal chains linking an initial faulty reaction and a subsequent pathological habit were discernible to the trained psychopathologist.[94]

Meyer thought that a person's "constitutional make-up" represented another important and multifaceted group of factors. He described it as the product of parental influence in the broadest sense, the sum of three factors: "(1) genuine heredity, that which comes with the germ cells and is itself inherited—a property of the chromosomes; (2) early growth and nutrition; (3) early training and habit formation. It is impossible to separate these three factors in man owing to the long period of gestation and infancy during which the nutrition and training problems are combined." At the time of conception, germ cells might be damaged (for example, by fever or alcohol); such an event would be part of what Meyer described as true heredity and could produce defects in subsequent generations. "Individual injuries or experiences," he added, speaking about the transmission of so-called acquired characteristics, "do not influence the stock." As he emphasized, he considered those events that took place after conception greater determinants of what became the mature adult's relatively fixed constitutional make-up.[95] He conceptualized the person's constitution and its causal role in terms borrowed from immunology and bacteriology. He described a hypothetical case to colleagues in 1908: A woman of "somewhat restricted capacity" was forced by unpleasant circumstances to move on two occasions. Each time, she "worked herself into a depression." She was overwhelmed by the prospect of accomplishing the task. "Instead of doing the best she could," he recounted, "she dropped into a state of evil anticipation, lamentation, perplexity—a typical depression of several months' duration." (I suspect, but cannot prove, that when Meyer offered this example he had in mind his own mother's illness, which developed under similar circumstances.) On no other occasion had the patient in his example shown signs of mental illness. "Where others get along with fair balance, this patient is apt to react with a peculiar depressive reaction." It was the result, he proposed, of a lack of psychobiological immunity. He argued that the etiological constellation had two interdependent elements: the constitutional make-up and the precipitating factor. He contended that both facts were relevant but that the latter was of greater importance to pathology and therapy: "It alone gives us an idea of the actual defect and a suggestion as to how to strengthen the person that he may become resistive."[96] The bacteriological metaphor worked to explain habit disorders and emotional disturbances. He described one patient's psychotic state as "a realm of unrealities—a sort of parasitic growth of pathological experience." Meyer made a metaphor of the ontological concept of disease that he so often rejected in order to emphasize that pathological processes existed not only in the organism's anatomy and physiology but also in its experiences.[97]

The tendency among physicians to pin mental disorders on defective hereditary constitutions frustrated Meyer as much as blaming them on hypothetical lesions or poisons. At the end of the nineteenth century, "heredity degeneration" was widely accepted as a scientific explanation for alcoholism, sexual perversion, "feeblemindedness" (what we refer to as an intellectual disability, previously, "retardation"), criminality, and insanity. Many Americans believed that these weaknesses were inheritable and progressively threatened the genetic integrity of the human species. In 1894, Meyer cautioned that conclusive data on degeneration theory were meager and its practical value overrated.[98] A few years later, he was more critical: "We certainly must do something to outgrow the stage at which 'degeneracy' is considered a sufficient verdict instead of being shown up as a block in the way of much needed knowledge."[99] His mother's recurring mental illness shaped his attitude toward hereditary causes, as did his empiricism and his preoccupation with differentiation, therapeutics, and prevention. "I am one of those," he remarked, "who see too many explanations by more direct non-hereditary causes in many of the defects and disorders of total behavior which we meet in our practical work."[100] Heredity, he said, hung like a sword of Damocles above the heads of families; the "blind belief in predestination" led too many physicians to dismiss relevant and potentially modifiable medical factors.[101]

Meyer asserted that a psychobiological reaction was a single biological event constituted by a neural action and its functional expression, via mentation, as adaptive behavior. Accordingly, he implied, experience had the potential to alter brain activity. This conclusion was inherent in his conception of a plastic nervous system, which he viewed as the basis of mind's causal agency in biological adaptation. "Mind, like every other function," he began asserting around 1906, "can demoralize and undermine itself and its organ and the entire biological economy." In a discussion about the persistent misapplication of "habits of adjustment," he warned that "the delicate balance of mental adjustment and of its material substratum must largely depend on a maintenance of sound instinct."[102] In other words, a morbid psychobiological habit signaled altered neural patterns and impaired mentation. An assault on either front risked permanent impairment. "A fairly large number of our people are exposed," he suggested of asylum patients in 1909, "to conditions in which they jeopardize the health of their brain so as to endanger their individual and social plasticity."[103] Psychiatrists, including Smith Ely Jelliffe and Stanley Abbot, balked at Meyer's insinuation that mind caused material changes in the brain. In 1911, Abbot—whose biological views usually harmonized with those of Meyer—publicly reproached the new chief of psychiatry at Johns Hopkins. "He thus stands practically alone," Abbot remarked, "and reverses the general tendency of medical science to take morbid conditions out of the functional class and put them into the organic."[104] While the suggestion may have appeared exceptional within psychiatry, the hypothesis that habit modified nervous structures had been discussed by the German physiologist Johannes Müller and the Darwinian Herbert Spencer much earlier.[105] It does not appear that Meyer responded publicly to Abbot's critique. In 1921, Abbot organized a symposium at a national meeting and invited Meyer, who accepted, to explain his views. Meyer's assumption that life experiences altered neural patterns in the brain helps to explain his confidence that the environmental conditions of his new Phipps Clinic could be used effectively to study and to treat patients with mental disorders.[106]

Throughout the quarter century under study, 1893 to 1918, Meyer struggled to convince his peers that his biological working hypothesis obviated the need to choose between mind and brain. Neither radical materialism nor holism, moreover, was a prerequisite for a scientific study of mental disorders. "We must use terms of psychology," he clarified of mind's causal agency, "not of mysterious events, but of *actions* and *reactions* of which we know that they *do* things."[107] Nevertheless, he consistently encountered reductive and dualistic impulses among his fellow specialists. In 1906, an older colleague lauded Meyer's efforts to bring back "psychological medicine" (what the commentator regarded as the mainstay of the great alienists of the previous generation) now that "Virchow's cellular pathology has had its swing." In the discussion of his paper that concluded with Lady Macbeth's sleepwalking, Meyer was asked pointedly: "A woman hears bad news and drops dead. Does the mind kill the body or the body kill the mind?" And, in 1915, a colleague suggested to him that psychotic states might be caused when the dendrites of neurons interlaced themselves. "Who on earth ever saw those dendrites wiggle?" Meyer demanded rhetorically, knowing that the answer was, no one. "How much more important it is," he told his questioner, "to look for what we can see." What could be observed directly was "the conduct, its inner consistency and its consistency with the situation, showing the extent of adaptation [and] rapport with the persons and objects of the environment." A choice between mental and bodily phenomena was artificial and misleading, he argued, since the patient's total adaptive performance—his or her psychobiological reactions—was the proper subject matter of the new clinical psychiatry.[108]

Endlessly—albeit ineffectively, as a rule—he repeated his message that mentation was a biological function and subjective experience was an instrument of adaptation. Tirelessly—and, often, contemptuously—he insisted that the pragmatist epistemology of common sense made it possible to study adaptive behavior scientifically despite its immateriality. And, in overtly optimistic language—sometimes misconstrued by listeners as naïveté—he assured physicians and the public that academic psychiatry could do something constructive *today* to advance knowledge and ameliorate patients' symptoms.

Psychobiology was Meyer's attempt to place a complex phenomenon, human behavior, at the center of a specific method of scientific inquiry, general pathology. He was motivated by his confidence in clinical methods, by his conviction that medicine should always be a patient-centered endeavor, by the notion that social behavior was instrumental in biological adaptation, and by his personal experiences with mental illness. "The human mind is the triumph of biological evolution," he wrote in 1908, "but, like all the boons that come to us, our mental development will at times miscarry." Minor and severe forms of mental disorder jeopardized both the individual and society. "To have a [family] member alienated from the common ground of mental life for months, for years and perhaps for life," he testified, was "the saddest experience that can befall a family." Scientific medicine, Meyer believed, had a duty and a means to confront mental disorders as medical conditions. Psychobiology, he promised, could overcome the theoretical and methodological roadblocks that had prevented psychiatry from evolving into a clinical science, as had other medical disciplines during the nineteenth century. "Psychiatry deals with the social organ of man," he declared confidently when the Phipps Clinic opened in 1913, "and is bound to become a prominent part of the field of social medicine, which so far has limited itself mainly to the infectious diseases."[109]

The remaining chapters explore Meyer's strategies at Johns Hopkins Hospital in the prewar period for implementing a new clinical science of psychiatry based on psychobiology. Between 1920 and 1960, the term *psychobiology* became synonymous with Meyerian psychiatry. In the 1920s, Meyer appointed Curt Richter director of the psychobiology laboratory in the Phipps Clinic. Richter undertook pioneering research on circadian rhythms and animal behavior. In the 1930s, the Rockefeller Foundation funneled millions of dollars toward psychosomatic research based on psychobiological principles. Both psychoanalysis and behaviorism thrived under an ascendant psychobiology, and, eventually, Meyer was battling two new forms of reductionism. Thomas Rennie, an outspoken Meyerian, explained in 1943 that the principles of psychobiology were simple: to determine why and how the patient's mental disorder developed. "If this has become so evident that it prompts the listener to ask, *Is that all there is to psychobiology?*" he declared, "then Meyer's task has been well done."[110] In 1980, it was estimated that Meyer had trained over a hundred full-time teachers of psychiatry, most of whom occupied chairs of psychiatric departments throughout the English-speaking world.[111] By then, however, the term *psychobiology* referred to psychiatric pharmacology and genetics, and most of the methods and concepts introduced by Meyer had become so integral to American psychiatry that few suspected they were once controversial or innovative. These developments paralleled, as one observer phrased it, the strange disappearance of Adolf Meyer.[112]

The Meyer family, Zwinglian Protestants from the countryside near Zurich, Switzerland, c. 1885. A teenaged Adolf is standing at far right with his hand on his father's shoulder. His mother, who suffered with serious mental illness a decade later, is seated at far left.

Meyer in his quarters at Worcester Lunatic Asylum in Massachusetts, where he was chief pathologist, c. 1896. He considered his five years in the wards and laboratories of the hospital the most productive of his early career, and his off-duty time there the most lonesome.

Some members of the inaugural medical staff of the Phipps Clinic in 1914. Flanking Meyer (*center*) is D. K. Henderson (assistant director) on the right and Macfie Campbell (chief resident) at left. Both Henderson and Campbell were Scottish physicians who had trained with Meyer in New York. Meyer had an affinity for Scottish medicine and clinicians.

Meyer with his only child, Julia, in 1918 on the steps of their Baltimore home. "Marie is very much determined to move to Roland Park, a very attractive suburb," Meyer told his brother in Switzerland one month after the Phipps Clinic opened in 1913. He added, "I often wish I could live at the Clinic." (IV/3/149, AMC)

SURGEON FOR CRIPPLED SOULS

Dr. Adolf Meyer,
Professor of Psychiatry in Johns Hopkins University, Director of
Henry Phipps Psychiatric Clinic.

"A Surgeon for Crippled Souls"—that is how an editorial cartoonist from the *Baltimore American* newspaper described Meyer in 1922. Indeed, psychiatry had come of age as a clinical discipline by the 1920s, just as surgery had thirty years earlier.

Meyer socializing with University of Chicago neurophysiologist Heinrich Klüver in the 1930s. On Sundays, the Meyers entertained Johns Hopkins colleagues and visiting scientists and intellectuals at their home in Baltimore.

Each Saturday, Meyer held a teaching clinic for Johns Hopkins medical students and psychiatric residents in this lecture hall. The chief or a trainee interviewed a Phipps patient, and Meyer lectured on particular aspects of the case. The door to the left of the chalkboard at the back of the stage provided direct access from the wards.

The admission ward was the first stop for most new Phipps patients. It was designed to look like any other ward in a general hospital, in order to emphasize that mental disorder was a medical problem that required medical expertise.

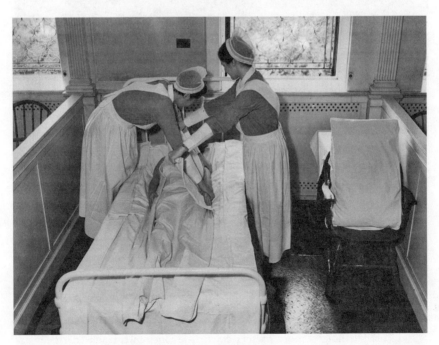

A staged demonstration by Phipps nurses in the 1920s of the wet or dry sheet pack, a technique used for both sedation and confinement in the Phipps Clinic.

Meyer at a conference around 1918. During the 1920s and '30s, he became the most authoritative and influential psychiatrist in the United States, imposing his vision for a clinical science of psychiatry and directing the general trajectory of the new discipline.

A banquet attended by staff and guests, held in the auditorium on the top floor of the clinic to celebrate the fifteenth anniversary of the Phipps Clinic. Phipps patients listened to concerts and theatricals, played school yard games, and exercised in the large room. Each Friday, female patients were induced to help prepare and enjoy a tea party with dancing and singing. The male patients gathered to play billiards down the hall.

CHAPTER 3

Unique Soil in Baltimore

The Phipps Psychiatric Clinic at
Johns Hopkins Hospital

On 16 April 1913, speaking to an audience of one thousand gathered to celebrate the opening of the Henry Phipps Psychiatric Clinic at Johns Hopkins Hospital, Adolf Meyer, the clinic's director, declared that those involved in its realization had together "created unique soil in Baltimore" by building a "an institute representing a true clinic, in other words, a hospital for practical work, research and teaching." The previous five years had been spent planning, budgeting, building, staffing, and equipping the new clinic. In the audience, alongside the social elite of Baltimore, sat its benefactor, the industrialist Henry Phipps, as well as the prominent Johns Hopkins faculty who had engineered the endowment: William Welch, Lewellys Barker, and Henry Hurd. Fittingly, former Hopkins luminaries Stewart Paton and William Osler—the earliest and most active lobbyists for a psychiatric clinic at the famous medical university—addressed the assembly. "If Dr. Adolf Meyer does not suffer from sore shoulders, hands and arms for the next few days," quipped the *Baltimore Sun*, "it will not be because of the lack of pump-handle shakes on the part of some of those who congratulated him." Three days of celebrations were filled with symposia, banquets, and outings arranged for the scientific dignitaries in attendance. All agreed that the new Phipps Psychiatric Clinic was an exceptional institution destined to play a role in the advancement of psychiatric knowledge and the training of future leaders in the field. While there existed pockets of trained researchers at private and state institutions who were producing (and continued to produce) meaningful scientific findings on psychopathology, "the Phipps," as it would quickly be dubbed, was the first academic institute of its kind in the English-speaking world. Among members of the American medical profession and the general public there was an overwhelming sentiment that the opening of the new department in the School of Medicine at Johns Hopkins University represented the first time mental disorders—both the debilitating nervousness that appeared endemic in America as well as the various forms of insanity seen in asylums—would become the subject of serious scientific study, medical training, and treatment efforts.[1]

As Meyer had discovered when he arrived in the United States in 1892, American mental asylums had not participated in the modernization general hospitals had undergone during the second half of the nineteenth century. Once a place of charity and convalescent care, hospitals had become increasingly associated with the optimism of scientific progress, effective technology-based diagnostics and therapies, and medical experts with specialized training. American asylums still loomed in the public imagination as archaic, unsafe, and terrifying. They had emerged earlier in the century as benevolent social institutions, but by the 1890s they were overcrowded and considered a last resort for individuals with a hopeless prognosis. They resembled prisons more than hospitals, and indeed, institutionalization in an asylum required a legal declaration of insanity. In addition, the shame of hereditary stigma accompanied a diagnosis of insanity, because much of middle-class America was horrified by any hint of "degeneration" in the familial line and its potential to weaken the genetic health of future generations. Unlike general medicine and surgery, moreover, psychiatry possessed no magic bullets, no technological marvels. As a result, the custodial character of mental institutions had remained unchanged as they continued to fill with people popularly perceived as hopeless cases.[2]

Concurrently, the ideas and methods of the German clinician Emil Kraepelin moved to the center of American psychiatry. Kraepelin had demonstrated that psychiatry was not only viable as a clinical science but could be fruitful in each of its three constitutive spheres: research, training, and practical innovation.[3] Calls sounded within the medical community at every opportunity for the creation of institutions modeled on the university clinics that Kraepelin had established in Heidelberg and later Munich. Throughout this period, the most vocal proponent for psychiatric clinics in America was the well-respected (and independently wealthy) Baltimore neurologist Stewart Paton. Paton had summoned the young Adolf Meyer to his Maryland estate in 1899 to hear about the "Kraepelinian views" that Meyer and August Hoch were implementing in Massachusetts. Paton became a friend and ally to Meyer and likely recommended him for both the New York and Johns Hopkins directorships. The State of Michigan and the City of Boston were the first jurisdictions to legislate funds for urban psychiatric hospitals, and in each case the institution was located near a university campus with well-founded intentions of stimulating a reciprocal relation-ship. When plans for a psychiatric clinic at Johns Hopkins were announced in 1908, commentary in both medical journals and newspapers reflected widespread approval and a relief that the United States would finally have an institution worthy of Kraepelinian psychiatry.[4]

The new clinic at Johns Hopkins brought publicity for what Meyer had in 1901 called the "new psychiatry."[5] He was part of a small cohort of elite neu-rologists and alienists in the United States who professed a newfound thera-peutic optimism. This group—which included August Hoch, Smith Ely Jel-liffe, Morton Prince, and William Alanson White—abandoned the absolute distinction between insanity and sanity and the traditional model of dis-ease upon which scientific medicine was based. Instead, they embraced new models based on evolutionary principles in which normal and abnormal mental states resided on a common spectrum. As historians have demon-strated, this altered perspective expanded psychiatry's jurisdiction well be-yond the asylum and into the lives of many troubled individuals (both self-identified as mentally ill or deemed so by others).[6] This conceptual shift—from insanity to maladjustment—resonated loudly in an age inspired by hopes of social progress and deeply concerned about genetic and moral decay. It also dovetailed with Americans' increasing fascination with personal enrich-ment, healthy living regimens, and talking cures.[7] By virtue of its well-known benefactor and parent institution, the opening of the Phipps Clinic shone a spotlight on this psychiatric reform movement, already under way when construction of the clinic was announced in 1908. Between 1913 and 1917, the Phipps Clinic attracted and admitted a socially diverse group of patients with a wide range of symptoms and motivations for using this new type of psychiatric hospital.

From his powerful institutional position at Hopkins, Meyer had the op-portunity to shape the new psychiatry around his clinical vision and his particular evolutionary model of psychopathology, psychobiology. The po-sition also enabled him to influence medical and public attitudes about psy-chiatric illnesses and institutions, which was critical to his professional, in-vestigative, and therapeutic strategy to bring prospective patients into contact with psychiatry long before a legal declaration of insanity became necessary. His design and organization of the clinic were equally integral to his methods of inquiry and treatment (explored in the final three chapters of this book). As they relied, conceptually, on the realignment of disordered thoughts and behaviors with the common sense consensus, he deemed it important that the patient perceive his or her condition as an illness requiring modern hospital methods and medical expertise. The Henry Phipps Psychiatric Clinic was established to fulfill institutional requirements at Johns Hopkins Hospital, to meet the medical needs of people perceived as sick (rather than "nervous" or "insane"), and to serve Adolf Meyer's psychobiological agenda.

A Psychiatric Clinic for Johns Hopkins

Five years earlier, on 15 June 1908, the front page of the *New York Times* announced that Henry Phipps, Pittsburgh steel magnate turned New York philanthropist, had endowed a psychiatric clinic at the Johns Hopkins Hospital in Baltimore. The famous university's newest addition would be "the first of its kind" in the United States, with "modern apparatus for use in the treatment of patients, and laboratories for the scientific investigation of mental abnormalities by pathological, chemical and psychological methods." Medical periodicals reflected the consensus that such an institution was much needed in the United States. The endowment was "[s]ignificant, as well as munificent," cooed a writer in the *Journal of the American Medical Association.* The *Boston Medical and Surgical Journal* declared, "[T]ime is certainly ripe for the establishment of such a clinic" and called for more philanthropists and medical schools to follow suit. Meyer would discover that making psychiatry operational as a clinical practice was exceedingly difficult. In principle, however, the idea that specialized medical training and clinical research were necessary to advance psychiatric knowledge was self-evident, especially if undertaken at Johns Hopkins.[8]

Johns Hopkins University had been established in 1876 in accordance with the wishes of Johns Hopkins, a Quaker merchant in Baltimore, with the fortune he had bequeathed for the purpose. The institution was organized around Germanic academic principles and was the first American university to place an emphasis on graduate education and scientific research.[9] When its hospital opened in 1889 and its school of medicine began classes in the autumn of 1893 (Meyer's first full year in the United States), they too were modeled on German medical universities and were heralded by educators and scientists as the best medical education available in the United States. The medical school's inaugural faculty wielded unparalleled influence within American medicine. William Henry Welch, pathologist-in-chief, had trained and worked extensively under Europe's leading university pathologists and bacteriologists. His authority was a prevailing influence at Johns Hopkins and he upheld its German emphasis on laboratory teaching and research. William Osler, a Canadian, was appointed first chief of medicine. Osler was already a recognized professor of medicine, first at McGill University and later at the University of Pennsylvania in Philadelphia. A talented diagnostician and pathologist, his clinical skill at the bedside made him a legendary and beloved figure with students and patients alike. Osler infused Johns Hopkins with an English tradition of clinical medicine. It is not difficult to see why leaders with such mindsets, when presented with the opportunity to establish the country's first academic department of psychiatry, found Meyer's well-known reform program to reunite laboratory and clinical work in mental asylums appealing. Osler was pleased Adolf Meyer had been selected to oversee the psychiatric clinic, he told his friend Henry Phipps in 1908, because of his "admirable training" and special gift for cultivating productive subordinates. "I have known him for many years and have the highest appreciation of his ability."[10]

One week after the news of the Pittsburgh millionaire's bequest, the *New York Times* announced that Adolf Meyer, then head of the New York Pathological Institute on Ward's Island, would become the first chief of psychiatry at Johns Hopkins. Reporters, who would describe the alienist as distinguished both in America and abroad, promptly called on Meyer at his Manhattan address. "The general plan," he informed them, "is to provide under the Phipps donation a clinic for the treatment of those afflicted with mental disorders, just as clinics for the treatment of those needing surgical attention have already been provided at this university." The comparison between surgery and psychiatric treatment was an obvious one. Once considered a last resort with limited scope, surgery had been revolutionized in the 1880s by the use of anesthesia (which allowed surgeons to open the viscera and to perfect more ambitious techniques) and by aseptic procedures (which radically reduced deaths by infection). As a result, surgery had recently undergone a fundamental cultural transformation within the public imagination as a synthesis of craft, science, and healing. For most Americans, the hereditary taint of insanity and the irrevocability of an asylum commitment remained as terrifying as the old amputation saw before the advent of hospital operations. Meyer maintained that the work, "a great one," would open up "almost boundless opportunities for the treatment of curable mental disorders." The reading public had reason to be intrigued. The plans for the psychiatric clinic and the statements of those connected with it were infused with the progressivism and optimism that pervaded American life at the beginning of the twentieth century.[11]

It is not surprising that Henry Phipps offered to finance a psychiatric hospital at Johns Hopkins. Discussions had begun inconspicuously one month earlier. "When I had the pleasure of lunching with you in Baltimore," Phipps wrote to William Welch, "you dropped a remark that set me to thinking in regard to the treatment of the insane." In fact, the philanthropist had long been interested in the subject. In the reportage on the endowment, one journalist claimed that Phipps had paid nurses at a Pennsylvania asylum to provide him with clandestine reports of conditions and treatment of inmates. Also, he was well-acquainted with Harry Thaw, the notoriously licentious son of another wealthy Pittsburgh family, who was on trial for the public murder of his wife's lover in New York City. Thaw remained in New York mental asylums for years and during several sensationalized trials; there he was examined by Meyer and dozens of other insanity specialists who provided legal testimony about Thaw's mental state. Phipps had maintained a strong friendship with William Osler after the opening of the Phipps Tuberculosis Dispensary at Johns Hopkins in 1905. In Osler's widely publicized farewell address to the university the same year, he decried the poor standing of academic psychiatry in the United States in comparison to the "great psychopathic clinics connected with each university" in Germany, and he had urged Johns Hopkins to take the lead in America. After the two lunched in May 1908, Welch sent Phipps a recent memoir by a man named Clifford Beers. Beers was a Yale graduate who described in writing his experience of severe mental illness, his detainment in a mental asylum and misuse there, and his inspiring personal efforts to regain his reason and life. Reading Beers's *A Mind That Found Itself* reinforced for Henry Phipps his decision to endow a psychiatric clinic at Johns Hopkins.[12]

A few weeks later, Welch, accompanied by Henry Hurd and Lewellys Barker—the university's longtime hospital superintendent and the new chief of medicine, respectively—met Phipps in New York City. Hurd, an alienist himself, had lobbied persistently for a psychiatry department at Hopkins. Barker, a Canadian neurologist, was equally enthusiastic. Phipps committed $500,000 and asked to see a proposal containing detailed engineering and budget specifications before he sailed for Europe in ten days. Building plans, operating estimates, and administrative approvals were compiled in haste, and they met with the benefactor's approval. On 12 June, Welch wrote to Hurd from New York: "Mr. Phipps gives us the psychiatric clinic and everything asked for." Phipps asked Welch to delay the announcement until after his departure for Europe. Welch's lunchtime remark to Phipps had culminated, in a period of less than a month, in a firm pledge of over half a million dollars. As Welch informed Hurd, however, he intended to obtain one more guarantee before returning to Baltimore, in order to ensure the success of the newest Hopkins venture: "I hope that we can secure Meyer." Later that afternoon, Barker joined Welch to extend the wholly unexpected job offer to the forty-one-year-old director of the New York Pathological Institute. Welch reported to Phipps that the interview had gone well, and the latter was pleased. Two days later, on 14 June, Welch released a statement to the Associated Press proclaiming the philanthropic windfall. No potential director was named, because Meyer, astounded by the offer, had agreed only to think about it.[13]

Meyer wrote immediately to his mother and brother in Zurich. "Prof. Welch and Prof. Barker of Johns Hopkins were just here," he informed them. "They offered me the professorship of the newly founded clinic." He relayed the astonishing details: a $5,000 salary (equivalent to that of other Hopkins chiefs), the "first clinic of its kind," and "ideal conditions" for scientific work. "I am deeply stirred," he confided. Nevertheless, as a lover of opera, classical theater, and orchestral music, he lamented the thought of leaving the cultural metropolis. He was forced to grapple, too, with personal principles regarding the appropriate oversight of institutions. His partiality for Switzerland's system of direct democracy, as well as his ringside seat to the petty politics of American asylums, had convinced Meyer that hospitals and medical training belonged under national control, to ensure their integrity.[14] Thirty years later he recalled having embraced the job opportunity, since Johns Hopkins represented "the fruits of capitalism at its best in its use of its accumulation of wealth for the benefit of science." Back in 1908, he marveled to his loved ones in Zurich that Johns Hopkins attracted the best students and most stimulating colleagues in the country. He wired Welch in the affirmative, even as he confessed a private doubt to his family: "The most important professorship in the English-speaking domain—if only I were a better speaker." Whereas Meyer was always sure of his abilities as a pathologist and clinician, and resolute as to the validity of his concept of psychobiology, he remained ever-conscious of his shortcomings as a communicator.[15]

As psychiatrist-in-chief at Johns Hopkins, Meyer would gain an extraordinary opportunity to establish, finally, a clinical science of psychiatry based on his psychobiological conception of mental disorders. He was pensive about the significance of the appointment. "The strikingly unique spirit of Johns Hopkins," he effused in an early—unsent—draft of his acceptance letter to Welch, "is in itself the highest inspiration one can meet in the academic and practical field of medicine." There was no comparable opportunity in an Anglo-American country, he added, and he felt keenly the honor and responsibility. In the end, he sent a less exuberant response stating that the best way to express his respect for Johns Hopkins was to accept the post and "do it as full justice as I can." Four separate executive bodies of the university and medical school had to approve the appointment, but within a few days everything was finalized. "I have never known such unanimity of opinion in the selection of a professor and head of a department," Welch reported when he welcomed Meyer officially to the faculty. He also assured Meyer that individuals in the neurology department were "sympathetic with you and the new movement." Meyer had insisted upon remaining at the Pathological Institute for another year, to conclude work with trainees in more than a dozen state asylums and in the laboratories on Ward's Island. His final directive—to rename the institution the New York Psychiatric Institute—reflected his clinical vision for American psychiatry. His successor in New York was his foremost professional ally, August Hoch. In 1909, he became chief of psychiatry of a department that existed only on paper. He moved to Baltimore in 1910 to oversee the clinic's construction, hire staff, attend meetings, and begin seeing patients in the general dispensary and hospital at Johns Hopkins.[16]

Johns Hopkins had hired yet another scientific star whose impressive credentials marked him for membership in America's citadel of modern medicine. The *Baltimore Sun* reported that "the opportunity to develop a strong department of psychiatry in connection with the Johns Hopkins Hospital and Medical School along lines little cultivated in this country" fulfilled Meyer's "well-known ideals." He was a skilled neuropathologist able to supervise laboratory research, the foundations upon which Welch believed the institution rested, as well as an experienced clinician who embodied William Osler's commitment to hospital medicine and teaching. The *Johns Hopkins Hospital Bulletin* published a short biography of Meyer in August 1908 and reported, accurately, "[H]e has always been a voluminous and forceful writer." The titles of his sixty-four scientific papers were appended to the article for the perusal of Johns Hopkins faculty, students, and alumni. Beginning with his doctoral thesis on reptilian forebrains, the list arrayed his work on comparative neurology, dementia praecox, mental hygiene, pathology, psychotherapy, epilepsy, brain anatomy, teaching methods, aphasia, childhood education, psychiatric epidemiology, and psychoanalysis. His resume and bibliography were also printed in the *American Journal of Insanity*, the specialty's foremost publication. This body of work confirmed the criteria that had earned him the job offer: his elite European training and proficiency in laboratory methods and his allegiance to the ideals of scientific research, clinical medicine, and community service upon which Johns Hopkins was founded. Asked for a comment on the choice of director for the new psychiatric clinic, Welch assured journalists that Meyer's qualifications were unmatched.[17]

Meyer understood what his newfound institutional power offered him. "I do not see how I could look forward to any greater opportunity," he wrote in 1908, "of spreading the kind of influence on the growing generation which I have so much at heart." As psychiatrist-in-chief at the country's foremost research university and professor of psychiatry to its elite medical recruits, Meyer acquired the opportunity to direct the general orientation of the specialty, both methodologically—toward general pathology—and conceptually—toward his model of psychopathology as failed psychobiological adaptation.[18]

A Psychiatric Clinic for Adolf Meyer

Both the Phipps Clinic and its new director embodied the hopes of the Progressive Age for a scientific solution to what Osler described at the clinic's opening as "the saddest chapter in the history of disease—insanity, the greatest curse of civilized life." When it was completed in 1913, there was little doubt that the handsome, five-storied, Queen Anne–style structure housed all the provisions necessary to pursue psychiatric research, training, and treatment according to the principles of scientific medicine. D. K. Henderson, the clinic's first resident psychiatrist, recalled that Meyer and the clinic's architect, Grosvenor Atterbury, "had planned every part of it" with tremendous thought and care to the "clinical, teaching, research and treatment facilities which set a standard for all to follow." It occupied the southeast corner of the medical campus, behind the hospital's domed administration building and across from the ward pavilions, which stretched east toward Jefferson Street. According to the hospital's annual report in 1913, Henry Phipps had paid more than one million dollars to build and subsequently equip and staff the clinic for the next ten years.[19]

Beyond the ornate lobby and waiting rooms just inside the front doors, an open courtyard and garden separated two wings—the East Wing for women and the West Wing for men—each of which comprised four floors of wards (the wings' internal divisions are discussed in Chapter 5). Teaching, research, and treatment spaces filled the main edifice. The second floor contained a library stocked with the fruits of Meyer's European book-buying missions, as well as rooms for the director, for transcribing and storing patients' medical records, and a lecture hall with one entrance for students via the street and another leading from the wards to fetch patients for clinical lectures. The third floor was occupied by a histological and a psychological laboratory with ten additional research rooms and two dark rooms. It was here that Curt Richter, the experimental psychobiologist appointed by Meyer in 1922, pursued his many well-known discoveries for over five decades. Provisions on the fifth floor for an animal hospital and surgery also became Richter's domain.[20]

An outpatient dispensary (which would be called a walk-in clinic today) was located in the basement of the Phipps Clinic. It resembled a general dispensary, with a waiting room, long wooden benches, and several examination rooms. The basement was also devoted to therapeutics (the focus of Chapter 5). There was a hydrotherapy suite with state-of-the-art baths and showers, and a gymnasium for calisthenics, weight lifting, and basketball. Most of the top floor was occupied by a spacious recreation hall equipped with a stage at one end and room to seat a hundred people on folding chairs. The space was used for concerts and lectures organized for the benefit of patients, as well as for dance and exercise classes. At the end of this auditorium there was a large pipe organ. The decision to install this expensive instrument might have been linked to the need to provide religious services to patients who were unable to leave the hospital; at the same time, being an accomplished organist whose socialist politics left him disinclined to join a congregation, it is possible that Meyer requisitioned the organ hoping to play it himself on occasion.[21] There were roof-top terraces for leisure activities, a billiards room and wood shop for male patients, and a women's activity room for sewing and crafts.

Given Meyer's longstanding and outspoken insistence that American psychiatry adopt the clinical model of Emil Kraepelin, observers might have assumed that his clinic would replicate Kraepelin's in Munich. Kraepelin—who had been slated to speak at the opening celebrations before the date was changed to accommodate the only-slightly-more-important

William Osler—had reviewed plans for the Phipps Clinic with Meyer in 1910, and his recommendations were incorporated.[22] The clinical agendas of these two friends were very different, however. In addition to experimental psychology, Kraepelin's research aims were to develop reliable diagnostic techniques based on a stable classification system, something he pursued by charting patterns of periodicity in his patients' symptoms and remissions over extended periods. This longitudinal approach was available to Kraepelin because his university clinic and regional asylums were both regulated by the German state, giving him continued access to patients over long periods. His Munich clinic became what one historian characterizes as a transit station for expeditiously examining, observing, and documenting high volumes of patients before redistributing them, based on prognosis, to secondary institutions (such as a state asylum or one of Kraepelin's private hospitals for treatment). In the United States, nonacademic institutions called reception hospitals emerged which similarly diagnosed and dispersed patients. That is not what Meyer intended for the Hopkins clinic.[23]

The Phipps Clinic reflected Meyer's goal to mobilize a clinical science of psychiatry based on psychobiology—a space to study the conditions under which mental disorders developed and might be modified. To facilitate his objectives, he required a clinic that was integrated into the community it treated, advised, and studied. Like Kraepelin, Meyer wanted voluminous clinical observations that were meticulously documented for future indexing and comparative analyses. And, as Kraepelin did, he hoped to study the presymptomatic and remissive phases of mental disorders, as well as periods of acute illness. Meyer wanted to attract patients long before they required a diagnosis and, especially, a legal declaration of insanity. In 1913, he explained that psychiatry and psychopathology must become "preinstitutional," playing a productive role in workplaces, schools, and in families facing the "possible disastrous experiences" of managing mental illness. "This decentralization and refocusing of psychiatry in the centers of community life and research and teaching," he considered "the most important new aim of the new type of institutions."[24] In 1915, he compared a psychiatric clinic to an agricultural station and called for the erasure of the boundary between psychiatric treatment and mental hygiene (a term that, for him, was synonymous with mental health). "A small community," was how he described such a psychiatric research station, "trying to determine in a quiet unostentatious way the value of the various claims in vogue." In other words, testing various scientific claims in psychiatry by applying them to patients and observing and documenting the results.[25] Similarly, Meyer spoke of establishing "psychiatric missionary settlements" modeled on settlement houses like Hull House in Chicago. The Phipps Clinic was therefore designed as a living field laboratory for exploring the uses and limits of psychobiology. Far from transiting patients expeditiously, he often negotiated with referring physicians to keep patients for an observation period of two to four weeks.[26]

Early and continued contact with all types of cases in every stage of illness, he predicted, would finally generate for American psychiatrists the kind of longitudinal data that led Kraepelin to his groundbreaking conclusions. In the asylum, cases were too numerous to document adequately, and mild or preliminary forms of psychopathology were not represented in that population, since legal commitment was considered a last resort. Indeed, in 1896 Kraepelin had warned Meyer that his clinical research model could not be implemented effectively in a large institution (which did not stop Meyer from trying).[27] "The peculiar asylum habit of dealing with the insane in hordes," the optimistic Meyer finally admitted, once he had his own clinic, "has retarded the application of what conclusions we are able to derive from the experience." In light of Kraepelin's successes comparing case histories and outcomes, however, he always regretted that more data were not collected in asylums.[28] Meyer's amendment to a description of the Phipps Clinic, prepared by someone else for publicity purposes around 1909, encapsulated his ultimate aim to establish a clinical science of psychiatry at Johns Hopkins. He expunged the unrealistic claim made by the original writer that the clinic would "open new gates of hope of permanent recovery from mental troubles," and in its place he inserted "will put mental disorders on the same ground as every other problem of human health and disease."[29] As a physician, Meyer wanted to ameliorate and prevent illness (though he did not anticipate curing high numbers of cases). As a pathologist, he sought to identify abnormality and to study the factors that caused or perpetuated it. The purpose of his clinic, he told colleagues in 1914, was to ensure that all phases of mental disorders received the medical attention and scientific scrutiny accorded to any other disease or medical condition.[30]

Readjusting faulty psychobiological habits and pursuing clinical research using Kraepelin's methods both depended upon gaining access to individuals before and after the symptomatic stage of mental disorder. On the therapeutic side, Meyer considered a person's willingness to accept medical help for mental troubles indicative of a phase in which mental abnormalities might be modified or prevented. It was a disgrace, he charged, when families squandered extravagant sums on private hospitals and medical consulting fees while the patient was so agitated, delusional, or depressed that he or she was unable to benefit from the interaction with the doctor or hospital environment.[31] Given the proximity of the Phipps Clinic to the communities it served, he wrote in 1913, he could establish "a relation between the hospital and the home, along the line of prevention and aftercare." On the investigative side, the patient's ability and willingness to provide a medical history, recount life experiences, and collaborate in talk therapy would furnish clinical data that Meyer considered essential to a general pathology of psychobiological reactions.[32]

As he attempted to expand public and medical conceptions of which symptoms and behaviors warranted the use of a psychiatric hospital, he also had to publicize the message that, as in a general hospital, all admissions were at the discretion of physicians. Disturbed individuals were not to be brought to the clinic unannounced, as was customary at a state asylum. Unlike Kraepelin's clinic and the psychopathic hospitals in Ann Arbor and Boston, the Phipps Clinic did not accept individuals delivered to its doorstep by city or state authorities. This autonomy, according to Meyer, made it unique. "It is the first university clinic, in this country at least, free to receive patients according to the judgment of the staff." Admissions were "facilitated by the ambulance of the Johns Hopkins Hospital, consisting of a limousine motor car, provided with nurses," he explained to colleagues in 1913. Individuals in police custody were transported from the station house either by ambulance or by relatives. Meyer's capacity to select which cases to admit or refuse was integral to the investigative and therapeutic strategies he was able to implement at the Phipps Clinic in the years before World War I.[33]

He considered it important to both his research goals and the therapeutic environment to build a patient population that represented a broad range of complaints and disorders. To discern meaningful patterns of pathological reactions using Kraepelin's clinical method, Meyer needed to compare sizeable numbers of all forms of psychopathology. In terms of therapeutic advantages, he reasoned, in a hospital setting "the patient sees others who may have seemed worse off get well, and he can get an inspiration from that and give an inspiration to others."[34] In 1914, he advertised to colleagues that the clinic would accept "any type of case for study, observation, diagnosis, and the outlining of the plan of treatment." The same year D. K. Henderson also instructed local Baltimore doctors in what kind of case they should refer to the Phipps Clinic: "any patient in whom a mental break-down is induced by either mental or organic factors, from so-called nervousness and instability to the most pronounced cases of mental disorder." The new clinic, Meyer told the *Baltimore Sun* in 1912, would be prepared to deal with a wide range of conditions: fears, obsessions, suspicions, peculiar experiences, queer notions, hallucinations, moodiness and irritability, difficulty of attention and memory, fear of insanity, depression, nervous spells, hysteria, fits which affect the mind, multiple personalities, alcoholism, cravings for drugs, and "the more clearly recognized insanities." Quantitative analyses of the diagnostic data in case histories produced between 1913 and 1917 confirm that the patient population of the Phipps Clinic in this period reflected the diversity Meyer sought.[35]

Citizens of Baltimore and elsewhere in Maryland were encouraged to consult a psychiatrist without an appointment at the Phipps outpatient dispensary regarding any unusual mental condition. General dispensaries, such as the large Monument Street Dispensary at Johns Hopkins, were common in American cities at the end of the nineteenth century. Divided into specialized departments, many general dispensaries had a neurology section like the one Meyer had supervised in Chicago twenty years earlier. It was through the Phipps dispensary, as well as the clinic's social work department, that Meyer hoped psychiatrists and patients might establish ongoing and reciprocal relationships with each other on an outpatient basis—his staff dispensing medical advice and patients providing new data to be recorded for future analyses. "The facts can be collected," he explained, "and where possible a close acquaintance between the patient and physician can be established, pending a careful investigation and possibly adjustment of the home environment."[36]

In order to attract and monitor the wide cross-section of cases for which he hoped, Meyer had to reconfigure the public's negative perceptions of psychiatric disorders and institutions. He set out to challenge common beliefs about hereditary stigma, the incurability of insanity, and state institutions as horrific almshouses for the wretched and destitute (much as he knew such perceptions contained elements of truth). As always, he was motivated equally by his quest to inaugurate a general pathology of psychobiological reactions (for which he required clinical observations of living patients) and his professional and personal experiences of the struggles faced by mentally disturbed persons and their families. Within the medical academy generally, derisive and nihilistic attitudes toward psychiatry limited the resources available to its clinicians. It was a discipline "hampered on all sides, always with half-officered hospitals, half-trained staffs, half-equipped laboratories," Meyer despaired in 1915. "When will our intelligent leaders begin to realize the agony to which we psychiatrists are put when we have to watch, too often helpless, the long illness and frequently the irresistible fading out or perversion of human minds, sometimes in the wives and husbands and children of our nearest friends and sometimes members of our own family?" Misplaced fears of mental illness and its institutions, Meyer and his staff believed, fueled apprehension among the public. While training and undergoing psychoanalysis with Carl Jung at the Burghölzli in 1910, Phipps psychiatrist Trigant Burrow sent his mother important instructions: "Mother, *please never* say to any one that I am interested in the *insane* or visiting an insane asylum." Rather, he was interested in correlations between minor neuroses, such as neurasthenia, and psychoses. "No one, *especially no one in America*," he advised his mother, "wishes to believe that there is any analogy, however remote, between a neurasthenia and a true insanity." Macfie Campbell remarked dramatically about hereditary stigma, "[W]e are not free from the trammels of medieval thought." Outmoded and distorted notions about chronic or inherited insanity, Meyer persisted, must be replaced with a more optimistic view of mental disorders as potentially correctable maladjustments.[37]

He regarded the obviation of asylum commitment, when possible, as a key function of the Phipps Clinic. "There is no doubt," he maintained, "that with proper facilities many a patient can be tided over difficult periods or be started on adequate treatment *without an official declaration of insanity*." A potential patient's willingness to undergo voluntary hospitalization was an important criterion in admission decisions at Phipps. "If we are able to get access to the cases," he proposed, "at a time when they are still in a condition of cooperation, the chance for recovery must be very much better."[38] The typical Phipps patient signed a promissory note upon admission by which he or she agreed to "abide by the rules and regulations of the Institution and the detention thereby enforced" and to provide "at least three days' notice" of their "desire to leave the hospital." In fact, the right to request discharge from a mental institution was already a law in Maryland. In 1914, Meyer boasted to colleagues that "only twelve out of 370 patients were held under commitment" at Phipps during its first year in operation. The importance of avoiding involuntary hospitalization was also shaped by his conviction that mental disorders were medical conditions and should be treated as such.[39]

Meyer recognized, as did other experienced alienists and neurologists, that psychiatric disorders carried fundamentally unique challenges for medicine and that psychiatric patients were not typical convalescents. He explained this reality in terms of psychobiology. Medical and nursing procedures in "the average general hospital" were based on "the supposition that any normal person has internal resources with which to stand forced rest or restrictions imposed by disease."[40] For example, under the healthy regulating function of mentation, a surgical or tubercular patient—even a youngster whose broken limb must be set in a cast—inhibited the instantaneous gratification of primal impulses (for example, to escape physical discomfort or avoid the trepidation that instinctively accompanies forced confinement) in favor of a more beneficial psychobiological response (the benefits of medical intervention). The human ability to delay gratification in such a way required the synthesis of past experience, present demands, and future rewards. "When we suffer from digestive or cardiac or infectious disorders," he pointed out, "knowledge of evil consequences makes us go to the physician and to hospitals, and we behave in a manner more or less receptive towards helpful advice and treatment."[41] Psychobiological explanation or no, it was (and is) a conundrum faced by all societies in which madness is equated with sickness: how to treat medically those deemed sick but who refuse medical help? In his private scientific notes, Meyer reflected that this was the most challenging problem in psychiatry.[42]

Meyer believed that a psychiatric hospital must provide for both cases of nervousness, typically involving anxiety or obsessive thoughts and compulsions, and active psychoses, which often induced erratic or violent behavior. "Neither type has so far been welcome in general hospitals," he conceded, "partly on account of certain risks, like risks of suicide, and partly on account of the difficulty of management and the occasional transgressions of hospital rules by noisiness." General hospitals, he acknowledged, routinely discharged patients with symptoms of insanity who did not cooperate, but Meyer reasoned that this was nonsensical. Many psychiatric patients experienced a "noisy and objectionable" phase when acutely ill but eventually recovered. "If we made our criterion of admission absolute submissiveness," he argued, "we should have to eliminate many of those cases who today fare very poorly, but who are not at all less hopeful and grateful patients in the end." He likened his authority to confine patients involuntarily to that of the physician's or health officer's responsibility to quarantine individuals with infectious diseases: "Nobody has a right to go where he pleases with diphtheria or small-pox. Nobody has a right to display himself in the street in a delirium. Nobody has a right to attempt or to commit suicide." Most importantly, according to Meyer, the decision was a medical one. "Danger and threat of suicide can be foreseen by any experienced physician," he wrote, "but a judge may want evidence of unmistakable threats and attempts made." Only a physician should rule on whether to hospitalize a patient involuntarily—not a judge, not a committee, and not a family member, Meyer insisted. He felt strongly that physicians needed to decide who occupied hospital beds.[43]

Though Meyer certainly anticipated a transfer of professional power within American psychiatry from asylums and superintendents to medical universities and clinicians, he did not suppose that psychiatric clinics would replace state and private institutions. Traditional asylums and sanitariums were essential, he maintained, due to the unpredictable and protracted nature of mental disorders. When his clinical judgment indicated that a Phipps patient required temporary or long-term institutionalization elsewhere, he bristled when he met with resistance from the family. Indeed, families, having internalized cultural fears of hereditary insanity and asylum horrors, were often shocked by this recommendation. Some offered to double his fees if he would extend the patient's Phipps admission indefinitely. One concerned mother questioned the necessity of committing her twenty-year-old daughter to the Bloomingdale Asylum in New York State. "I hate to put her among the lunatics," she fretted. Meyer's rebuke was uncharacteristically severe (perhaps because his own mother was again very disturbed and hospitalized in the Burghölzli). The term *lunatic*, he told the woman, was a harsh and indiscriminate slur. "They are all patients for whom the physicians are trying to do their best," he continued, "and your daughter is decidedly one of the more difficult and most profoundly affected patients they could want to admit." The young woman was in a "manifest phase of her illness" and needed "medical attention by experienced people," he would report to the physician at Bloomingdale. Meyer urged hospitalization. He told the mother to put her energies into helping "eradicate the erroneous conception concerning mental diseases and their successful treatment." In the end, the woman did commit her daughter. Moreover, Meyer's tirade had its intended effect. Two months later, she was gushing about the hospital's beautiful grounds and concerned staff. "I feel very differently since I have been up at Bloomingdale so often and seen the patients," she told him, "and could find nothing shocking about them." Echoing his directive, she pledged to do her best "to eradicate the stupid and wrong idea" regarding asylums.[44] "One of the best contributions I can make," Meyer announced at the annual meeting of the American Medico-Psychological Association in 1914, "is to draw our state institutions from their isolation in the eyes of the public and the profession." He was confident that the Phipps Clinic, especially by virtue of its affiliation with Johns Hopkins, would alter public and medical perceptions about psychiatry, its methods, and the medical problems it could address.[45]

On 1 May 1913, the Henry Phipps Psychiatric Clinic at the Johns Hopkins Hospital was at last ready to receive patients. After five years of theorizing and envisioning, planning and building, organizing and hiring, Adolf Meyer had his psychiatric clinic. As the *Baltimore Sun* reported, that day a single patient was admitted—a fifty-one-year-old man from the city whose wife brought him to the hospital. He had lost consciousness and awoken confused on several occasions of late, including once on the streetcar on his way to his work. He had the undivided attentions of eight head nurses and several psychiatrists that night in the admission ward, but more patients followed, and by summer's end most of the clinic's eighty-eight beds were filled.[46]

A Psychiatric Clinic for Sick People

The Phipps Clinic was anticipated by potential patients, their families, and their local physicians. In January of 1913, with the clinic's opening still months away, Meyer received a letter from a man saying, "I have been waiting for four and half years for the Phipps Psychiatric Clinic to be open, so that I might take treatment."[47] A year after its opening, readers of the *Baltimore Sun* saw the headline "Big Demand On Phipps Clinic," and during the next year, Meyer reported to the hospital's superintendent that the clinic could not accommodate all applications for treatment.[48] Because of its small size, urban location, well-appointed décor, and modern hospital facilities—traits that contrasted with those of traditional asylums—one might infer that the Phipps Clinic catered to moneyed invalids desiring treatment from an eminent nerve doctor, rather than to ordinary families or seriously disturbed individuals. Indeed, Meyer revealed in 1916 that the trustees of Johns Hopkins had originally instructed him to admit "cases in the mild stages" and not to accept "really insane" patients. He ignored the directive.[49]

Between 1913 and 1917 the Phipps Clinic attracted and admitted a socially diverse group of patients with a wide range of complaints and symptoms and of motivations for using the new institution. Of the 1,897 admissions during this period, 47 percent were women and 53 percent were men. Since the clinic's eighty-eight beds were rarely empty before World War I, this reveals little about relationships between gender and admission decisions. The average admission was fifty-six days, and the longest stay was that of a thirty-nine-year-old taxi driver from Maryland hospitalized for 1,015 days. Trades people and clerical workers accounted for 17 percent of admissions, laborers 8 percent, and 9 percent for those in occupations that had acquired professional status by the nineteenth century (for example, medicine, law, nursing, dentistry, engineering, architecture, accounting, or teaching). The remaining 29 percent of income-earning patients included merchants, farmers, clergy, and commercial and government service workers. Thirty percent of all admissions were female patients without an income, a group that also reflected social diversity. Information about nationality, religion, and ethnicity was not documented systematically. In recorded cases, patients described as Protestant, Methodist, or Episcopal constituted approximately half, with the remaining half split evenly between Catholic and Jewish. Represented sparsely were Greek Orthodox, Dunkard, Christian Scientist, Quaker, and atheist. Of those patients for whom a "race" was recorded, most were described as white, a significant number as Hebrew, and some as Russian, Bohemian, or Irish. Unlike the Johns Hopkins Hospital, the Phipps Clinic did not admit so-called colored patients. It is unknown whether this was due to budgetary restrictions (segregated wards would have had to be provided), the personal wishes of Meyer or of Henry Phipps, or some other factor.[50]

The national prominence of the new clinic meant that Meyer received requests for advice and pleas for help from all over the country. After seeing his name in the evening newspaper, a Baltimore man wrote to Meyer asking for assistance. He confessed that, due to his insufferable nervous trouble, "life has no charms for me." He lived with his widowed mother and, he told Meyer, it was "only for her sake that I am trying hard to forget my suffering." Within the medical profession, Meyer was, like his fellow Johns Hopkins specialists, an obvious authority, to be consulted regarding a difficult case or to whom to refer a patient. One man claimed that his doctor told him "that the Phipps Clinic is the only place for him, and that he himself (the local doctor) could not diagnose the case." In one instance, a physician of national repute traveled from the Midwest with his daughter (who was in a violent psychotic state) to admit her to Phipps as a private patient of Meyer's.[51]

Referrals to the clinic came from various sources. Sometimes an individual's appreciation of his or her own odd thinking or uncontrollable conduct triggered a personal decision to contact Meyer or visit the Phipps dispensary. Other times someone close to an individual—a family member, employer, or physician—contacted the clinic after having a peculiar conversation with the person or witnessing disconcerting behavior. Twenty-six-year-old Steffi, for example, was fired from her manufacturing job because of her odd conduct and boisterous outbursts. According to her Phipps case record, she had informed the mayor of Baltimore and the governor of Maryland in writing that "a spell had been cast on her by the machinery of a noisy factory across the street" and it had "interfered with her thoughts and compelled her to do things over which she had no control." Steffi's behavior had come to the attention of neighborhood police on several occasions. When her mother, with whom she lived, kept her indoors, however, "she sat and stared into space, brooding, or laughing foolishly and unaccountably to herself." Steffi's mother worried that local authorities would insist that her daughter be placed in an asylum and hoped that doctors at the new "nervous hospital" at Johns Hopkins might cure her. She and Steffi walked to the clinic, which was only a few blocks away from their house.[52]

Local use of the clinic was facilitated by the outpatient dispensary. Of the 1,897 admissions during the first five years of its operation, 49 percent were from the City of Baltimore and 11 percent were Marylanders from outside the city. This emphasis on local use was intentional, not only in terms of Meyer's vision of a psychiatric settlement house, but according to the wishes of Johns Hopkins himself. Like its parent institution, the Phipps Clinic was expected to serve the urban, often impoverished population surrounding it. In 1914, Meyer reported to colleagues that more than one hundred of the 370 admissions were local "free patients" and 25 percent were patients from Maryland who paid subsidized rates of less than $10 per week. After the first year of operations, Henderson reported as "a matter of great satisfaction" that "no patient from Baltimore has as yet been refused admission because he could not afford to pay anything." For families like Steffi's who were confronting the vicissitudes of mental illness, the opening of the

Phipps dispensary in 1913 created an accessible alternative to a traditional choice: enduring either the disruptions (and sometimes dangers) of severe mental disorders or a legal declaration of insanity and asylum commitment.[53]

With Baltimore and Maryland residents constituting 60 percent of the admissions, the remaining 40 percent were primarily from neighboring states Virginia, Pennsylvania, West Virginia, New York, North Carolina, and Washington, D.C. Patients from outside Maryland were charged $25 per week and, along with free and subsidized local patients, were admitted to one of two communal public wards (by comparison, nearby Mount Hope Retreat operated by the Sisters of Charity charged between $8 and $15 per week).[54] Meyer was sufficiently proud of the public wards to assure a concerned husband, worried about the cost of treatment for his wife, "I would not hesitate much in recommending it to any member of my family." He would not have dared to make such a claim about any of his former places of employment—three of the largest overcrowded public asylums in the country.[55] Eighteen of the clinic's eighty-eight beds were allocated for the private patients of Meyer and the assistant director, Macfie Campbell. Private rooms with en suite baths were located on the fourth floor. Like the other departments at Johns Hopkins Hospital, revenue from private patients constituted a significant part of the clinic's operating budget and ranged from $50 to $100 per week.[56]

A decision to utilize this unfamiliar type of psychiatric institution might be made by an individual, family member, general physician or community agency. "My sister has just called me up informing me that she has received a letter from the Phipps Institute, advising her of a vacancy," one man wrote to Meyer "[S]he is very anxious to enter the Institute and I believe that a stay there will have beneficial results." A fifty-six-year-old cobbler, convinced that his wife and son were plotting to shoot him, awoke another son at 5 a.m. and "begged for help—that he was sick and something should be done for him." He appeared at the dispensary unaccompanied a few hours later.[57]

The majority of patients decided to enter the hospital in consultation with their relatives. One Midwestern couple came with their son, Beryl, to consult Meyer about the young man's compulsive and debilitating delusion that a hazardous odor emanated from his genitals. After the meeting, the three returned to their Baltimore hotel to discuss Meyer's recommendation that the boy enter the hospital. A few hours later, Beryl bade farewell to his parents and took a taxi alone to the clinic, and he was admitted. John, a fifty-four-year-old minister and farmer from a southern state, arrived at the clinic accompanied by his uncle. On the way, John's uncle had harangued him about his many shortcomings and, once there, declared to the admitting psychiatrist that "if he had had more whippings as a young boy, he wouldn't be here." The uncle also made clear that he resented his familial obligation to pay for his nephew's hospital treatment. A woman who had worked in several nearby asylums took the streetcar to the clinic accompanied by her sister, but once there she sobbed for fear that she would be institutionalized herself, after which everyone would know about her trouble. Asked what that trouble was, she explained, "I've had it pretty rough—everything I've undertaken has failed." Her sister convinced her of the value of hospitalization at Phipps.[58]

Members of kinship and community networks beyond the immediate family also opted to utilize the Phipps Clinic as a resource. The manager of a utilities company summoned the brother and cousin of one of his employees, instructing them "to bring him to the hospital" after "the sudden falling off of his work and the appearance of odd behavior." A twenty-year-old telegraph operator who suspected his coworkers of misdoing barged "into the main office and said that the boys were after him," after which his manager recommended he be escorted to the Phipps Clinic. Another man, a carpenter from Baltimore, was ordered by a social service agency to report to the Phipps dispensary "because he behaved queerly in their office." Just because a community member persuaded or instructed an individual to appear at the clinic, however, did not necessarily mean the person in question was eager to see the doctors there. For example, a twenty-seven-year-old artisan from Switzerland who worked for his uncle in Baltimore reported to the Phipps dispensary when "his dentist noticed that he talked peculiarly about religious matters and sent him to the hospital." The young craftsman, Kasper, made a noteworthy impression on the staff when he arrived. "The manner of his entrance," the psychiatrist observed, "was very dramatic." Kasper wore a hat, the rim pulled low, and had bound a handkerchief over his eyes as a mask. "He later explained this by saying that his eyes had been hypnotized and he wore the mask to prevent others from knowing it." He reported that his dentist "gave me the address of a hospital" and said "you better take care of yourself." He carried a letter from his dentist, who wrote that he would telephone Dr. Meyer later to discuss the situation. Kasper had reported dutifully, as instructed, but he paced in front of the clinic for some time, perhaps contemplating the implication of the stone tablet above its door: *Henry Phipps Psychiatric Clinic.* "After some argument," the psychiatrist recorded in Kasper's case history, "he was persuaded to come into the Dispensary, and about 4 p.m. was admitted." It required some effort, evidently, to convince Kasper of the safety or value (or both) of speaking to the kind of doctor that might practice in such a hospital.[59]

Individuals who exhibited psychosis or paranoia also made choices about using the clinic. In some cases, like Kasper's, it was by simply tolerating hospitalization; in other instances, the patient actively sought and participated in medical treatment. Joe, a thirty-four-year-old bookkeeper from Atlanta, spent several years traveling from city to city in an attempt to outrun a secret society whose members tormented him wherever he went. When this conspiratorial surveillance continued in Baltimore, he sought assistance at the Phipps dispensary. An account that he wrote of his experiences, entitled "Statement to the Public," was appended to his case history. In it, Joe described how one day he was standing on a corner in St. Louis and noticed that the street sweepers continually spat or coughed—deliberately to unnerve him. "They were put up to it by the boss who I found out was a member of this lodge," he explained. "Now at first this may seem absurd," he acknowledged, "but let it go on for months and months and you would find it surprising the way it will affect one." Joe wished the public to know that he was "willing to admit that I have always been very nervous and have been treated for it at Johns Hopkins Hospital." He conceded that people probably attributed his worries to "the imagination of a disordered mind" and, he reassured his readers, "that's what I am afraid of, so am writing this before I do lose my mind through worry." Concerned that the stressful effects of constant surveillance were threatening his nervous health, he sought medical treatment. He was seen periodically by psychiatrist Trigant Burrow in the Phipps dispensary for eighteen months, where (according to both Joe and Burrow) he improved. When Joe's paranoid delusions intensified, however, Burrow convinced him to admit himself to the hospital for inpatient treatment.

From his bed, Joe explained to Meyer, "I was one year with Dr. Burrow, which nearly cured me," but he reported that he rarely slept, due to the torment of the secret society and its cronies. The Phipps staff noted that "most of his time on the ward was spent in bed with the bedclothes pulled over his head." Seven days after he was admitted, Joe penned a simple but unequivocal request on Johns Hopkins Hospital note paper and addressed to no one in particular: "You told me when I entered this Hospital I could leave if I would give three days notice. I, herewith, wish to give this notice." He was discharged and it is unknown whether he continued to use the dispensary.[60]

There are several important points to be gleaned from this snapshot about how mentally disordered individuals brought themselves to the Phipps Clinic. First, believing that at every turn a secret society conspired against him, Joe looked to the psychiatric dispensary and its physician as legitimate sources of help. He may have been directed there after seeking aid elsewhere—at a police station or general dispensary—but *he* decided that a psychiatrist might be of some assistance. Second, not only did Joe conceive of his predicament as amenable to medical intervention; he clearly valued the outpatient treatment he received, because he utilized the dispensary for over a year and asserted that Burrow had nearly cured him. Despite Meyer's strong recommendation that he remain in the hospital, however, he tried but decided against inpatient treatment. In various ways, then, Joe shaped the manner in which he utilized this new kind of psychiatric institution.

Another young man struggling with paranoid thinking was Saul, a twenty-one-year-old bank teller working in New York City, who took the train to Baltimore to consult doctors at the new psychiatric clinic at Johns Hopkins. Of late, he had become convinced that his mother was poisoning him and carrying on an illicit affair with his uncle. He also suspected that his superiors at work, as well as several well-known tycoons, were conspiring to prevent his advancement in the banking world. The dispensary psychiatrist noted, "patient feels that he is confused—perhaps over-suspicious, and feels extremely worried and muddled—he has come to know whether he is crazy." Saul returned a few days later and admitted himself to the clinic.[61]

General hospitals in Baltimore City, as well as private and state asylums within Maryland, referred patients to the university clinic for diagnostic evaluation and specialized care. Richard, the nineteen-year-old son of a hardware store owner, spent ten days at a local sanitarium after becoming abusive and profane with his father and the customers at the family shop. Although Richard "improved nicely" under care, the owner of the private asylum recommended that Richard be examined by Meyer. The physician traveled with Richard and the father to the Phipps Clinic to make arrangements for admission. The family of another patient, Ruth, took her to Women's Hospital in Baltimore because "she frequently aroused the neighbors by screaming during the night." "Contrary to advice, the family refused to commit her to any mental institution." The unmarried, middle-aged daughter of a prosperous Jewish family in Maryland, Ruth rarely ate and was emaciated. Her longtime family doctor described her typical demeanor as "erratic, inclined to be grandiose," and he informed Phipps staff that years earlier she had been successfully treated for a "morphine habit." Women's Hospital was not equipped to cope with agitated patients and she was "so noisy and hard to manage" the hospital transferred her to the Phipps Clinic.[62]

Patients were referred to Meyer or to the Phipps dispensary by their family doctor or local specialist. One Baltimore physician explained to Meyer that a thirty-year-old patient was "very agitated mentally and a violent love affair seems to be the exciting factor." Referring the man to the Phipps Clinic, the doctor said, "I felt that he needed to be taken actively in hand." It was also common for a family physician to make arrangements on behalf of the family, including the travel or services of a private nurse, as well as making all financial and medical decisions.[63]

Individuals, families, physicians, and community agencies all turned to the Phipps Clinic inpatient, outpatient, and consulting services for assistance in dealing with mental illness. As Meyer had hoped, the university psychiatric clinic—emulating, as it did, the respectable general hospital and obviating the requirement of a legal declaration of insanity—evidently represented an acceptable or advantageous option for many of those faced with the vicissitudes of mental disorder.

A Psychiatric Clinic for Psychobiology

The Phipps Clinic also served psychobiology. It was crucial to Meyer's investigative and therapeutic schemes, relying as they did upon the common sense consensus, that the patient perceive his or her condition as an illness treatable by medical experts working in a modern hospital. Meyer reckoned that the high esteem in which the public held the modern general hospital was invaluable to his efforts to study and modify individual social function. Once inside the clinic, the hospital setting and the medical authority embodied by Meyer and his staff became tools of clinical investigation and therapy. For centuries, so-called mad doctors had utilized personal authority and environmental conditions to influence their charges. In Meyer's case, according to the principles of psychobiology, he would attempt to induce the patient to modify his or her psychobiological habits to realign with the social reality of a modern hospital. The day-to-day machinations of this strategy are explored in the final three chapters of this book.[64]

From the perspective of psychobiology, it was critical that severely disordered individuals be seduced, not forced, to enter the hospital. Meyer traded on his conviction that a hospital environment—with its scientific experts, trained nurses, and modern technological therapies—would appeal as genuine medical help even to patients experiencing delusions or hallucinations. He told fellow physicians, "[W]hen we meet the feeling of exaggerated well-being of an expansive patient, or the cocksureness of a person with a beginning paranoic development, cooperation becomes clearly possible only if we succeed in making our help attractive and acceptable even to the shaken or twisted confidence of our patients." He suggested asking the psychotic or delusional patient what prompted the medical consultation. If the patient disclaimed the notion that he or she was ill, the psychiatrist might try: "Your friends are anxious about you because you are nervous or changed—will you tell me how this situation developed?" According to Meyer, it was the task of the psychiatrist to inspire "a spirit of hopefulness and helpfulness" by providing an attractive "open door" to hospital conditions. "I realize the difficulty," Meyer conceded, "but I rather think it is best that physicians should have to exert some efforts and exercise their imagination." He felt strongly that resourceful psychiatrists, nurses, and attendants could make their help appear advantageous to most patients.[65]

Having agreed to enter the hospital for study and treatment, an individual was formally admitted to the Phipps Clinic as a psychiatric patient. An important ritual of transformation—from troubled or nervous person to hospital patient—was enacted (or at least attempted) at the beginning of every admission. An admission ward was located on the ground floor of the East Wing (for women) and of the West Wing (for men). This ward could be accessed from the main corridor, nearby the waiting rooms in the lobby, and also from the rear of the clinic to accommodate the arrival of noisy, resistant, or infectious patients. For most patients, hospitalization at the Phipps Clinic began here, although some private patients were probably admitted directly to their rooms. Unpopulated, the admission ward was indistinguishable from one of the general wards of Johns Hopkins Hospital. It contained eight beds in two rows of four on each side of the room with the head of each below a window. Scattered about were folding screens and metal bedside tables on wheels, and everywhere there was crisp, white fabric—on the beds, the pillows, folded on the shelves and, importantly, adorning the nurses and doctors that awaited each new patient.[66] Like their colleagues in the general hospital, Phipps nurses donned floor-length white

aprons and starched caps, and clinicians sported the white coat that signi-
fied their profession, embroidered with the identifier "Phipps Clinic, J.H.H.
Staff." A dream recounted by a patient confirms that the modern hospital
environment and its attendant personnel indeed carried meaning for patients. "I dreamed Dr. Hammers came in the ward downstairs—dressed in
white—to perform some operation on me," she recalled, "and Dr. Hammers
had on [night nurse] Mrs. Beiter's cap."[67]

Accompanying the ascent of the modern general hospital and the revolution in patient care was an army of professional nurses with training comparable and complementary to that of scientific medicine in the period. Effie J. Taylor was in charge of nursing at the Phipps Clinic and answered only
to Meyer and the hospital's superintendent of nurses. Born in Canada and a
graduate of the Nurses Training School at Johns Hopkins, Taylor played a
pivotal role in the day-to-day management of the clinic, as she supervised
the work of all nurses, attendants, occupation instructors, and cleaning
staff. She also worked closely with Meyer before the clinic opened, furnish-
ing wards, equipping occupations rooms, and hiring staff.[68]

The imperative to medicalize the patient's conception of his or her condition imbued the Phipps admission process with social meaning and therapeutic function. Each new patient—paying or free, calm or raving, male or
female—was asked (or persuaded or coerced) to undress, take a warm
"cleansing bath" and to surrender his or her clothes in exchange for a hospital gown (stamped with the words *Phipps Psychiatric Clinic, Johns Hopkins
Hospital*), a robe, and slippers. This technique was called "bed therapy" and
had emerged in German psychiatric clinics in the 1880s. Meyer may have
observed it first in Kraepelin's clinic in 1896. The aim was to convince the
mentally disturbed individual of his or her illness, to produce beneficial
physiological and psychological effects. This strategy resonated harmoni-
ously with the epistemological and therapeutic goals of psychobiology.
When he reported on the work of his new clinic in 1914, Meyer explained
the importance "of helping the patients reach a correct understanding of the
meaning of such a hospital."[69] Each Phipps patient was prescribed three days
of bed rest. During the admission period, a Phipps patient was expected to
behave as any other patient at Johns Hopkins Hospital. This initial period
stood in stark contrast to the remainder of the hospitalization, in which the
therapeutic goal was to have the patient reenact everyday situations outside
the hospital. The linoleum floors, smell of disinfectant, and the nurses and
doctors milling about in their starched whites impressed upon the new

patient that he or she was not "insane" or "nervous" but sick and receiving medical care from experts. Some patients were familiar with general hospital care while for others it was a new experience; some eagerly embraced its practices and some repudiated them.[70]

Every reaction of the new patient offered potentially important data. "Admission notes on chart should show whether patient came in walking, in wheel-chair or on stretcher," Phipps nurses were instructed, and should document the patient's "mood, reaction to admission routine, reaction to nurse, to other patients, and any remarks indicative of his state of mind or feeling." A patient's utterances were to be recorded verbatim and special attention given "to allay the apprehension or dread which usually is present on admission." When Meyer asked one patient to recount the events of her admission, she told him, "I came in here and my husband kissed me when he left—then a doctor and a hefty Irish girl forced me into the bath tub." When he asked her if a bath was not necessary when entering a hospital, she promptly informed him that "I've been in four hospitals and I'll take orders when they are right." Another patient "made herself thoroughly at home, was elated, and talked and laughed constantly." She was delighted with everything and constantly referred to the spirits surrounding and protecting her.[71]

Other aspects of the admission routine were practical. During the first twenty-four hours in the clinic, every patient was monitored closely to determine the risk of suicide. An employee of Johns Hopkins, a Greek man, was brought to Phipps after he attempted to hang himself on the massive marble statue of Jesus Christ located under the hospital's hallmark dome. In the admission ward, he was "restless, getting out of bed constantly," repeating again and again "me no speak English—you good, America good, Greece good—I am no good, give me knife." On admission, each patient's personal clothing was stored in a locked room on the ward and access to it by the patient was not guaranteed. According to the nursing manual, baggage was searched for "razors, knives, nail files, scissors, sharp instruments of any kind, mirrors, glasses, spectacles, bottles, drugs, medicines, bathrobe cords, belts of all kinds," all of which were inventoried and "kept under lock and key" to prevent injuries and self-harm. Articles of clothing were marked with the patient's name, logged into the clothes book, and stored in a locked dressing room. Luggage was stored in the trunk room on the fifth floor. Eventually, most patients donned hospital-issued clothing for daily use (khaki pants and shirt for men, and plain percale shifts for women).[72]

Once nurses and attendants had bathed, gowned, fed, and put a patient to bed, the next scene of the admission unfolded: the first bedside visit with a Phipps psychiatrist. Every patient—private, house, dispensary—was asked to state his or her medical complaint. Meyer considered this a vital step in the transformation of troubled person to hospital patient, and in later years he described his brand of psychiatry as "complaint-centered." He commented, "[I]t is very striking how much more composed and amenable to explanation and conversation many patients are just at the time of admission, than only a few hours or a day later." He advised taking full advantage of this window. Asking the psychiatric patient to describe his or her medical complaint for the psychiatrist mirrored the traditional medical encounter in which a patient retains a healer to effect a cure. It thus reinforced assumptions that in the conventional patient-healer dyad were implicit: the patient's volunteerism and agency; his or her conscientious participation in helping to bring about a cure; the hope and expectation of relief; and the healer's expertise, skill, and responsibility to the patient.[73]

Meyer placed great emphasis on the rapport between psychiatrist and patient. In order to wield the common sense consensus as an investigational and therapeutic tool, the patient must be treated "as a sensible man or woman," to establish trust and win confidence. "We introduce ourselves by a few questions establishing the mutual clearness of the situation," he wrote in his textbook on examination techniques in 1918. Questions were to be considered carefully "so that they do not suggest too low an opinion of the patient's capacity." Ruses and trickery were to be avoided. The procedure called "fagging" in which an agitated patient was deliberately excited by taunting, to exhaust him or her, Meyer called "as contemptible as it is harmful." He also warned that the "approach to those topics to which the patient is most sensitive must be made with the greatest amount of tact and mutual understanding." A colleague once reflected that "to watch the skill with which [Meyer] developed a relationship with patients was to learn by example the art of interviewing and history-taking." Meyer maintained, "[W]e would be unfair to the patient if we did not allow him to present his own story and interpretation" and that the clinician should avoid interrupting the account "so as not to raise any doubt in fair play."[74]

Most patients were able to describe the experiences and events that preceded their hospitalization. Some were desolate as they did so and had to be encouraged; others became eager, even excited or agitated, when asked to explain their present situations. Sometimes they were unable to express themselves in terms understood by the staff because they were impaired by hallucinations, delusions, alcohol, or neurological damage. This was the exception, however. When asked for their medical complaint, most patients had a medical response. A typesetter from Baltimore was admitted because of his erratic behavior and strange declarations about being dead, depleted of blood, poisoned, and his windpipe being clogged. In the admission ward, he stated that his complaint was "this stuffiness in my throat." The psychiatrist asked him to explain in what way he was not well. "In a mental way," the man answered, "my mind's all upset—I'm afraid something is going to happen." A woman from West Virginia explained that she had "heat in her veins" and "auto-intoxication from the bowel" and "suicide mania." Then she cried and lamented that things would be much easier if she were "really insane." This initial interaction, in which the physician asked for the psychiatric patient's "complaint," served to medicalize the individual's conception of his or her mental condition and paved the way for the clinical investigation and hospital treatment to come.[75]

In contrast to popular conceptions of mental asylums, the new clinic at Johns Hopkins—with its associations with modern hospital methods, technology, and scientific expertise—often represented an acceptable or appealing source of medical help for individuals faced with managing mental disorder in the decade before World War I. The reconfiguration of mental disorder as a potentially correctable maladjustment, as advertised by Meyer and a small vanguard of specialists, also fit contemporary Americans' preoccupation with self-improvement and the pervasive perception that the modern fast-paced environment was dangerous to one's health. Unlike the general hospital when it rose to prominence a generation earlier, however, the urban psychiatric clinic never supplanted its predecessor, the mental asylum, in terms of patient care. Nor did it instigate the widespread deinstitutionalization of asylum patients that occurred a few generations later. Rather, it became an integral part of an expanding network of social and medical institutions that managed and treated mental illness during the twentieth century.

At the end of the nineteenth century, the social obligation of the alienist to protect the insane was considered incompatible with scientific research, and the asylum itself incongruous with hospital medicine, clinical teaching, or therapeutic advance. The founding of a department of psychiatry at Johns Hopkins around 1908 was a commanding reification of Meyer's insistence that psychiatry could operate as a clinical science and academic discipline. After World War I, more and more clinics and university departments of psychiatry were established in North American cities, many of them staffed by clinicians trained at the Phipps Clinic by Meyer.

The clinic was very intentionally exemplary, down to the smallest detail. In addition to myriad practical considerations such as decorative window bars and special door handles that could not be used for self-strangulation, Meyer designed the Phipps Clinic to meet two important objectives: to apply the methods of general pathology to psychiatric disorders and to practice psychiatry using his instrumentalist concept of psychobiological reactions. As the remaining chapters show, for Meyer the physical and social spaces of the clinic constituted an environment that became tools of clinical investigation and medical intervention.

CHAPTER 4

The Baptismal Child
of American Psychiatry

The Meyerian Case History

"We *have* our opinions and we *do* act," Adolf Meyer declared when explaining his rationale for clinical research to alienists in New York in 1903. "All I
recommend is to record our reasons so that we can control them in the light
of events, formulate them into useful principles, and get from them suggestions for research and for future observations and deductions." For Meyer,
this was a straightforward explanation of how wisdom was transformed
into knowledge by the scientific method.[1] In 1912, shortly before the Phipps
Clinic opened, he mused that when it came to explaining and curing insanity, ancient Hindu and Greek healers certainly thought and acted no differently than medical practitioners in the present: based on reasonable speculation as to the causes of mental disorders, they developed and applied
treatments without ever confirming the relationship of theory to practice.
Meyer was adamant that at the beginning of the twentieth century a continued reliance on this speculative approach was preventing the progress of
modern psychiatry and that pathology, the proven scientific method of medicine, was the solution. "Speculation too easily solves many puzzles which it
takes many years of experimental and clinical work to put on a safe working
basis," he campaigned, equating "safe" and "scientific" knowledge. "It is, nevertheless, concrete work that has to be done and will prove the soundest
ground for stimulating the interest of the physician in his work of understanding and modifying cases." If psychiatrists did not begin to systematically record and evaluate their practices and results, Meyer predicted, failures would be repeated and successes lost to history.[2]

The clinical system Meyer devised and mandated at the Phipps Clinic
was the product of his resolve to harmonize the practices and expectations
of clinical medicine with his psychobiological conception of mind. Psychobiology was premised equally on the idea of a "plastic" nervous apparatus
capable of endlessly adaptive sensory-motor responses and a pragmatic and

instrumentalist interpretation of the common sense epistemology. This merger made a person's experience and interpretation of his or her social environment a focal point of Meyerian investigational strategies. Meyer utilized the case history as a device by which to convert ephemeral experience into fixed clinical data that could produce new disciplinary knowledge. All clinical activities at Phipps revolved around the construction of the case history: interviewing and examination, laboratory testing, record keeping, staff conferences, diagnosis, teaching, and therapies. Ubiquitous concepts within Meyerian psychiatry were the "experiment in nature" and the discovery of "facts." Often dismissed as merely rhetorical idiosyncrasies, both the experimental metaphor and fact collecting structured Meyer's retooling of the methods of general pathology in order to study and differentiate psychobiological reactions. The case history performed a crucial clinical function for Meyer: it objectified the patient's unique life history and social behaviors. "These charts must naturally be the result of many clean observations," he said in his earliest formulation of psychobiology in 1898. "They will become the expression of definite clinical data comparable with anatomical data in a dissection." By objectifying unique experiences to create a standardized unit of study, and by embodying medical authority and scientific precision, the case history enabled Meyer to operationalize his psychobiological psychiatry as a clinical science.[3]

Opening Up Some Lives

In the second half of the nineteenth century, Wilhelm Griesinger inspired a whole generation of pathologists to adopt his dictum that "mental diseases are brain diseases." The expectation that a unique histological basis would be discovered for every form of insanity was a commonplace one within medicine at that time. By the time Meyer emigrated to the United States in 1892, however, no such causes had been demonstrated. Many suspected a causal link between syphilis and insanity, but no conclusive pathological process would be discerned until after the turn of the century. Neither anatomical nor cellular pathology had yielded conclusive results comparable to those of the Paris clinicians in pulmonary diseases or to the remarkable discoveries of bacteriologists. Theoretical models of brain disease devised by Theodor Meynert and Karl Wernicke became vehicles for important research in neuropathology, but they were also circulated in textbooks as proven fact. In time, a new generation of researchers that included Meyer's mentors August Forel and Emil Kraepelin condemned these theories as "brain mythology." Meyer, too, advertised himself as allergic to conjecture and abstract explanations of psychopathology. He became an outspoken critic of traditional mechanical conceptions of the nervous system and was convinced that psychopathology represented a human organism's failure to adapt to its complex environment. Nevertheless, trained and skilled in pathology, his confidence in the pathological methods of pioneers such as Griesinger, Meynert, and Wernicke remained unshaken.[4]

When he had chosen neurology as a field of study, only a few years earlier, he had assumed that it dealt with well-defined diseases discernible by characteristic pathological processes. "It was not long, however, partly from experience in my own family," he recalled in 1916, that he faced "the realization that we do not deal with such a clean-cut group of cases." A combination of experiences initiated Meyer's reorientation away from pathological processes in brain tissue and toward those that interfered with the perception of subjective experience and its use by the individual in adaptive behavior. Meyer identified one of his first autopsies at the Kankakee asylum in 1893 as having been particularly transformative. The patient had died suddenly, and an autopsy was performed for legal purposes. A jury of citizens was convened in the morgue to observe first-hand the pathological lesion that led the pathologist to conclude that the heart had ruptured. Satisfied with the anatomical evidence for the cause of death, the jury's foreman, a country doctor, could not resist asking to see what had caused the patient to go insane: "Now can you show us what you find in the mind, I mean, in the brain?" Meyer had dissected the brain of the asylum patient and found nothing noteworthy. "When you want to look into the mind in this case," he heard himself explaining to the doctor, "you have to go to the history—you have to go to the life." He remembered this moment as significant in igniting his earliest deliberations about a compelling potential source of research material: the biographical history of the asylum patient.[5]

As he reconceptualized mental disease as failed psychobiological adaptation rather than merely a symptom of anatomical or physiological failures, Meyer never doubted that general pathology was the means of studying medical conditions scientifically. He affirmed that the systematic collection and correlation of clinical, laboratory, and autopsy data must continue—with the addition of mental reactions—alongside chemical and cellular reactions, as pathological material. Nevertheless, he shifted his attention away from the customary organic matter to collect, compare, and test and divested himself of the techniques and instruments essential to the clinical-pathological method. Neither his dexterity with the scalpel nor his mastery of the microtome and microscope would help him to look into the mind. In the early nineteenth century, the Paris clinician Xavier Bichat had famously decreed that the only way to dissipate the obscurity of disease was to "open a few corpses." Meyer's instinct as a pathologist, mediated by his conceptualization of mind as an instrument of adaptation, convinced him that psychiatry might achieve comparable strides in the next century by opening up some lives.[6]

Emil Kraepelin, the Heidelberg clinician whose reclassification of the psychoses transformed clinical psychiatry in the 1890s, showed Meyer how to go about doing just that. Ten years Meyer's senior, Kraepelin was part of the first wave in the generational backlash against the hegemony of brain mythology in the late nineteenth century. Echoing the views of his mentor, the experimental psychologist Wilhelm Wundt, Kraepelin insisted that psycho-physical functioning should replace anatomical mechanics as the subject matter of university psychiatry. He insisted that, without explaining psychological functions, pathological anatomy alone could make no claim to scientific legitimacy; such an explanation had to be supplied by clinical studies of patients. When he became director of the psychiatric clinic at Heidelberg in 1891, Kraepelin centralized patient records, opened surveillance wards for observing and documenting patients, and set up psychological laboratories to study the effects of exhaustion and intoxicants on mental performance. Liberal admission criteria and expeditious discharge procedures transformed the clinic into a transit station through which the greatest possible number of patients were admitted, diagnosed, and distributed to state asylums. There, patients' conditions continued to be monitored regularly by Kraepelin and his staff. These institutional circumstances allowed Kraepelin to carry out a specific research agenda: developing reliable diagnostic techniques to expedite prognosis. Entering his clinical data into blank forms derived from those used by the imperial state for census taking, Kraepelin constructed a meticulous medical and biographical history for each patient. He was convinced that studying comparatively the longitudinal development of symptoms would reveal patterns that would correlate to distinct disease forms.

Kraepelin was well known for his obsessive determination to collect enough histories to generate statistically significant nosological differences, and his quest revealed key prognostic differences among the most common psychoses. For instance, what he termed *dementia praecox* was deteriorative and incurable, while *manic-depressive insanity* was periodic and curable.[7] His findings at once imposed order on a previously amorphous group of cases that shared similar symptoms but among which no distinction traditionally had been drawn. The majority of American psychiatrists, including Meyer, considered this reclassification based on clinical studies and quantitative analyses to be revolutionary.[8]

Meyer spent the summer of 1896 in Heidelberg working under the master clinician, an apprenticeship that was coincident with the publication of the fifth edition of Kraepelin's textbook introducing the reclassified psychoses. The techniques he learned—the use of biographical data, surveillance wards, comparative analyses, and longitudinal studies—made a deep impression on Meyer and returned with him to the United States. Meyer's ally at McLean Hospital, August Hoch, had also trained with Kraepelin and was also instrumental in importing Kraepelin's techniques of psychological experimentation. Hoch and Meyer campaigned to have demanding standards of clinical examination, observation, and history taking implemented in asylums in Massachusetts and New York. Owing to the overwhelmingly positive reception of the new diagnostic categories in the United States, and to a regular stream of apprentices like Meyer and Hoch, it was Kraepelin's ideas and techniques that brought the psychiatric patient's biography to the forefront of American psychiatry at the end of the nineteenth century.[9]

Meyer considered Kraepelin's delineation of dementia praecox and manic-depressive insanity a great advance for psychiatry. Yet, he grew increasingly incensed that American physicians implemented Kraepelin's new nosology with speed, zeal, and a complete absence of its innovator's clinical acuity. Kraepelin was a gifted clinician and therapist whom Meyer respected. His research agenda, however, was to develop stable nosological categories of mental disease and the diagnostic techniques by which to distinguish them. In 1896, Kraepelin described this as psychiatry's most important practical problem. On this point, the pathologist in Meyer persistently disagreed. He lamented the ensuing Kraepelinian craze in the United States to diagnose, classify, and, especially, to generate statistics. These pursuits had little value, he insisted, because the majority of subjects were asylum patients, who tended to be so-called borderland cases that could not be matched to any single characteristic type. The primary task of clinical psychiatry, Meyer argued, was to collect and derive correlations from the data of general pathology, to discern *developmental* patterns. The pathologist's goal, he agreed, was to identify distinct forms of psychopathology—not for classification, however, but for the purposes of understanding, treating, and preventing illness. While Kraepelin focused on the results of pathological processes, Meyer hoped to clarify and modify their development.[10]

Meyer thought that his rationale for a clinical science of psychiatry was straightforward. He found, though, that his peers did not, despite the growing acceptance of Kraepelinian psychiatry. "Certain discussions at the meeting of the Medico-Psychological Association in New York," he mused in 1899, "showed that it is exceedingly difficult to make it plain what clinical psychiatry purports to be." He was motivated by a desire to ensure that mental disorders were subjected to the same scientific scrutiny as any other medical condition encountered in specialty clinics. When he returned to Worcester from Heidelberg he explained that clinical psychiatry "implies nothing but a careful examination of the previous history of the patient and of the condition from day to day, such as every painstaking physician makes on his patients." Even though it was standard procedure in general hospitals and private medical practice, it was uncommon to make detailed notations about patients in American mental asylums at the end of the nineteenth century. Meyer urged that the documentation of systematic medical examinations, laboratory tests, and autopsies for every patient was "the means of modern medicine" by which the seasoned alienist's wisdom was transformed into scientific knowledge, and brain mythology could finally be tested. "To keep such records of our observations," he campaigned, "will show us our shortcomings and also help us to make a truer picture of the disease forms than superficial observation and the faithful following of second-hand text-book opinions." Working without records and comparative data, he often remarked, was merely the "impressionist method"—the only technique available to physicians before the emergence of scientific medicine. Hypotheses were useful interpretive tools, but they were not data and they were not facts.[11]

Psychiatrists derived factual data, Meyer insisted, from observations of patients, results of neurological and laboratory examinations, and autopsy findings. They also harvested critical information from the patient's medical and biographical history. All of this was to be brought together in the case history, in the pursuit of a general pathology of mental disorders. "Records should be made the criteria of our medical work," he concluded. "This is the only means for the accumulation of a sound stock of information." Whereas Kraepelin had scoured the patient's biography for occurrences and remissions of disease (disregarding unnecessary personal information), Meyer—in an equally compulsive quest for data—spent the rest of his career mining the life histories of patients to detect abnormal adaptive patterns that corresponded to specific types of pathological reactions.[12] In his 1896 review of Kraepelin's textbook, Meyer described the author's conclusions as "dogmatic at first sight," but he immediately withdrew the comment. "The conscientious critic must refrain from comparisons," he acknowledged, "unless he have as many or more records of patients collected with the principles in view which Kraepelin has brought forth for the first time." One of Meyer's central objectives for the Phipps Clinic, then, was to amass enough case histories to produce medical knowledge equal to that of Bichat or Kraepelin.[13]

The Experiment in Nature

Twenty-five years after he told the country doctor that to discern pathology in the mind "you have to go to the history—you have to go to the life," Meyer published *Outlines of Examinations*. It was a comprehensive guide to psychiatric interviewing and history taking, examination and testing, and clinical record keeping. This manual emphasized that the case history must document observations of psychobiological reactions with the same degree of exacting clinical rigor and descriptive accuracy found in a pathologist's autopsy report. "The ideal which Virchow has promulgated concerning autopsy reports—the description of just what one does and sees—should also be the ideal of clinical notes." In such form, Meyer declared, the case history became a "permanent demonstration of the data." The case history, in other words, rendered the immaterial, material—the ephemeral, fixed—and it embodied the disciplinary authority of no less than Rudolf Virchow, the revered pioneer of modern cellular pathology. In Worcester and New York State, Meyer had urged physicians to compile an exhaustive medical history, and he had met with moderate success. His newfound autonomy at Johns Hopkins, beginning in 1913, allowed him to mandate that his psychiatrists dissect the patient's life history and construct their own Virchowian narrative in the form of a comprehensive case history for every patient. Psychiatry had its corpse.[14]

Like his fellow pragmatists, Meyer rejected abstractionism and embraced the instrumentalist view that history can explain the present. To validate the reasoning and methods of this new mode of scientific inquiry for psychiatry, Meyer relied on a powerful conceptual device. He suggested that every form of psychopathology was an "experiment in nature." This was a familiar concept to him and his peers, associated with the so-called genetic methods of natural history. The German anatomist Jacob Henle advocated that studying the results of disease—what he called "natural experiments"—elucidated normal physiological functioning otherwise unobservable to the anatomist. Later, both John Hughlings Jackson and Karl Wernicke reiterated this view in neurological terms, emphasizing that brain lesions and nervous disorders were experiments in nature that revealed valuable information to the neurologist.[15] "Diseases are the most crucial experiments in man," Meyer similarly reminded psychiatrists in 1909; "here the momentous things occur in a way which might well supplement the man-made experiments of our laboratories." Conceptualizing each incident of mental disorder in this way, he suggested, highlighted avenues of inquiry related to the causal relationships within Nature's experiment.[16]

In clinical practice and teaching, Meyer employed the concept of an experiment in nature (sometimes "of" nature) to emphasize that each form of psychopathology—even if unique to each patient—had a natural history subject to scientific scrutiny. "To do justice to the principles of an experiment, which is the fundamental requirement of pathology, and of science in general," he explained, "we must give each fact its true value in the chain of cause and effect." He understood, of course, that the methods of pathology and the observational sciences were quasi-experimental and that the factors directing causal chains in these sciences were not controlled variables as in a laboratory experiment. The metaphor of experiment, however, was more than scientific window dressing for psychiatry, it was an epistemological stance. As opposed to simply pondering the unknowable unity of Kantian noumena, Meyer declared, scientists could use the experimental approach to study actively the multiconditional character of natural phenomena.[17] "The fundamental syllogism of the modern mind," he wrote in 1905, was "the formula of an experiment." In terms of its use in pathology, he defined the experimental formula in the following way:

> Given a definite constellation of factors, what will be the result? And given a modification of the definitely tried and established constellation, what will be the modification of the results? We have, indeed, the rudiments of a general pathology which puts together the best-tried constellations and their results as standard guides in line with this simple rule of thought. In our

clinical work the equation is naturally turned around. The question is: given the abnormal condition, what is the constellation leading to it, and how can it be modified?[18]

In his private scientific notes, he ruminated that progress resulted from "the spirit of experimentation with limited hypotheses rather than the assumption of a universal principle." For him, an experiment was defined by the application of scientific rigor to explaining causal relationships between events. The ideal to which psychiatry must strive for a general pathology of insanity, he advised, was "to understand all the happenings in mental pathology with principles of thought which come up to the accuracy of those of experiment." Meyer's ubiquitous reminders to view each patient's mental disorder as an experiment in nature reflected his conviction that the isolation of causal relationships, not the proving of physical laws or mathematical formulae, was the essence of science.[19]

Meyer needed to mediate the tension between the infinite number of variables of his experiment in nature—a uniquely disordered personality—and perceptions of scientific precision. By using the experimental metaphor to structure the narratives in the case history, he hoped to capture the fluidity and subjectivity of psychobiological reactions while still casting them as a series of causal events with a natural historical development. He explained to colleagues in 1912, "The first task is to describe critically the plain events of abnormal reactions" as "experiments of nature." Viewing pathology as a natural experiment also framed the patient's adaptive behaviors as potentially modifiable experimental phenomena. Conceptually and rhetorically, this traditional metaphorical device aligned the clinical methods of psychobiology with both the genetic approach of the observational sciences, especially evolutionary biology, and the methodological rigor of the experimental sciences. "The soundest instinct in medicine," he told his readers, "is grouped around the interest in causes and effects."[20]

Meyer's rhetoric of an experiment in nature implied, quite deliberately, a social experiment without an end, one that encompassed all the variables of individual behavior that could be comprehended and redirected by scientific experts. It resonated with one of the Progressive Era's most dominant cultural notions: that scientific measures could be used to guide human social and biological development in profitable directions.

Meyer utilized the case history to discipline unwieldy but indispensable clinical material. "Chains of mental happenings," could be reconstructed by combining critical common sense and the rigor of pathology. According to Meyer, the "fundamental law and aim of medical science" was to pursue two tasks: first, to translate the "essential facts" of a patient's illness into the terms of a natural experiment (also called a disease); second, to determine which new "variables" modified how the harmful "experiment" progressed. This is how pathologists and bacteriologists had identified in organs and cells the characteristic progressions of several diseases and, in some cases, discovered effective therapies or vaccines. Meyer maintained that psychopathologists could use the same approach to study the "psychobiologic assets," with the goal of identifying distinctive pathological processes that undermine them. The psychobiological reaction, moreover, served as a unifying frame for plotting anatomical, physiological, instinctual, emotional, and intellectual processes on a single chain of causality. "The task of a general and specialized pathology of insanity," he wrote in 1902, was "to reduce these heterogeneous data to some sort of useful and practical order and especially into strings of causal connection." Meyer employed the standardized form and language of the case history to objectify highly contextualized thinking and behaviors and to integrate disparate streams of data. He reasoned that doing so increased the likelihood that unique forms of psychopathology could be compared, tested, discussed, taught, and modified.[21]

A conscientiously crafted case history objectified the patient's life history so that "every step is like an experiment telling us the story." To begin reconstructing causal chains that led to a person's mental disorder, the psychiatrist questioned the patient extensively about the circumstances that preceded hospitalization. He or she also asked about childhood experiences, schooling, family environment, friendships, religious views, work life, and hobbies. Whether there had been contact with poisons, drug or alcohol use, and what the patient's sexual experiences were, especially "perversion, masturbation, irregular intercourse or infections" had to be ascertained and noted, as well as evidence of an "irregular or unhygienic existence." Such factors might be contributory or incidental to the development of the disorder, or indicate the patient's current capacities for making adjustments to environmental demands that were advantageous to him or her. It was important, consequently, for the psychiatrist to observe and document the situations in which experiences arose, in the past and in the clinic, as well as the patient's interpretations and reactions in relation to them.[22]

Once biographical details were elicited from the patient, the psychiatrist interviewed the patient's family, friends, and other members of his or her community network, including employers and landlords. A full statement of symptoms and previous treatment was requested from all referring physicians. Relatives were asked for their accounts of the events surrounding the onset of the illness and, importantly, a picture of "the make-up of the patient at his or her best," in order to establish the comparative measure of normal that would be necessary to pathological study. "Wherever possible," Meyer suggested, "it is well to have some member of the family write in detail the history of the abnormal developments and the idiosyncrasies likely to be met." He and his psychiatrists did not rely on these accounts exclusively, but they considered this data important both for verifying the accuracy of the patient's report and for evaluating the degree of deviation from his or her usual disposition.[23]

Visiting informants at their homes or places of employment also illuminated the environmental conditions that might have impinged on the development of the disorder. At the New York Pathological Institute, Meyer, his assistants, and his wife, Mary Potter Brooks Meyer, all visited the families of patients "to understand the conditions of the home and to get a chance to inquire on the spot as to the early course of the disorder and the precipitating factors." He discovered that this supplemental field knowledge consistently benefited his work with patients. Information about the conditions in which abnormal adaptive reactions developed, he insisted, was imperative for reconstructing causal links in the experiment in nature. "A reaction," the pragmatist stated more pointedly in 1918, "cannot be judged without a knowledge of the stimulus."[24] If no informants accompanied or visited the Phipps patient, they were sought out. Phone calls were made to employers or other institutions, and letters sent cross country and ocean soliciting information about the family, medical, or developmental history of the patient. The female staff of the Social Services Department trekked to Baltimore neighborhoods to talk to neighbors or even instruct individuals to report to the clinic and provide testimony. Three days after a young craftsman named Kasper was admitted to the clinic, his landlord was summoned to Phipps. His tenant chattered absurdities, he reported, claiming everyone in Baltimore gossiped about him. The night before his admission, Kasper had wakened the entire boarding house and "insisted on having milk and raw eggs, and when they could not be obtained he rushed out into the street to find them." He called a policeman to report that he would die without milk and eggs. The landlord's testimony proved significant to the ongoing investigation of Kasper's case.[25]

From patient and informant accounts, the psychiatrist compiled a detailed family history that documented the patient's lineage and fraternity. It described the health and occupations of family members, as well as any medical, social, or moral abnormalities. The information was often portrayed visually as a family tree with symbols identifying those members who had died of disease, those with mental disorders or cognitive defects (so-called feeblemindedness or idiocy), and instances of serious illness, alcoholism, or suicide. For example, one patient's mother was identified as "feeble minded" and from a "bad family." She also had a "twin sister who seems normal, a brother who is able to make a poor living for himself, and a sister who is feeble minded." Harriet, a wealthy widow, was also descended from what was considered a deficient familial line: "two maternal aunts who died in asylums, two maternal uncles who were eccentric [and] eighteen aunts and uncles, twelve of whom reportedly died in infancy." The information elicited from Kasper's uncle, by contrast, revealed that the family tree contained six healthy brothers and "no cases of insanity, epilepsy, criminality, feeble-mindedness, suicide, cancer, tuberculosis, heart or kidney trouble, or goiter." Despite their apparently very different pedigrees, both Kasper and Harriet suffered severe psychotic disturbances, and both spent the rest of their lives in asylums.[26] Meyer had confronted similar discrepancies in his contact with thousands of patients and their families during the first twenty-five years of his career, while working in large asylums. Because of his conviction that psychopathology had a natural history that was multicausal—not to mention the instance of mental disorder in his own family—he tended to marginalize the role of heredity. One patient believed that a faulty inheritance was responsible for his difficulties. "I am especially anxious to make him realize," Meyer told the man's father, "that it is not so much any vague idea of heredity but experiences in personal life that combine in these upheavals."[27] In *Outlines of Examinations* he noted that the "precaution" of charting family histories methodically, as he instructed his clinicians to do for each patient, was the only way to establish or disprove the role of heredity scientifically.[28]

For Meyer, the family history provided information about more significant causal factors: the physical and social environments in which the incipient stage of derangement evolved. A thorough life history, he wrote, identified "determining factors" such as "any undermining conditions due to race, urban or rural community life, or abnormal surroundings—lack of home, bad company, or the other extreme—a tendency to choose over-religious or emotional associates." A psychiatrist's familiarity with the patient's "various conditions of life and ways of succeeding and failing in psychobiological adaptation" was essential. "The *final problem*," asserted Meyer, was always to determine "the conditions which the patient had to meet"; in other words, the clinician needed to identify discrepancies between environmental demands and the patient's normal adaptive resources for meeting them. "Every patient has his thoroughly normal assets, and others not so well-managed," he explained to one patient. "They are sorted out in terms of events of the past and the way they were handled." When they compiled genealogical histories, Phipps psychiatrists placed a greater emphasis on the patient as a product of his or her early environment.[29]

The metaphor of an experiment helped him emphasize to his staff, students, and readers that abnormal psychobiological reactions could be described in historical terms and tested like other natural phenomena. The most important "control" Meyer thought he could impose on Nature's experiment was the Phipps environment itself. "In the hospital the patient is reduced to a common level with all the others," he explained, "he is put upon a strange experimental field" in which "the rules of the institution thwart many sallies of individuality and create uniform conditions." For the psychiatrist observing the patient's adaptive behavior, the environment became his or her fixed experimental apparatus. "It eliminates the factor of variation of the external conditions to a great extent," Meyer reasoned, "and fulfills the most fundamental requirements for comparable tests of abnormality." This was one way, as he claimed in this instance, that psychiatrists might manipulate and test clinical data, and a way to discern inefficient from efficient psychobiological reactions.[30]

A standardized set of physical, mental, and laboratory examinations performed during the patient's first week in the clinic also facilitated the uniformity Meyer sought to impose and produced sets of common data. These examinations, too, he characterized as means of testing variables in the experiment in nature. According to his instructions, the standard medical examination of each new psychiatric patient was to "follow the rules of clinical observation of general medicine." This was a basic procedure learned as a medical student and refined and supplemented during specialization. By traditional means of palpation, auscultation, and percussion—as well as direct questioning and the use of diagnostic techniques borrowed from other specialties—the Phipps psychiatrist evaluated the overall physical health of each patient and carefully recorded significant findings.[31]

A systematic neurological exam was performed to evaluate reflexes, vasomotor and nutritive function, and motility (for example, gait, grip, and balance). The psychiatrist used established diagnostic techniques to detect known organic pathological processes. For example, if the patient repeatedly stumbled when asked to stand with eyes closed (Romberg's sign), it indicated a problem of the sensory nerves, not the brain or mentation. In another instance, an asymmetrical smile, countenance, or protruded tongue might reveal that a paralysis being displayed by the patient was induced by spinal, rather than emotional, trauma.[32] The neurological examination also tested subjective sensations such as pain, dizziness, coordination, or weakness. Pricking the tongue with a pin, for example, tested for desensitization to pain (characteristic of some psychoses). The normal functioning of the peripheral nerves was assessed by having the blindfolded patient identify various smells, such as vinegar, talcum powder, or an empty container (to detect olfactory hallucinations); or different tastes, such as molasses, whiskey, or peppermint syrup; or the sounds of a ticking watch or tuning fork placed at various distances from the ear. While still blindfolded, the patient's sensory perception was further evaluated by instructing him or her to describe various stimuli applied to the body (heat, cold, or pinpricks) and to identify everyday objects by touch.[33]

These examinations were not novel, but Meyer's use of them was distinctive. Meyer had dual expectations of these routine procedures. First, they generated data for later comparative analyses—of the same patient over time and across multiple cases for purposes of comparative research. Second, they operated as a medium of social interaction between physician and patient during which the act of the examination itself was conceived as another controlled variable imposed on the natural experiment. By performing this examination the same way on a hundred patients, so his theory went, the clinical psychiatrist could discern instructive differences and learn to perceive these same clinical signs in isolation. The process was similar to how a medical student learned to detect a heart arrhythmia, for example.

Histological and chemical analyses clarified the parameters of the experiment in nature. Nurses collected and measured the quantity of each patient's urine, after which it was analyzed chemically to test specific-gravity, concentrations of albumin and sugars, and microscopic bodies unique to urine (each of which might indicate known organic pathological processes). Blood and spinal fluid were drawn from every patient and tested for a Wasserman reaction, a serological test for syphilis. Various organs could be tested for the Abderhalden reaction, an analysis that purportedly distinguished histological markers indicative of dementia praecox. Several studies that evaluated the effectiveness of this test were inconclusive. "I do not allow myself to be influenced by the Abderhalden results," Meyer told a colleague, "at the same time, I do not wish to push them aside." Characteristically, despite his skepticism, he felt strongly that all knowable data—including every success or failure of a laboratory test done at the Phipps Clinic—must be entered into the case history for comparative and quantitative analysis.[34]

These examinations were no different from those employed in the other clinics of the Johns Hopkins Hospital, but attempts to conduct exams on and collect specimens from psychiatric patients could generate unpredictable behavior. Clarence Neymann, the physician-in-charge of the Phipps histology laboratory, discovered this when he was summoned to draw blood from Kasper. Kasper had already declared that he refused to surrender his blood. When Neymann entered the room (presumably with equipment for drawing blood), "the patient immediately sprang from the bed and made an unprovoked attack on him, tearing his lip badly with his finger nail." The ward psychiatrist insisted forcefully that Kasper submit to the procedure, which he eventually did. He then harbored a grudge against Neymann for the duration of his lengthy hospitalization.[35]

The mental status exam, as it was called, was yet another standardized test to which every patient was subjected systematically. It was designed to confirm normal modes of mental functioning for the individual patient and to expose abnormalities. Its lengthy series of questions and tasks isolated and tested specific mental abilities, such as abstract thought, reasoning, learning, judgment, planning, problem solving, and the comprehension and expression of language. Some exercises helped to differentiate between pathological mental states that were often confused—for example, to distinguish psychosis from mania, or depression from unresponsiveness. The psychiatrist also inquired about preoccupations, dominant ideas, delusions, hallucinations, obsessions, or otherwise unusual experiences. Meyer warned that the psychiatrist's "approach to those topics to which the patient is most sensitive must be made with the greatest amount of tact and mutual understanding" in order to engender trust. The psychiatrist's rapport with the patient was established during these initial comprehensive exams, which facilitated access to the crucial data of the patient's life history and personal experiences.[36]

Historians have been critical of mental status exams, citing inherent socioeconomic biases as obstacles to any meaningful assessment.[37] Transcripts of these exams in the Phipps case histories indicate that Meyer and his staff were attentive to some of these factors. When Phipps psychiatrist Edward Kempf evaluated one patient, he commented that the man "received a meager education, but is a successful butcher and handles his business well—his education should, therefore, give him an average basis for the following tests." In *Outlines of Examinations*, Meyer advised that "the examination must furnish the essential facts as to the lasting stock of mental resources (or at least those as yet unaffected), as well as any chance for comparisons of any temporary or progressive changes." The exam, he insisted, had to remain fluid in order to reach these clinical objectives. For example, when Kempf examined another patient, a young man who could not name the longest river in Africa, he suggested to his patient that perhaps it was the Mississippi River. "No, the Mississippi is in the United States," the patient chided Kempf skeptically, "you know that as well as I do." Indeed he did—they had established a point of consensus and at least one marker of normal abilities. "Perhaps it is the River Nice?" Kempf then asked, and immediately the patient remembered that the longest river in Africa was the Nile. Responses to the mental status exam were relative, not absolute, and Meyer expected examiners to modify their technique to discriminate between ignorance (a normal phenomenon, relative to a patient's life experiences) and psychobiological impairment.[38]

Like their neurological counterparts, results of standardized mental status exams were intended to bring differences between cases into relief. Some patients reacted with hostility to this extensive and seemingly bizarre slew of tasks (such as being asked to name the animal imitated by a psychiatrist who barked, chirped, or mooed). Others reveled in being under the intense scrutiny of medical experts. Asked to name the last five American presidents, or to explain the significance of the Titanic's maiden voyage or the Suffragette Movement (all standard questions), one patient might do so politely, the next might claim to have drowned in the famous shipwreck, while another might refuse to answer. During every exam, Meyer explained, the psychiatrist must note "the conduct, its inner consistency and its consistency with the situation, showing the extent of adaptation" and "the extent of rapport with the persons and objects of the environment." Every reaction was important and recorded in the case history. After naming correctly the Kaiser of Germany, capital of Austria, and several American presidents, a fifty-six-year-old cabinetmaker became irritated. He yelled in a thick European accent that "naturally anybody would be sick and confused in the head after being asked so many questions!" The psychiatrist noted diligently the man's opinion that "the doctors are much better able to tell than him whether he will get better, and be able to get back to work." When requested to describe her thoughts, one woman retorted, "that this must be purgatory and that you are all detectives." Seeking to differentiate pathological delusion from healthy disdain, the Phipps clinician asked if she really thought the assembled doctors were detectives. "No," she clarified, "but you certainly work like detectives." That they did.[39]

Interpreting Data and Transmitting Psychiatric Knowledge

The detective work of making observations and collecting information during interviews and examinations was the first phase of compiling the case history. Meyer insisted that raw data be interpreted and reinterpreted as the experiment in nature—the patient's illness—progressed according to the physical and social environment of the Phipps Clinic. He described this process in one instance as "the sifting of the facts and their grouping into chains of cause and effect." It was not the mere assemblage of data that illuminated patterns of psychopathology, he emphasized, but the psychiatrist's interpretation, organization, and presentation of them. "To expect that a history will be good enough if it simply records the facts mechanically as they are observed, without a continuous digestion of the material," Meyer chided, "is absurd." The importance of continual analysis and reinterpretation, according to both psychiatrist's and patient's changing perceptions, reflected the pragmatism inherent in Meyer's approach.[40]

Meyer proposed that data became factual when the question of their presence or absence represented an observable differential. In his private scientific notes, he mused that a psychiatrist was more of a rag-sorter than a dissector. To the seasoned pathologist, relationships between the fixed elements in the autopsied body were relatively predictable. Understanding the multifaceted relationships among elements in psychobiological reactions, he reflected, was to sort out ratios. As one of his former psychiatrists clarified, "to him, a fact might be a brain tumor, broken leg or a dream or a hallucination or a nagging mother-in-law."[41] His conversation with one Phipps patient about the number thirteen exemplifies this principle in action. A few weeks after giving birth to her first child, the thirty-nine-year-old wife of an insurance salesman was admitted to the Phipps in a paranoid state. "I say I'm unlucky because I was born on the thirteenth," she stated firmly to Meyer, to which he countered, "I too was born on a thirteenth, but I have no reason to think of it as being unlucky." The woman considered this discrepancy carefully, musing that it was undeniably strange that she was unlucky yet her examiner was not. "How can you account for that?" she asked him, "have you not had a lot of changes in your life? Have you not been unhappy?" Refracted through Meyer's experimental lens, she was asking him why this single variable had not produced a universal outcome. In answer, he redirected her focus to the search for additional factors that might explain the *difference* between their respective responses to their shared birth date: "the question is whether it began to play any *special* role, *your* being born on the thirteenth?" The patient could not trace the origins of her conviction, an insight that Meyer deemed important to explaining her atypical reaction to the number thirteen. So, the psychiatrist continued to scrutinize the data and discern more facts.[42]

In order to establish a clinical science of psychiatry based on psychobiology, Meyer needed facts to perform a clinical function for him. His fealty to the "fact" was well known. He sermonized incessantly about the importance of identifying facts in psychiatry, instead of relying on the "impressionist methods" of what he called prescientific medicine. Some historians have been quick to characterize his fact collecting as unproductive, arbitrary, audacious, equivocal, or defeatist.[43] Historian Ruth Leys, by contrast, deals substantively with Meyer's rhetoric. She argues that, despite his professed pragmatism (in which facts were always situated within a particular interpretive context relative to specific human goals), he failed to enforce theoretical distinctions between subjective and objective facts.[44] From his perspective, I suggest, practical needs trumped theoretical consistency, and his goal was to render subjective experience and situated knowledge *objective enough* to serve as material for collective clinical practices. He thought that this was achieved by converting psychobiological reactions into sharable facts. He declared that collective scrutiny and consensus—"a presentation of the facts under the critical eye of all the physicians of the hospital"— played a critical role in establishing (psychiatrists') subjective observations of (patients') subjective behaviors as objective and factual enough for clinical investigation, research, and teaching. "Inconsistencies of flawed observation, indifference about the procuring of information, looseness of diagnostic reasoning," he concluded, "are thus eliminated in a way that is bound to make the work interesting and profitable to patient and physician alike." His intellectual position on the status of scientific facts aligned with the emerging philosophy of phenomenology, especially Edmund Husserl's notion of a "presuppositionless" methodology in which efforts to clarify presuppositions and, subsequently, to diminish the importance of what remained unclarified, were considered productive. Knowledge acquired by this method was to be considered "as certain as possible," not absolute or unquestionable, and the language used to describe it was to be as "presuppositionless" as possible. Meyer's claims that a psychopathologist was trained to convert subjective interpretations into "positive facts" did not imply that it was actually possible to eliminate all presuppositions and achieve truly objective knowledge, as positivists had claimed—only that the attempt to do so generated valuable perspectives. Moreover, he did not advocate collecting and recording facts arbitrarily. "The ideal is a maximum efficiency in the differential estimation of the events," he explained cryptically in 1909, and not "photographic accuracy of really subordinate incidentals." In other words, for Meyer a fact was constituted as a fact—worthy of documentation in the case history—by its mutually acknowledged status as a differential in the experiment of nature.[45]

Meyer was not unique in his reverence for facts or his devotion to documenting them methodically. By the 1870s, aphasia researchers in London had introduced standardized formats and language for clinical recording. Personnel at every level of the medical hierarchy were tasked with compiling scrupulous reports of observations and examinations. By the time Meyer visited in 1890, these clinical protocols were sacrosanct. John Hughlings Jackson, who exercised such influence over Meyer's clinical sensibilities, insisted that the clinical record reflect an observation and not an inference. "Better to record the fact that the patient does not put out his tongue when told," Jackson counseled in 1864, "than the inference that it is paralyzed." The record-based and fact-based methodology that Meyer instituted at the Phipps Clinic bore a family resemblance to both the English clinical model and the one that led Emil Kraepelin to his striking correlations. Eventually, it would draw criticism for being protracted and unfocused. "I consider it a justified indulgence of my own scientific curiosity," he declared, ever-confident in 1928, "to penetrate into all kinds of promising details." By the turn of the twentieth century, however, the utility of methodical observation, fact-collecting, and documentation in generating scientific knowledge had become self-evident.[46]

Others adopted his techniques and found them fruitful. In 1903, a New York asylum physician thanked Meyer for "showing me how to find out all there was to know about a patient," something that the writer claimed allowed him to reduce significantly his use of drugs and restraints, because "my judgment was then clear to know what to do for them."[47] Clinical thinking, this alienist suggested, had transformed his practices from custodial to therapeutic. Psychiatrist John Oliver was on the Phipps staff from 1914 to 1917. A few years after his departure, he professed deep faith in the value of Meyer's methodology when he described his favorite room in the Phipps Clinic:

> In that large complex building, with all its wards, laboratories, offices, and workshops, there is a quiet little room, off the library, which was always to me the sanctum sanctorum of the whole institution. This was the history room. Here were filed the thousands of case histories of all patients ever treated at the clinic since its foundation; and each history, a small volume in itself, with all the details of the patient's family, personal history, physical and mental status, type of mental reaction and hospital progress, was a mine of wealth, gathered with infinite patience—a mine that future psychiatrists might use . . . for the working out of the great basic principles of psychiatry.[48]

Oliver's description echoed Meyer's own confidence in the methods he promoted for a clinical science of psychiatry at Johns Hopkins.

One Phipps psychiatrist recalled that, rather than any particular system of thought, the chief demanded "scientific thinking." For Meyer, that meant collectively identifying causal chains in the development of psychopathology as rigorously as experimentalists studied those in a controlled experiment.[49] Unlike controlled variables in a laboratory experiment, however, unknown factors were a constant reality in psychiatry, one that Meyer was determined to integrate into his approach. This tolerance for the unknown drew criticism from some of his peers, but it was something his adherents appreciated about the Meyerian approach. "Dr. Meyer, you represent to me Courage," wrote Jacob Conn in 1948 in honor of his mentor's eighty-second birthday, "the courage to admit there are discontinuities—gaps between sets of facts." The interpretation and ordering of facts using devices such as the case history and experimental metaphor helped navigate this pervasive reality in psychiatry—that symptom pictures were amorphous and changeable and that psychiatrists' interpretations of them were provisional and incomplete.[50]

Where Meyer departed from research traditions that used standardized records to discipline data was in his quest to contextualize the content of case histories. Clinicians such as Kraepelin and Jackson utilized clinical records to universalize symptoms, disease, and medical interventions. This process involved both selection and interpretation on the part of the writer, as historians have shown, in order to convert complex medical perceptions of illness into written narratives. This was certainly true for the Meyerian case history, a product of "rag-sorting," as Meyer described it. Unlike the writers of traditional clinical records, however, who endeavored to depict an acontextual snapshot of a disease, the pragmatist Meyer argued that abnormal psychobiological responses were incomprehensible to other readers without conveying the highly contextualized settings in which they arose.[51] "Dependence on merely momentary statements and reactions" of the patient during a "detached examination" in the clinic, he asserted in *Outlines of Examinations*, was wholly inadequate. Elsewhere he explained that identical symptoms regularly appeared under very different circumstances, so that conclusions drawn from a temporary picture were apt to be guesswork. "The most valuable determining feature is, as a rule, the form of evolution of the complex, the time and duration of circumstances of its development, and the character of the possible transformations of the picture." He wanted the psychiatric case history to "read like a continuous, consistent story" that depicted the person's dynamic interaction with his or her environment. Due to the instrumental nature of human mental functioning, he maintained, achieving this goal must include an attempt to document causal relationships between the patient's perception of past, present, and prospective experiences.[52]

Never a snapshot, the Meyerian case history was supposed to capture cinematically the dynamic developments in the ongoing experiment in nature. The Phipps staff recorded copious notations throughout a patient's hospital admission—not only the raw data collected by clinicians, but also social interactions, utterances, overt and covert behavioral observations, disturbances, and emotional displays. Nurses maintained their own hourly logs, from which the psychiatrists harvested information. Meyer, moreover, rarely traversed the clinic unaccompanied by a stenographer. Even his impromptu encounters with patients in the garden or elevator were recorded and transcribed into the history. During the prewar period, Meyer was developing another technique for recording and standardizing psychobiological data; he called it the Life Chart. After World War I, this device became well known beyond the Phipps Clinic.[53]

Meyer's cinematic adaptation of the traditional medical narrative was possible because of new record-keeping technologies adopted by modern general hospitals at the end of the nineteenth century. Like the rest of Johns Hopkins Hospital, the Phipps Clinic operated on a progressive model of business management for processing medical records and generating invoices for large numbers of paying patients. Because data collection was distributed throughout the clinic and its staff, the centralized record-keeping system was essential not only for billing and the management of lab results but for synchronizing chronologically the narrative case history. The Phipps Clinic maintained a busy stenographic pool larger than most small specialty clinics: four stenographers, two secretaries, and one librarian. Phipps psychiatrists, when not on duty in the wards or dispensary, organized and interpreted their clinical observations. They then recited their notations into a new device called a Dicta-type machine, in the record room. The following day, a typist transcribed each notation onto 8½-by-11-inch blank chart paper (with the letterhead "Johns Hopkins Hospital") and marked it with the date and the clinician's initials. All notes relating to a particular patient were inserted into the individual's case history in a way that sustained its chronological integrity and cinematic sensibility.[54]

Patients could not help but notice that, while being asked to talk about their troubles—often rooted in intimate or embarrassing situations—someone was zealously taking notes. During an admission interview with psychiatrist Ralph Truitt, one woman evidently found it amusing that someone would want to record her every word: "I hope you are not writing down everything I say!" she reportedly scoffed. Within two weeks, however, she was worried about the note taking and told Truitt that she did "not want anything written down about her sister's husband." Another patient was known especially for her vitriolic attacks on the staff (she loudly referred to Meyer as the "common man" and "infamous Jew" who was her jailor). She was asked if she knew the date. Characteristically, she ignored her questioner but harped at the stenographer, "you must be Jim the Penman—I never saw anybody write so much." It was not uncommon for note taking, like other elements of the Phipps environment, to become implicated in a patient's psychopathology. "I'm sick of this place—you can put that down if you please," one patient remarked. He then commenced a lengthy monologue that a stenographer attempted to record verbatim. Lost in a stream of seemingly unconnected details, the psychiatrist finally stopped him to ask the meaning of his story. "Oh it doesn't mean anything," the patient explained, "if you don't want to write it down, it doesn't matter."[55] Although the presence of a stenographer during a medical exam or a discussion with a doctor was novel and sometimes elicited antagonism, Meyer contended that "in the hospital it is practically always possible to make the patient see the importance of notes." Indeed, more of his time was spent convincing his peers of the necessity of astute observations and careful note taking—especially before diagnosing and classifying mental disorders.[56]

Reaction-Types

Both his own experiences and his confidence in psychobiology convinced Meyer that the diagnosis and, especially, the classification of mental disorders needed to remain secondary to therapy, given the dismal state of psychiatric knowledge at the turn of the twentieth century. He persistently swam against the rising tide of Kraepelin's descriptive psychiatry and tried to counter demands for standardized terminology. In *Outlines of Examinations*, he warned readers that "without some qualifications, more than one half of the cases are not sufficiently clean-cut to represent a type." No experienced alienist or psychiatrist, he argued, would be satisfied with a one-word diagnosis. "What is the faulty reaction? What are the conditions that led to it? What can we expect to achieve, and what are the steps to be taken?" These were the questions that should guide clinical practice, he told alienists and trainees in 1908, not the search for the right diagnosis. In too many cases, he assured them, there was not one. In the Phipps Clinic, only after the case history was scrupulously compiled did he expect his clinicians— indeed, permit them—to venture a possible diagnosis. This directive is reflected in the Phipps case histories, in which expressive "diagnostic formulations," as Meyer called them, typically appear only on the discharge summary created at the end of the patient's hospitalization. In most instances, the diagnosis is the lone entry on the form written in Meyer's hand. This suggests that the act of diagnosis at the Phipps Clinic was reserved, at least officially, for the director and senior staff. [57]

Medical practice had long revolved around a correct diagnosis, however, and improving and ensuring diagnostic accuracy was a central concern of pathology. At the end of the nineteenth century, discourses surrounding the differentiation and classification of mental disorders were characterized by the same uncertainty and debate that imbued the theoretical chaos in neurology and psychiatry. All the same, consistent diagnostic categories were necessary for statistical analyses, teaching, and research. Kraepelin's prognostic rationale for distinguishing deteriorative dementia praecox from recoverable manic-depressive insanity was considered revolutionary because of its immediate utility in medical practice. Based on the patient's symptomatic history, a physician could render an expeditious diagnosis that facilitated family decisions about hospitalization.[58] Meyer comprehended the great significance of Kraepelin's findings (achieved by comparative analyses of case histories, he pointed out, and not the discovery of a histological marker). Nevertheless, he observed that, when consulting on a case of severe mental disorder, too many general physicians and alienists viewed making the diagnostic choice between dementia praecox and manic-depressive insanity as their sole medical duty. Between these two characteristic *types*, he insisted, there were countless cases that fit neither diagnosis (for which he encouraged the term "allied to" a particular type). Such patients still required medical attention and, as Kraepelin had established, their case histories teemed with valuable clinical data. "If the facts do not constitute a diagnosis we nevertheless must act on the facts," he railed.[59]

While he recognized the practical need for diagnostic categories, he agonized over the issue of a fixed classification system. Seeing the spontaneous cessation of symptoms in many patients—including those who had exhibited typical signs of dementia praecox—convinced Meyer to take a cautious approach. "The illness of one of the local patients disappeared quasi all at once," he wrote to his brother in 1912, admitting that "in the end, it is always difficult to tell how those transformations come about." For Meyer the pathologist, differentiating causal from incidental factors in the development of mental disorder—for the purposes of taking medical action—trumped the act of naming the patient's illness. "Do not think that the name embodies the knowledge and definition," he cautioned students in 1905; "you have to know your cases, and if you do, the name will be a secondary matter."[60] Psychiatrists must be trained to do more than juggle terminology, he clarified in 1913; they must learn "to understand each case of mental defect or perversion as an experiment of nature." The psychiatrist's task was "to get hold of it as near the point of morbid departure as possible, and to determine the factors which must and can be modified to restore the individual and social balance." The patent inconsistencies between fixed diagnostic categories and so-called borderland cases in everyday practice demonstrated the impracticality of classification.[61]

Meyer's views were formed according to his personal and professional experiences before coming to Hopkins. In 1905, he promised that there would be no imminent "nosological glories" in psychiatry comparable to those in bacteriology, only decades of conscientious clinical work, as Kraepelin's findings, he felt, had also demonstrated.[62] Yet, Kraepelin had a personal predilection for Linnaean classification that Meyer found repellent because of his own idiosyncratic distrust of closed theoretical systems. "Types of persons are difficult to define," he put it plainly; "once and for all we should give up the idea of classifying them as we classify plants." Other leading figures in American psychiatry, however, were pushing for a standardized classification system that would unite diagnostic practices. On the one hand, Meyer continued to deride the notion. It was absurd that "a systematic scheme of nomenclature and a classification of all the cases should first be demanded as an ideal achievement," he sneered in 1904, "while hardly any adequate effort is made to ensure sufficient value and accuracy of the facts to be classified."[63] Pressure from state authorities in New York, on the other hand, forced Meyer to impose consistent categories for generating quantitative data about asylum patients. The following year, 1905, he begrudgingly introduced a system, based on Kraepelin's categories, for use throughout New York State. During his sixteen years of service in large asylums, he had witnessed the manipulation of statistical data by superintendents and trustees in order to ensure funding and public support, a practice that soured Meyer's taste for counting disease forms and left him resistant to the notion of imposing a single system upon individual physicians working within specific contexts.[64]

By the time the Phipps Clinic opened, he was a well-known opponent of systems that forced all cases into fixed categories. A year later, in 1914, he began summarizing the work of the clinic. "It is wonderful material," he told his brother, "but so difficult to get into terms of a system—Kraepelinian psychiatry is a very poor mirror for this world of facts."[65] The Kraepelinian psychiatry for which he and Hoch had lobbied so strenuously less than two decades earlier—a methodological system for clinical practice—now heralded what Meyer saw as a pointless quest to describe mental disorders rather than study and treat them (even as he knew Kraepelin himself to be an able investigator and therapist). Ironically, then, it was the system he had devised for New York State that the American Medico-Psychological Association adopted in 1917 to standardize diagnostic categories across the profession.[66] With clarity that was usually limited to his informal statements, Meyer presented his views on "the meaning of psychiatric diagnosis" at the annual meeting that year. He protested that the organization's Committee on Statistical Classification had "sworn allegiance to the German dogma without provision for mixed and merely allied types." Afterward, committee member Stanley Abbot explained that cases deemed "allied to" a particular class were equivalent to "undiagnosed and it seemed better to call them so frankly." Meyer bristled at the conflation of diagnosis and classification. "To speak of the case as undiagnosticated because it contains more factors than the pigeon hole scheme contains," he hissed, "would be both disastrous to the attitude and conviction of the physician and ultimately to the welfare of the patient." He delivered the same message the following month with similar clarity to members of the American Medical Association, emphasizing again that a Linnaean classification was unattainable for psychiatric disorders without verifiable histological or clinical markers.[67]

Based on the data in his case histories and his concept of psychobiological reactions, Meyer concluded that cases could be *grouped* based on direct observations of overall adaptive performance—"efficiency in those mental adaptations which we know to become actually deranged, the emotional sphere, the equilibrium of reason."[68] The best medical standard for evaluating and sorting cases, he proposed, was that of "adequate or efficient function." In 1902 he observed that "the analysis of a large number of faithful records of cases of insanity furnishes certain natural groups of almost identical conditions." Similarities might cohere around causal factors, symptomatology, the course of disease, or pathological experiences. "Where there is a coincidence of the main points in all four directions we have every reason to surmise a definite law of development," he concluded, "especially if the type occurs often enough to free one of the impression of chance." He derived these conclusions from comparative analyses of 3,000 case histories compiled from clinical observations of asylum patients between 1896 and 1902 (abstracted and indexed on note cards, as he had learned to do from Osler and Kraepelin). He deduced that there was, in fact, a small number of typical abnormal reactions behind the infinite variety of clinical pictures presented by patients.[69]

Meyer called these natural groups "reaction-types." Within the psychobiological paradigm of adaptive reactions, psychopathology was something a person did. Like his fellow pragmatists, Meyer made action and activity his primary categories of analysis in his approach to grouping mental disorders. A patient did not contract a disease called schizophrenia any more than he or she became a schizophrenic, both of which implied that the disorder was an ontological entity distinct from the person. He urged that many prevalent forms of psychopathology were not the result of an attack on the organism (like syphilitic insanity, for example) but developed as part of its adaptive performance. "Every individual is capable of reacting to a very great variety of situations by [adopting] a limited number of reaction types," Meyer proposed in 1906, and he deemed this true of both healthy and pathological reactions. Schizophrenia, to continue with the example above, described a particular type of maladaptation.[70] Rather than a nosological term, then, he suggested a diagnostic adjective that would precede the term *reaction-type*. Instead of schizophrenia, he proposed "schizophrenic reaction-type." Meyer's concept of a psychobiological reaction-type emerged within the context of widespread interest in typology among European intellectuals and the extensive body of literature they had produced.[71]

Typing was consistent with Meyer's objective to remake psychiatry as a clinical science. "The conditions which we meet in psychopathology are more or less abnormal reaction-types," he explained, "which we want to learn to distinguish from one another, trace to the situation or conditions under which they arise, and study for their modifiability."[72] It was imperative for Meyer that any data used to distinguish reaction-types or make a diagnosis meet the criterion of being "sharable facts." He maintained that diagnostic practices based on typical maladaptive reactions could be approached "directly from a simple empirical standpoint," unlike anatomical lesions, toxins, hereditary degeneration, or, for that matter, the activity of neurons underlying every psychobiological reaction. Meyer hoped that typing, rather than classifying, would satisfy medical expectations regarding diagnosis without sacrificing clinical observation to guesswork about essential cause or prognosis. A superficial glance into the future suggests that it did: the first edition of the well-known *Diagnostic and Statistical Manual*, issued to psychiatrists across the country in 1952, two years after Meyer's death, was a catalogue of Meyerian reaction-types.[73]

In 1908, Meyer described three general reaction-types—organic, affective, and substitute—each comprising several subtypes based on well-known diagnoses.[74] The category "organic reaction-type" included cognitive or motor impairments resulting from a head injury, brain tumor or abscess, Huntington's chorea, arteriosclerosis (softening of the brain tissue), aphasic disorders (loss of language and perception), and senile psychosis. Syphilitic insanity was also an organic reaction, in which a sufferer typically expressed delusions of great wealth or fame and general paralysis (a traditional term for the condition); cognitive and physical deterioration continued, culminating in derangement and death. A subtype was "toxic reaction-type." Commonly called auto-intoxication, examples included the delirium that accompanied exhaustion or fever, psychosis resulting from thyroid or diabetic disorders, postpartum psychosis, and some forms of epilepsy. Prolonged alcohol or drug use also induced a toxic reaction. Delirium tremens—in which the patient typically imagined rats, snakes, or bugs crawling everywhere—were common in chronic alcoholism. Drug "habitués" addicted to heroin, bromides, or morphine experienced hallucinatory psychoses and paranoia. In the case of an organic reaction, the principal factors driving the causal chain were assumed to be nonmental (anatomical or physiological) while the abnormal mental states that often accompanied an organic reaction—acute sadness, elation, jealousy, hallucinations, or delusions—were considered only incidental to its development.[75]

The second of Meyer's broad diagnostic groups, "affective reaction-type," he described as a disorder of extreme emotion. Pathological despair, elation, or fear was categorized as depressive, manic, or paranoid reactions, respectively. Adopting Kraepelin's differentiation, Meyer explained that a manic-depressive reaction was marked by limited episodes of hopelessness or frenzied excitation punctuated by periods of relative health. A paranoid reaction might comprise various degrees of disordered thinking, from unwarranted suspiciousness to systematized delusions. A sufferer might suggest that the conversation of strangers or the flickering of street lamps referred to his or her own person. Driven by pathological emotional cues to hide, commit suicide, or attack, patients exhibiting an affective reaction engaged in unaccountable, unpredictable, and often reckless behavior. Meyer speculated that, unlike the organic reaction, mental factors probably contributed to the development of an affective reaction.

Meyer's organic and affective reactions articulated with traditional diagnostic descriptions that were considered commonplace by his peers. His third category, "substitute reaction-type," however, depended upon the concepts of dissolution, faulty reactions, and psychobiological habits. By the 1890s, disorders such as hysteria, neurasthenia, invalidism, and psychasthenia represented a nebulous group of conditions with overlapping symptoms and a common feature: a bodily or mental disturbance that could not be linked to an organic disease process using laboratory or diagnostic tests (though many diseases could be ruled out using these means). Sufferers exhibited symptoms such as persistent pain, physical or verbal tics, bowel disturbances, difficulty breathing, convulsions, choking, paralysis, or selective amnesia. They might develop debilitating phobias or bizarre preoccupations that made everyday tasks impossible. Those physicians who did not simply dismiss these patients as malingerers or hypochondriacs termed their disorders "functional" (as opposed to organic). Such a dualistic distinction, of course, was antithetical to psychobiology. According to Meyer, these symptomatic behaviors, commonly diagnosed as hysteria or neurasthenia, advertised, to be more accurate, a substitute reaction. When dissolution impaired higher nervous functions and left lower emotional and instinctual responses unregulated, he hypothesized, the person necessarily substituted an insufficient adaptive measure that became progressively more harmful and automatic. The substitute reaction typically originated in a "pathological experience" that sustained the formation of an inefficient or morbid psychobiological habit. The principal causal factors of so-called functional disorders were mental, according to Meyer, and any anatomical and physiological abnormalities were probably incidental to the developmental chain. In 1906, he proposed that dementia praecox (and, eventually, the schizophrenic reaction) was a subtype of the substitute reaction-type.[76] (His approach to this category of disorders is the focus of Chapter 6.)

Meyer described a final category of mental abnormalities not as a type of reaction but as "constitutional inferiority" in which "there is an inability of adaptation to the demands and existing order of society." Such patients were often labeled by family members as the bad egg or black sheep because, despite promising circumstances (for example, a loving family and good schooling and opportunities), they appeared incapable or unwilling to behave appropriately. In some cases, there was a predominance of vicious, antisocial, immoral, or criminal compulsions. "The most frequent types," Meyer explained in 1905, were "sexual immorality, prostitution, alcoholism, disregard of family or other duties, a fiendish passion for making trouble, stealing and forging and lying." Because of the ways in which this disorder manifested itself, it had previously been called moral insanity. By the 1930s, he and other colleagues referred also to a "psychopathic personality," the origin of the current noun *psychopath*.[77]

The paradigm of reaction-types kept in view the specificity of anatomical, physiological, and mental factors in the causal chain, as well as the patient's subjective experiences and the interpretive contexts that shaped adaptive reactions. Meyer confidently believed that this approach also accommodated the nebulous character of patient's symptoms and the reality that "in a large number of our cases the organo-genic, the neuro-genic, and the psycho-genic disorders play into each other's hands."[78] In 1919, Meyer claimed that the principle of reaction-types was "finally accepted by Kraepelin" as well as several other European clinicians who expressed similar views.[79]

He used the concept of reaction-types to teach the principles of psychobiology and clinical techniques for disciplining disparate data. According to the *Johns Hopkins Circular*, the university's academic register, from 1913 through 1917 Meyer conducted a psychiatric clinic each Saturday afternoon at 3:30. Its purpose, like the medical school's other teaching clinics, was to provide students the opportunity to observe patients and discuss cases. It took place in the clinic's demonstration room, which had at its front a black-board, projection screen, and a stage that could be accessed by a door leading from the wards. Its lecture-hall seating was filled by residents, staff psychiatrists, house officers, and medical students. The gathering has associations to what is today called grand rounds.[80]

By employing the standardized case history as a common disciplinary object, Meyer was able to execute the collaborative practices of scientific medicine and clinical teaching. The case history supported group research and peer evaluation. It also allowed him to teach psychiatry the way other medical disciplines trained students and future specialists—by demonstration. "The ideal of any modern medical school," Meyer told colleagues at the International Congress of Medicine in 1913, "is to furnish comprehensive representation of the practical work and research in all those directions in which a physician should have first-hand experience." As was the case in departments of surgery or obstetrics in medical schools, training at Phipps was "a chance for practical work and investigation." For Meyer, however, this meant a continued campaign for eradication of the reductionism and dualism commonplace in American medicine. "Procedures cultivated by ordinary medical education must be avoided or adjusted, such as the inveterate simplism that real pathology can only be anatomy and histology," he lobbied, "and that the more complex events, the psychobiological reactions, are not worth the physician's attention before they are reduced to cerebral localizations or to hormones." Much as he griped about it, Meyer understood that this state of affairs was perpetuated, in part, because professors of anatomy and physiology and histology demonstrated the conceptual and technical knowledge of their respective disciplines using tangible, sharable, stable objects of study. By adopting the case history as a common unit of study, psychiatry could partake in what Meyer considered the defining practices and expectations of a clinical science: comparative studies of normal and pathological processes, systematic testing and evaluation of clinical data, the accumulation and transmission of new medical knowledge, and, above all, a persistent awareness that the purpose of correlating clinical and pathological findings was to direct therapeutic innovation.[81]

The clinical practices inaugurated by Meyer at the Phipps Clinic between 1913 and 1917 emerged from his longstanding ambition to harmonize his conception of mind as an instrument of biological adaption with the methods and expectations of clinical medicine. "Records are not the last step, but the means for the most advanced efforts, the collection of data of experience," he reflected in 1898, the year he formulated his concept of biology. Thirty years later, he could proclaim that "the clinical case today is the baptismal child itself, the acceptance of psychiatry as a basic and natural part of the training and practice of medicine, in other words, pathology and therapy."[82] So integral was the case history to the Phipps environment and its procedures, that even patients quickly appreciated its significance. One morning in 1914, during his weekly teaching clinic, Meyer asked a thirty-eight-year-old housewife to describe for those assembled the troubles that had brought her to the hospital. She hesitated, then offered the chief of psychiatry of Johns Hopkins Hospital some advice: "I think you can find out better by consulting the history." Undoubtedly, the students and staff in attendance exchanged a few knowing glances and amused expressions to see Professor Meyer's clinical demonstration momentarily derailed by his patient's perceptive acknowledgement of the fundamental truth upon which he had devised his methodology for the new clinical science of psychiatry.[83]

The Henry Phipps Psychiatric Clinic, and its donor and namesake, on a postcard in 1913. The clinic was designed by the architect Grosvenor Atterbury according to Meyer's specifications. Atterbury had worked with the philanthropist on the well-known Phipps Houses, model tenement housing in New York City.

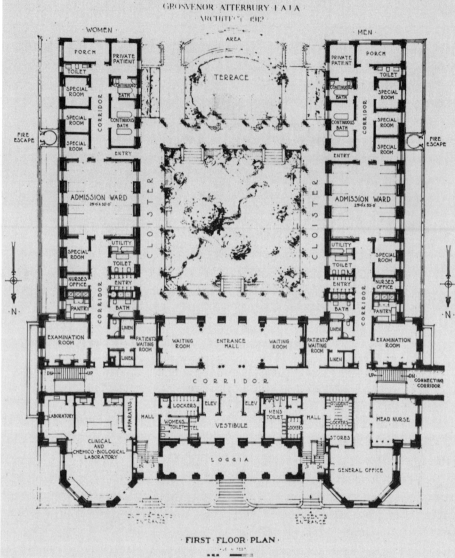

Plans for the ground floor of the Henry Phipps Psychiatric Clinic designed by Grosvenor Atterbury according to Meyer's specifications. Patients and visitors entered a well-appointed lobby through the front doors to see an interior courtyard and garden beyond. A door at the rear of the building led directly to the admission ward, for arrival of patients who were in an agitated state.

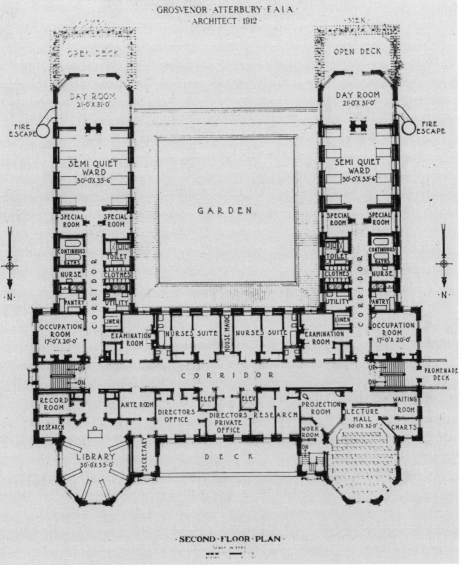

THE
HENRY·PHIPPS·PSYCHIATRIC·CLINIC·
FOR
THE·JOHNS·HOPKINS·HOSPITAL·BALTIMORE·
GROSVENOR·ATTERBURY·F.A.I.A.·
·ARCHITECT·1912·

·SECOND·FLOOR·PLAN·

Plans for the Phipps Clinic, second floor. The wards for women were in the East Wing and those for men in the West Wing, each with a day parlor and veranda. Every morning, Meyer entered the library directly from his office to discuss patients and their cases with his staff of psychiatrists.

·THE·
· HENRY· PHIPPS· PSYCHIATRIC· CLINIC ·
· FOR ·
· THE ·JOHNS· HOPKINS· HOSPITAL· BALTIMORE ·
· GROSVENOR· ATTERBURY· F.A.I.A. ·
· ARCHITECT · 1912 ·

· WOMEN ·

DAY ROOM
21'-0" X 31'-0"

· MEN ·

DAY ROOM
21'-0" X 31'-0"

FIRE ESCAPE

FIRE ESCAPE

QUIET WARD
30'-0" X 33'-0"

QUIET WARD
30'-0" X 33'-0"

SPECIAL ROOM   SPECIAL ROOM

SPECIAL ROOM   SPECIAL ROOM

SPECIAL ROOM   TOILET

TOILET   SPECIAL ROOM

NURSE   CLOTHES

CLOTHES   NURSE

PANTRY   UTILITY

UTILITY   PANTRY

·N·   CORRIDOR   CORRIDOR   ·N·

LINEN   LINEN

DINING ROOM
17'-6" X 20'-0"

SPECIAL ROOM   BATH   SPECIAL WARD   HOUSE MAID   SPECIAL WARD   BATH   SPECIAL ROOM

DINING ROOM
17'-6" X 20'-0"

UP   UP
DN   DN

CORRIDOR

ELEV   ELEV

RESEARCH DARK ROOM   DARK ROOM   RESEARCH   RESEARCH   RESEARCH   RESEARCH RESEARCH   RESEARCH   RESEARCH RESEARCH   ASSISTANT

PSYCHOLOGICAL LABORATORY
25'-0" X 30'-0"

HISTOLOGICAL LABORATORY
30'-0" X 41'-6"

· THIRD · FLOOR · PLAN ·
SCALE IN FEET

Plans for the Phipps Clinic, third floor. In the dining rooms, patients and staff dined
together. Laboratories and research rooms occupied the north side.

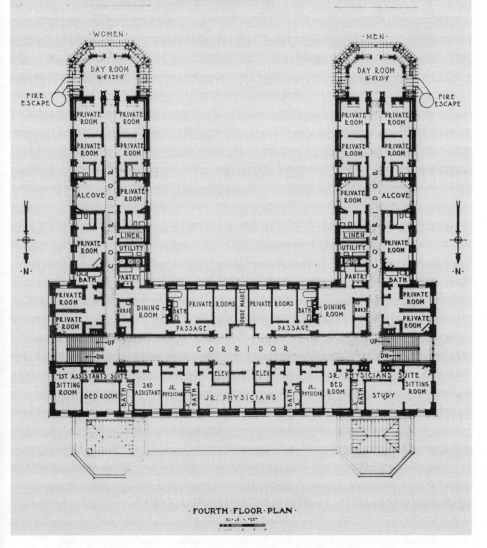

· THE ·
· HENRY · PHIPPS · PSYCHIATRIC · CLINIC ·
· FOR ·
· THE · JOHNS · HOPKINS · HOSPITAL · BALTIMORE ·
· GROSVENOR · ATTERBURY · F.A.I.A. ·
· ARCHITECT · 1912 ·

· WOMEN ·

DAY ROOM
16'-0" X 23'-0"

· MEN ·

DAY ROOM
16'-0" X 23'-0"

FIRE
ESCAPE

FIRE
ESCAPE

PRIVATE
ROOM

PRIVATE
ROOM

PRIVATE
ROOM

PRIVATE
ROOM

PRIVATE
ROOM

PRIVATE
ROOM

PRIVATE
ROOM

PRIVATE
ROOM

ALCOVE

PRIVATE
ROOM

PRIVATE
ROOM

ALCOVE

C O R R I D O R

C O R R I D O R

PRIVATE
ROOM

LINEN

UTILITY

LINEN

UTILITY

PRIVATE
ROOM

· N ·

· N ·

BATH

PANTRY

PANTRY

BATH

PRIVATE
ROOM

NURSE

DINING
ROOM

BATH

PRIVATE ROOMS

HOUSE MAID

PRIVATE ROOMS

BATH

DINING
ROOM

NURSE

PRIVATE
ROOM

PRIVATE
ROOM

PASSAGE

PASSAGE

PRIVATE
ROOM

UP

DN

C O R R I D O R

UP

DN

1ST ASSISTANT'S SUITE

SITTING
ROOM

BED ROOM

BATH

2ND
ASSISTANT

JR.
PHYSICIAN

BATH

ELEV

JR. PHYSICIANS

ELEV

BATH

JR.
PHYSICIAN

SR. PHYSICIANS SUITE

BED
ROOM

BATH

STUDY

SITTING
ROOM

· FOURTH · FLOOR · PLAN ·

Plans for the Phipps Clinic, fourth floor. Accommodations for private patients and resident physicians shared the fourth floor. As in the Johns Hopkins Hospital, the fees of private patients were an indispensable source of income, subsidizing the public wards and other expenses.

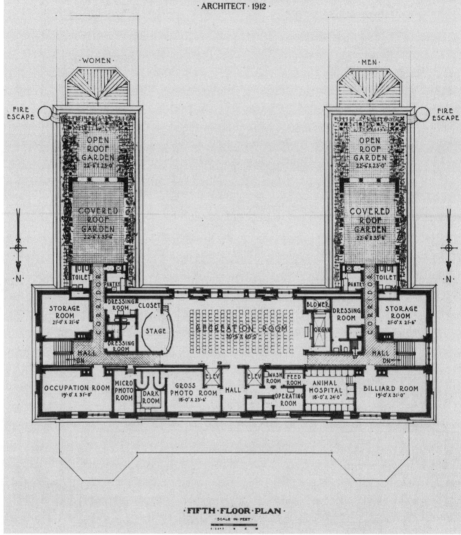

Plans for the Phipps Clinic, fifth floor. On the top floor of the clinic there was a large auditorium and rooftop gardens. An occupational therapy room and billiard room for patients flank research facilities, including a hospital and surgery for experimental animals.

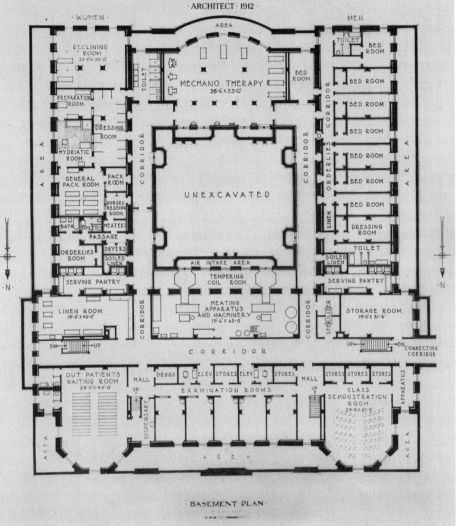

THE·
· HENRY· PHIPPS· PSYCHIATRIC· CLINIC·
· FOR·
· THE ·JOHNS· HOPKINS· HOSPITAL· BALTIMORE·
· GROSVENOR· ATTERBURY· F.A.I.A·
· ARCHITECT· 1912·

· WOMEN ·

MEN

AREA

RECLINING
ROOM
25·0"X50·0"

TOILET

BED
ROOM

PREPARATION
ROOM

TOILET

MECHANO THERAPY
38·6"X53·0"

BED
ROOM

BED ROOM

DRESSING
ROOM

BED ROOM

HYDRIATIC
ROOM

CORRIDOR

CORRIDOR

ORDERLIES

CORRIDOR

BED ROOM

GENERAL
PACK ROOM

PACK
ROOM

UNEXCAVATED

BED ROOM

NURSES
DRESSING
ROOM

BED ROOM

BATH TOILET HEATERS

LINEN

BED ROOM

PASSAGE

DRESSING
ROOM

ORDERLIES
ROOM

DRYERS

TOILET

SOILED
LINEN

SOILED
LINEN

SERVING PANTRY

AIR INTAKE AREA

SERVING PANTRY

TEMPERING
COIL ROOM

· N ·

· N ·

CORRIDOR

HEATING
APPARATUS
AND MACHINERY
19·6"X65·0"

STERILIZER

CORRIDOR

STORAGE ROOM
19·0"X 31·6"

LINEN ROOM
19·0"X42·0"

DN

UP

UP

DN
CONNECTING
CORRIDOR

C O R R I D O R

OUT PATIENTS
WAITING ROOM
29·0"X44·0"

HALL

DRUGS

ELEV

STORES

ELEV

STORES

HALL

STORES

STORES

STORES

APPARATUS

UP

UP

DISPENSARY

EXAMINATION ROOMS

CLASS
DEMONSTRATION
ROOM
29·0"X37·0"

AREA

AREA

A R E A

· BASEMENT· PLAN·

Plans for the Phipps Clinic, basement. In the dispensary, on the north side of the
building, psychiatrists interviewed outpatients who sought help without an
appointment. The basement also housed the hydrotherapy facilities,
gymnasium, and orderlies' quarters.

The waiting room of the Phipps dispensary, one of the first psychiatric walk-in clinics in the United States. Part of the space is set up for an occupation class in basketry. Meyer and many of his counterparts in Maryland considered basketweaving ideally suited to the goals of occupational therapy.

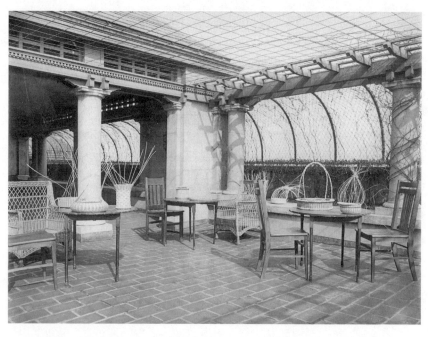

Baskets in various states of completion on a rooftop garden. Meyer went to great lengths to disguise or prettify necessary security measures such as window bars and fencing.

Hydrotherapy—involving high-pressured and gentle specialized showers (seen here) as well as friction rubs and steam baths—became common in psychiatric hospitals at the beginning of the twentieth century. In an era that embraced extreme health regimens, most patients considered these procedures central to their treatment at the Phipps Clinic. Meyer was ambivalent about them.

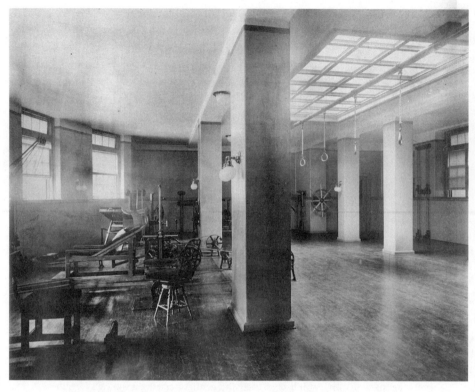

The gymnasium in the basement of the clinic. Exercises, including yoga and basketball, were part of the therapeutic regimen of the Phipps Clinic.

CHAPTER 5

A *Wonderful Center for*
Mental Orthopedics

Adolf Meyer's Therapeutic Experiment

"At last," Adolf Meyer could finally declare at the opening of the Phipps Clinic in 1913, "science begins to take up with new and forceful methods the great problem of mental life."[1] He was confident that at Johns Hopkins his concept of psychobiology, premised on the pragmatic notion that mind was an instrument of adaptation, would support a clinical science of psychiatry. One year later, he proudly told his peers in the American Medico-Psychological Association that the purpose of his clinic was "for study and for work with bed treatment, bath and hydrotherapy, occupation and mental readjustment."[2] Were these the new and forceful scientific methods to which he referred at the clinic's opening? A traditional hospital approach of proper nutrition and rest? Water treatments and a good day's work? Talking things over with a doctor? These remedies were hardly inventive, hardly new—they could all be found in the Hippocratic corpus. Even bringing them together in a kind of therapeutic commune or morally uplifting regime was suggestive of the so-called moral therapy introduced in various European and American retreats for the mentally ill dating from the eighteenth century. If there was anything innovative about the Phipps Psychiatric Clinic at Johns Hopkins—as Meyer and most of his contemporaries asserted—it was not in its offerings of convalescent care, baths, diversions, and good counsel.

He was describing the pioneering approach he had instituted at Johns Hopkins when he stated that the function of the Phipps Clinic was "for study and for work" with these traditional approaches. Recall that Meyer conceived of each individual form of psychopathology as an "experiment in nature," a perspective he considered important when applying traditional clinical methods to the subject matter of psychiatry. The metaphor of an experiment emphasized that the adaptive habits of each patient were observable, perhaps modifiable, natural phenomena with a developmental chain, and it reminded his clinicians that the data of adaptive behavior must be collected and recorded with the same degree of accuracy accorded laboratory notes or autopsy reports. Following this logic, a routine of rest, regular mealtimes, occupation, hydrotherapy, recreation, and talk therapy functioned as the chassis for the real engine of Meyerian psychiatry: a perpetual cycle of inquiry and rehabilitation that began and ended with the patient's psychobiological reactions.

Meyerian psychiatry was innovative because scientific investigation and therapeutics became two sides of a single coin. This was a reversal of the circumstances that had so surprised Meyer when he arrived in the United States twenty years earlier, when the supervision of insane individuals and scientific research were considered mutually exclusive. In 1913, he described the Phipps Clinic as a place to study "the *causes and modifiability* of harmful cravings and of poor mental habits." Three years later, he declared that "the great progress of psychiatry of today over the psychiatry of the past" is that "it tries to bring out what is sane and sound in every one of our patients, and to study even the morbid features for the sane and sound kernel they may present." And in his personal working notes, he mused, "[T]herapy is the pathology as far as it can be used—our task, thus, is to turn *pathos* into something manageable." Over and over, Meyer described the new psychiatry at Johns Hopkins by speaking of clinical investigation and medical treatment in the same breath.[3]

I have termed this pioneering approach the "therapeutic experiment." The environment of the Phipps Clinic was the confluence of its physical spaces, hospital routine, and social milieu. As the patient interacted with this environment, his or her psychobiological reactions were observed and recorded collectively and relentlessly by Phipps psychiatrists, nurses, students, attendants, social workers, and occupational therapists. These clinical data were deposited into a central record to be interpreted and reinterpreted daily by a ward psychiatrist in consultation with Meyer and his senior staff each morning at a meeting called a case conference. Meyer and his staff then applied the findings immediately to formal psychotherapeutic interviews and spontaneous interactions with the patient to modify his or her reactions—at which point the sequence began again with more observations, notations, case conferences, and interactions with the patient. This was the perpetual cycle of the therapeutic experiment introduced by Meyer at Johns Hopkins in 1913.

To facilitate both sides of the therapeutic experiment, the Phipps Clinic served simultaneously as a laboratory to identify pathological processes and as a medical instrument with which to modify them. As a pathology lab, it was a space to dissect the patient's life history and to discover factors that contributed to the development of the abnormality. The Phipps environment also acted as an experimental field against which idiosyncratic reactions were brought into relief by what Meyer described as "uniform conditions." He reasoned that how a patient adjusted to the hospital routine—every attempt to follow, conquer, subvert, or repudiate it—made manifest unique adaptive patterns not discernible within the uncontrolled conditions of everyday life. In devising the physical and social spaces of the clinic, he sought to distill the common sense consensus, the collective imperatives that prescribed so much of everyday life and his comparative standard of normal for psychobiological reactions.[4]

Meyer called his therapeutic strategy "habit-training." Adapting to life inside the Phipps Clinic *was* the treatment; it was less something done to the patient than something done by the patient, with the staff's facilitation. The hospital routine was supposed to artificially enforce what he considered to be psychobiological equilibrium. Just as a healthy heart must stay fit in order to perform efficiently during periods of rest and distress, the goal of Meyerian therapy was to restore the fitness of the personality. Meyer thought he could accomplish this—in many cases, not all—by overriding ineffective and deeply ingrained adaptive patterns and replacing them with habitual responses that were natural, wholesome, healthy, satisfying, and efficient— terms that, for him, were largely synonymous. "We naturally begin with a natural simple regime of *pleasurable ease*," he explained, "a sense of a day simply and naturally spent, perhaps with some music and restful dance and play, and with some glimpses of activities which anyone can hope to achieve and derive satisfaction from." Embedded in this description of what sounds like a delightful day at the Phipps Clinic was the pragmatic emphasis on the dynamic interaction between organism and environment and the assumption that mind serves as an instrument of biological adaptation. Meyer's aim was to help patients sharpen their mental tools. He endeavored to impose an environment that, first, obliged the patient to respond to physical and social demands and, second, incited spontaneous responses that, as William James had phrased it, moved him or her in the direction of survival. Meyer maintained that the only way to achieve psychobiological balance was "*actual doing, actual practice*, a program of wholesome living as the *basis* of wholesome feeling and thinking." Habit-training had more in common with physical rehabilitation therapies than surgery or drug regimens.[5]

This chapter analyzes Meyerian psychiatry in action as Meyer and his staff cycled through the therapeutic experiment of observing, recording, analyzing, and applying the data of psychobiological reactions. The narrative follows a day in the Phipps Clinic from morning to evening and describes institutional procedures related to risk management, patients' autonomy and leisure time, ensuring adequate nutrition and sleep, and the sedation and immobilization of agitated patients. Two elements of the hospital routine—occupational therapy and hydrotherapy—are examined closely. Throughout, the experiences of psychiatrists, nurses, and patients illuminate the realities of daily life in the clinic. This narration begins following a brief discussion of practical challenges facing Meyer and his decisions about how to organize the wards and staff.

Not surprisingly, least of all to Meyer, his idealized hospital routine was rarely routine in its execution. Putting the therapeutic experiment into practice with patients suffering a range of mild to severe psychiatric symptoms was not easy. The case histories show that staff were directed to modify intuitively aspects of the routine to suit a particular patient's state of mind or therapeutic needs. At the same time, with a premium placed on the observation of old and the inculcation of new habitual reactions, Phipps patients were granted a wide degree of latitude to deviate, ignore, or resist elements of the routine. The physical spaces of the Phipps Clinic were designed to avoid injury, minimize mischief, and prevent tragedy. Nevertheless, violence and lamentable incidents did occur, such as the one Meyer described to his brother only three weeks after the clinic opened. A patient jumped from the veranda on the third floor, escaped, and hid himself in a cellar, where he drank a good deal of petroleum. "We found him," Meyer recounted, "and he is now giving us considerable worry owing to the most remarkable blistering of the skin of the back." It was an inevitable and painful lesson, he conceded. He took heart in the reality that other patients "give one more pleasure and satisfaction." The erratic, sometimes volatile nature of psychiatric symptoms meant that patients and staff constantly engaged with and disengaged from both therapeutic tasks and each other as circumstances dictated.[6]

Uniquely, Meyer had at his disposal optimal human resources to implement this new approach to psychiatric inquiry and treatment (circumstances that would change after World War I). With a daily cohort of psychiatrists, nurses, student nurses, attendants, and occupation instructors numbering above fifty, the ratio of staff to the clinic's eighty-eight patients was better than 1:2. There was a senior staff of five clinicians (Meyer, assistant director Macfie Campbell, chief resident psychiatrist D. K. Henderson, and two assistant resident psychiatrists—ideally, one male and one female) plus a psychiatrist assigned to each of the eight wards. This put the doctor-patient ratio at roughly 1:7 (not including medical students also working on the wards). This was a striking departure from the traditional state asylums in which Meyer had worked during the previous two decades. In 1894, doctor-patient ratios in New York State hospitals were between 1:107 and 1:240, while the attendant-patient ratio was roughly 1:7. Nationwide, the proportion of attendants to patients living in public or private asylums was estimated to be 1:12. By any measure, the size of the Phipps staff was novel for American mental hospitals of the period. Factor in the Phipps cleaning and clerical staff, and the clinic hummed with more than sixty personnel and the eighty-eight patients.[7]

Meyer expected his staff to maintain and, if necessary, manipulate the Phipps environment to encourage patients' success in making efficient psychobiological adjustments. He defined adaptive efficiency as selecting and achieving self-interested goals. It was the responsibility of all Phipps staff members to create conditions for achievement by individualizing activities and social interactions. The prevailing objective was to cultivate in every patient some interests within the dominion of the common sense consensus and to help him or her experience satisfaction from pursuing and achieving them. For Meyer, this was the definition of psychobiological health.

Incidents frequently associated with psychiatric institutions, such as misconduct by staff or neglect of patients, are absent in the documentary evidence for the Phipps Clinic in this period. To maintain the integrity of what he perceived as his chief therapeutic tool—the social milieu and equanimity of the Phipps environment—Meyer demanded from his staff patience, flexibility, kindness, resilience, genuineness, and self-control. Indeed, the case histories portray the predictable skirmishes between hospital staff and some psychiatric patients as prosaic or unproblematic. Members of the Phipps staff undoubtedly found this degree of composure and forbearance difficult to maintain, even impossible or impracticable at times. Nevertheless, those who lost self-control risked dismissal. D. K. Henderson recalled that Meyer was never more angry than when he discovered that a patient had been misused. Letters of gratitude from patients are abundant in the case histories and depict the Phipps staff as largely benevolent, despite inevitable conflicts. The following passage from a letter to Meyer is typical: "I cannot close this letter without thanking you and Doctors Campbell, Hall, Oliver and other doctors, and all nurses (excepting one), and orderlies and attendants for unfailing kindness to me while there." A lack of documentary evidence does not mean that wrongdoing did not occur, although the extraordinary importance and wide sweep of record keeping at Phipps would make its documentation more to be expected there than in other institutions. The evidence we do have helps to explain why—within an oligarchy fueled by the rigorous documentation of all reactions and interactions—there are very few indications of misconduct by staff members in the case histories.[8]

The internal organization of the wards was designed to modulate the ideals of the therapeutic experiment with the realities of caring for patients with mental disorders. The women's wing of the clinic, on the east side, and its counterpart on the west side for men were each divided into four wards on four separate floors. Each of the wards on the second floor accommodated "ten semi-quiet cases," and the third-floor wards provided for "about sixteen quiet cases." These were communal wards, with all patients' beds occupying one large room. "The third floor is a ward for patients whose behavior is perfectly orderly," he told colleagues in 1914. "On the second floor we try to maintain, as far as possible, a similar regime, keeping the ward as free as possible of actively disturbing elements." Behavior, then, rather than diagnosis, dictated how patients were assigned to wards. In this respect the Phipps Clinic resembled the traditional asylum rather than the modern general hospital.[9]

Each ward had a parlor on its south side that was a vital common space for enacting the therapeutic experiment. Patients and staff often called it the sun room or day room because on most days it was bathed in sunlight. Each of the wards on the third floor had a vine-covered veranda that extended beyond the parlor and offered a view of Baltimore's downtown and harbor in the distance. On the communal wards there was a shared lavatory and dressing room, the nurse's office, an examination room, a linen room, and a pantry. The private patients of Meyer and Campbell occupied single or semiprivate rooms on the fourth floor connected by shared baths. The male and female fourth-floor wards could accommodate eight private patients each. With the director's permission, some private patients retained a personal nurse who lodged with them. Meyer occasionally deemed this arrangement counterproductive to therapeutic efforts. Finally, on the ground floor there were special provisions for those patients requiring continual supervision or isolation. Dubbed the "disturbed ward" by the staff, it is discussed in more detail below. In Phipps parlance, used by staff as well as savvy patients, wards were called East-1 or West-2 according to the corresponding division and floor.[10]

"Habit-training" and "Occupation Therapy"

Dressing, breakfast, and morning conversation began the day for most Phipps patients. Ambulatory patients shed hospital-issued nightgowns for bathrobes to bathe before breakfast. Those deemed not well enough to leave bed were bathed by nurses and changed into a clean gown. On the general wards there was a communal lavatory where patients performed their own ablutions or were assisted by attendants. Under the ward psychiatrist's direction, some patients dressed for the day in their own clothes while others wore identifiable hospital garments stamped with the words *Phipps Psychiatric Clinic.*

The frantic letter-writing of a twenty-two-year-old woman paints a rich though distorted picture of getting dressed on the ward. Due to her strange behavior, observed by her family and friends, she had been brought to the clinic by her sister and father. The latter, it was noted, was intoxicated. Although she had agreed to enter the hospital, she now claimed that she had been tricked into doing so. "Think of your child in a dirty ward," she wrote imploringly to her father, "bathing and dressing in a cold bath room with twelve other people, dressed in ward clothes, which consist of percale dress, drawers, shirt, short petticoat and stockings." The ward psychiatrist noted that she wrote her family several letters each day "denouncing everything and everyone in most severe terms and pleading to take her back home and let her explain it was all a put on." She tried to horrify her mother, writing that "if you could see your child going around in a short gown up to her knees, with a striped robe on, on this cold, cold, cold concrete floor you would get her home in some way." Another note warned: "My teeth are decaying, my hair coming out, and I will be wrecked for life if you don't come." When the woman realized psychiatrists were reading her correspondence, she left a "good letter" out for the doctor and surreptitiously asked another patient's visitor to mail her candid notes. The would-be accomplice dutifully surrendered these to the nurse. "I wish you could see your child bent down to her knees," she pleaded in yet another letter, "my hair down my back no powder on my face, nothing to brush my teeth with, no soap, nothing,

nothing." Though patients' possessions were regulated by staff, there is sufficient evidence that personal hygiene products like tooth powder and hair brushes were encouraged as part of healthy habits. All of her letters were transcribed into the case history and their contents became data in the therapeutic experiment.[11]

The capacity to get dressed and spend the day in one's own clothes—a psychobiological habit executed unthinkingly each day by millions of Americans outside the clinic—is a good illustration of how Meyer employed the common sense consensus as a comparative measure of normal adaptive behavior. Patients' behavior regarding dressing was recorded. For example, nurses reported that Percy had "been up and dressed several different times but never for any great length of time." Percy, who spent almost a year in the clinic, moved through periods of confused elation, unresponsiveness, and relative clarity—transitory states considered typical of a schizophrenic reaction. Five weeks later, nurses noted improvement: he "dressed himself without assistance [and] tries to feed himself." A notation about Richard (the nineteen-year-old son of a hardware store owner mentioned in Chapter 3) elucidates how a quotidian task like dressing was considered a variable in the therapeutic experiment. "Yesterday for the first time he consented to don his clothes and went down to the gymnasium," the psychiatrist noted, "he was rather timid and bashful at first, but became more confident." Richard wore his clothes all day and helped with jobs on the ward. "He is keen to go home," staff observed, "wants to rest a while there and then go to work." Richard's diagnosis was described as "manic-depressive insanity" on his discharge form. During his hospitalization at Phipps, his willingness to dress and go to the gym, his mood and demeanor, and his intention to return to normal life were interrelated and considered part of the same enterprise.[12]

In addition to being an experimental variable, dressing became a tool of institutional control and a form of personal expression for patients. Richard, who earlier had consented to wear his own clothes, became "very sullen, obstinate and resistive" when his father agreed with Meyer that Richard should remain in the clinic and refused to take him home. For several days, he refused to leave his bed because he was denied his own clothes. During this time, a nurse "removed a piece of sock and collar button from his mouth which he had been chewing for some time." While staff withheld street clothing to facilitate confinement, patients also exploited the act of dressing. Another patient walked "up and down the hall clad in a bath robe and his underwear, refusing to put on more of his clothes." On the one hand, Phipps psychiatrists noted it as a sign of improvement in Richard when he voluntarily donned his clothes. On the other hand, nurses and attendants were expected to persuade the underwear-clad patient to dress. In either case, from a psychobiological perspective, forcing the patient was considered pointless. The objective was not to dress Richard for the day—as was the responsibility of many an asylum attendant—rather, it was to induce him to react in a way that fit with the common sense consensus.[13]

Quotidian tasks provided opportunities for Phipps staff to observe manifestations of psychopathology that might elude the specialist in private practice who saw the patient only during consultations. Gladys sought treatment at the clinic because her compulsion to wash her hands too frequently interfered with her hopes and plans in daily life. In the clinic, staff observed that she sometimes spent all day dressing, because of fretting and indecision. One Saturday morning, Macfie Campbell interviewed Gladys in front of the weekly teaching clinic. He asked her about getting dressed: "When you first came to the clinic it took you a long time didn't it?" She assured him, however, that it took her ninety minutes, at the most, and that was because she was not accustomed to getting dressed in front of other people. Rather than challenging her directly and provoking an unwanted defensive reaction, Campbell hypothesized aloud to his students that perhaps she was uncomfortable due to their presence, since "she gives us her symptoms with a certain amount of modification—she would really begin dressing at 8 and would not finish until about 4 in the afternoon." The intensive scrutiny afforded by the therapeutic experiment generated a different set of observations about a case from those that might be collected by means of self-reporting in the neurologist's office or psychoanalyst's study.[14]

Once dressed, patients took the first meal of the day on the ward. The food for all meals arrived prepared from the central kitchens of the general hospital to two pantries in the basement of the clinic. It was then sorted and loaded onto one of eight dumbwaiters bound for the correct ward. There it was reheated and served in one of the dining rooms or delivered on trays to patients in bed. The standard "ward diet" was described as "any good ordinary food." Other diets were prescribed to suit particular cases. The alternative "light diet" (consistent with early twentieth-century notions of nutrition) included eggs, chicken, rare roast beef, mutton, lamb, sweetbreads, fish, oysters, potatoes, fruits, and desserts. A "limited light diet" was prescribed for patients with poor mastication, which often accompanied neurological disorders and recovery from alcoholism. It consisted of milk or juice, strained cooked cereal, soup, eggnog, eggs, minced meat, and mashed vegetables. Ideally, patients took breakfast together in the day room, where the ward psychiatrist, nurses, and attendants were expected to encourage conversation about items in the daily newspapers delivered to the ward.[15]

After digesting breakfast and the morning news, most patients began a busy day of activities, exercise, and hydrotherapy. Patients from the two communal wards traveled together to specially designed rooms throughout the clinic for "occupation therapy." These decampments were often hectic and precarious, and nurses were instructed to assemble their charges on the ward before unlocking the door. Patients could not be allowed to linger in the corridor waiting for the elevator "as the noise is very disturbing to people in the adjacent offices." Lining the main corridor were the psychological laboratories on the third floor and Meyer's office and research rooms on the second. The nursing manual for the Phipps Clinic in this period also advised that those male patients with the resident psychiatrist's permission to smoke were forbidden to do so outside the ward. Female patients were prohibited from smoking anywhere in the clinic. Alternate hours for occupation therapy were reserved for private patients, and the occupation instructor brought individualized activities to patients on the disturbed ward.[16]

The Occupation Department of the Phipps Clinic conducted classes for male and female patients in modeling of clay, painting, weaving, bookbinding, knitting, leatherworking, and, Meyer's perennial prescription, basketweaving. In most cases, Meyer regarded handwork as a tool for igniting impulses of self-interest and the cravings for satisfaction essential to efficient adaptation. In the carpentry shop, male patients made wooden trays, tables, and bookshelves, and they were encouraged to send their achievements home as gifts. In needlework and sewing classes, female patients made slippers, shawls, and tablecloths that they sent to family and friends. By 1919, the clinic also had pottery studios and a large loom.[17]

Manual occupation was an old form of treatment that had long capitalized on the therapeutic dividends of work. At the beginning of the twentieth century, it received new names that reflected new rationales: reeducation, reconstruction, and occupational therapy. In both Europe and North America, it played a central role in therapeutic regimes, not only for the mentally ill, but also for those with cognitive or physical handicaps or tuberculosis. In the United States the trend was underscored by a new cultural valuation of handmade objects, creativity, and manual work that culminated in the Arts and Crafts Movement. This handicraft craze gripped the American middle class, who bemoaned the loss of craftsmanship and individual accomplishment in an age when mass production was entering more aspects of the material culture. The assembly line of the Ford Motor Company came to life the same year as the clinic, producing an automobile every two hours, a revolutionary event.[18]

Concerns such as those about the decline of manual labor and individual craftsmanship were not solely aesthetic; they reflected widespread anxieties about the harmful effects of modern urban life on the nervous system. This notion was inherent in the ubiquitous diagnosis of neurasthenia in this period. Beginning in the 1880s, physicians prescribed rest cures to replenish the weakened nerves of neurasthenic patients, assumed to have been depleted of critical "nervous energy" by overwork or overstimulation. Medical explanations and cultural understandings of neurasthenia (literally, "nerve weakness") began to shift around 1900, away from the idea of exhausted nerves and toward the growing perception that mental exertion disrupted the natural balance between intellectual and physical functioning in the nervous system. Manual work, such as making traditional handicrafts or farming, was increasingly assumed to rebalance the brain. Therapeutic occupation and physical exercise gradually displaced rest and the replenishment of nerves in the treatment of Americans' nervousness.[19]

Within Maryland's mental institutions, including the new Phipps Clinic, occupation and recreation were considered active therapies, not merely distractions, and this attitude reflected national trends. In Maryland, superintendents of state mental hospitals and proprietors of private asylums were preoccupied with implementing therapeutic occupations and recreation in their institutions. In 1908, they formed the Maryland Psychiatric Society, and in 1911 (the year Meyer happened to join) the organization published the first issue of the *Maryland Psychiatric Quarterly*. Between 1911 and 1917, therapeutic occupation dominated the journal (far exceeding content on hospital management, histological laboratory techniques, or changes to Maryland's insanity laws). Advertisements for craft supplies and blurbs on Maryland's multi-institution craft sales or new basketweaving techniques also permeated their pages. Members of the society discussed the ethical implications of selling patient-manufactured goods and the use of occupation in managing excited patients or improving public relations. The journal also announced the appointments of young women, specially trained in occupational therapy, to the state's hospitals. Eleanor Clark Slagle, a dominant force within this emerging profession, was the inaugural occupation instructor at the Phipps Clinic. William Rush Dunton, another leader in the budding specialty, was assistant director of the Sheppard and Enoch Pratt Mental Hospital in Baltimore. Concurrently, he lectured and treated outpatients at the Phipps Clinic. Slagle, Dunton, and Meyer worked closely and shared a pragmatic orientation that shaped the trajectory of American occupational therapy in the twentieth century.[20]

Underwriting Meyer's use of occupational therapy was William James's instrumentalist notion that abstract thought activated evolutionary drives to achieve goals and satisfy self-interested needs. Occupational therapy at the Phipps Clinic also reflected Meyer's functionalist view of thought and action as a single continuum of mutually reinforcing sensory-motor activity. In 1916, he told a graduating class of nurses that action was both the foundation and the climax of mental life. "Talk and feeling and thinking," he said, "are but a way to action." When a person reacted to his or her thoughts and feelings with more thoughts and feelings, instead of productive activity, a potentially harmful autocatalytic process was begun. This mental feedback loop, as we might say now, could lead to "hypertrophy or drying up" of normal, useful functions like deliberation, confidence, imagination, or emotion. Cultivated exclusively, a commonplace instinctual, emotional, or intellectual impulse could become a "troubling fermentation of thought and fancy" manifested as symptoms, such as distraction, phobias, delusions, anxiety, or pathological emotional states (depression, elation, rage, or paranoia). Notionally, then, weaving a basket or constructing a wooden toy served the same therapeutic function as getting dressed or discussing the morning news: the activity was an (artificially produced) opportunity for the patient to select a self-interested goal within the external environment and execute the necessary psychobiological adjustments to achieve it.[21]

At stake, in Meyer's view, was the regulation of instinctual drives—the balancing of their inhibition and gratification and their integration with higher faculties. Occupation was intended to divert the patient "from the temptations or attractions alluring to his morbid appetite, longings, and fancies" and tap into his or her underutilized resources. Walter, a twenty-nine-year-old clerk from New York City, remained debilitated by anxiety after weeks in the clinic. When his brother inquired about his progress, Meyer said that the problem resided in Walter's reluctance to leave the internal world of his anxiety. "I am firmly convinced that thought without action and habit-training is of no earthly use," he explained to Walter's family. "I have insisted emphatically on him taking the occupations and all those things seriously, which he had been trying to subordinate to thinking about his problem." Included in the suite of adaptive resources regulated by mentation were, of course, social instincts. In a 1921 address in which he described his philosophy of occupation therapy, Meyer waxed cryptically that concrete performance was the means of "bringing the very soul of man out of dreams of eternity to the full sense and appreciation of actuality." Achieving goals through action, in other words, necessarily entailed meeting demands in the external environment, which was structured by common sense. Meyer surmised that feelings of inferiority, often sexual, were at the root of many patients' disorders and were reflected in an "inability to use the hands to produce things" that were respected by the patient or others.[22]

Meyer expected occupation instructors to arouse a spontaneous adaptive reaction in the patient, not simply plan and supervise activities. He described the special qualities of the ideal occupational therapist. He or she had to possess "freedom from premature meddling, and tact in avoiding false comparisons or undue expectations fostering disappointment, orderliness without pedantry, cheer and praise without sloppiness and without surrender of standard." In other words, the goal was to inspire social confidence with genuine encouragement, not hyperbole or pretense, while always upholding the tacit standards of the common sense consensus. Above all, he emphasized, he or she must value "the native *capacities and interests* of the patient." One wonders how many of his listeners inferred the missing pragmatic adjectives—*adaptive* capacities and *Jamesian* interests—that were, for Meyer, inherent in his choice of nouns. He imagined occupational therapy as a social exercise in which the patient learned to recognize his or her usefulness to the collective. In turn, this recognition fostered adaptive resources, such as empathy and altruism, that helped a person to thrive in a predominantly social environment.[23]

There are numerous examples of patients who responded to habit-training in the way Meyer hoped. Staff noted that Agnes "reads, plays games, works raffia, sews." One Sunday afternoon, however, she told the psychiatrist that "I've just got it into my head I'm not going to get better." Perhaps, she suggested, the thought she had expressed "was due to not being occupied," since it was Sunday. A patient named Edward, who was thirty-five years old and eventually diagnosed with "psychasthenic depression," was evidently productive during his time at Phipps despite the existential gloom that pervaded his thoughts and outlook. "I sent you by Father my first tray," he wrote to his wife; "you may keep it if you like, or later you may decide to give it to some one as a gift. I shall make some more in different styles." He also reminded her that his father had promised to send some wood: "Don't let him forget, because the bases are quite expensive here." One day, Edward and two other patients ventured to nearby woods with an attendant to gather materials for basket making. He told his father, "[E]ach of us had great bunches of mountain laurel." He enjoyed the occasion, he told the ward psychiatrist, but only "as much as my condition would permit." He boasted to a friend, "I have made many useful and beautiful things in both carpenter and basket line," including, he enthused, a mahogany book case that he shipped to his wife for her birthday. Many other patients complained about the activities and occupations, refused to participate, or questioned whether occupational therapy was a legitimate form of medical treatment.[24]

Because achieving satisfaction and pleasure was the operative principle in Meyerian occupation therapy, merely participating to satisfy or placate those supervising the work was insufficient. In a second letter, Meyer informed Walter's brother that the patient "must be able to make some plan of getting some concrete results from his occupations." Perhaps some Christmas gifts, Meyer had suggested to Walter, "instead of relatively aimless job-work." Meyer claimed that it was ten years earlier when he had begun to understand the difference between "busy work" and productive activity, on the wards of the massive Manhattan State Hospital on Ward's Island. "I was greatly assisted by the wholesome human understanding of my helpmate, Mrs. Meyer," he recollected in 1921. He credited his wife, Mary Potter Brooks Meyer, with introducing "a new systematized type of activity into the wards of a state institution." On Ward's Island the Meyers observed that productivity transformed the behavior of some asylum patients. Where there had once been "bored wall flowers and mischief makers," he marveled, there were patients doing basketry, leatherworking, and bookbinding. "Abstract exhortations to cheer up and to behave according to abstract or repressive rules" had limited therapeutic value he observed, but the pleasurable experiences that accompanied productive work appeared to act as an incentive that induced a patient to willfully adapt his or her behavior.[25]

Coercion interfered with spontaneous reactions, Meyer reasoned, and compromised the integrity and aims of the therapeutic experiment. Accordingly, patients were granted a wide degree of latitude in participating in occupations. For the first few weeks on the ward, one patient simply wandered around and "went into any room where the door was open and stood around." At night he slept very little and spent most of the night whispering and smiling. After a while, however, the patient's "behavior on the ward was fairly normal, he took an interest in the basket work, went to gymnasium and played ball in the yard." With personal satisfaction the ultimate goal, there was little to be gained from bullying a patient. This autonomy, none-theless, was not supposed to supersede the adequate protection of patients whose mental states presented grave dangers.[26]

Patients who were deemed at risk of suicidal behavior remained under "special observation," their activities were modified accordingly, and they were never to be left unattended. The following case demonstrates the inevi-table inconsistencies between the ideals and execution of the therapeutic experiment. Abigail was admitted to the Phipps Clinic after trying several times to kill herself, a desire prompted by a belief that she had wronged her husband. By all accounts, including her husband's, this appeared to be a delusion. In the clinic, she was restless, "unable to sit still to do fancy work," and spent "a great deal of her time lounging around on couch in day room wondering what her husband will do with her." The ward psychiatrist noted, "[S]he finds the routine irksome." One afternoon she was too restless to abide instruction in handicrafts, so a nurse walked with her around the ward. When Abigail needed to use the bathroom, the nurse did not accom-pany her. She was discovered ten minutes later hanging by a towel she had looped through the open transom window above the door leading to the tub room. Abigail had stopped breathing and her face was blue. "When she was cut down she sank limp to the floor," the psychiatrist noted in the case his-tory, and "artificial respiration was given and she began to breathe irregu-larly." The nurse had failed to heed the warning in her nursing manual: "If a patient is determined to end his life, he is extremely cunning and watchful for an opportunity. While you may feel that he is diverted and at ease, noth-ing can ever be taken for granted and the strictest sort of observation must be carried out." Abigail seemed wholly recovered from the event a few days later and said she remembered nothing of the incident.[27]

Immediately thereafter, the psychiatrists and nurses noticed a decided improvement in Abigail, one of them noting that "she became quite indus-trious and worked well at all occupations." They were concerned, however, that she might be doing only what was necessary to be discharged. "It is quite hard at times," ward psychiatrist Augusta Scott conceded, "to tell whether the patient really improved or whether she was just putting on a smooth exterior so that she would be taken home." Despite these doubts,

Abigail was discharged to an eager and hopeful husband, who reported that when they arrived home she was "perfectly calm and cheerful." Less than a week later, Scott received news from him that validated the staff's reservations. "During a brief absence of myself, she secured a rope and hanged herself." She had left several notes for her husband, one of which confirmed that she had intended to accomplish her purpose all along "and would have done so before but that she wanted to have a little time with me." He assured Scott that "in her normal life she was of a bright, sunny disposition, devoted to all of her loved ones and had an exceedingly tender conscience." In this case and others, Meyer's occupational intervention had failed to subordinate the thoughts and feelings of the patient's internal environment in favor of activity and productivity in the external social sphere.[28]

The staff's capacity to protect patients like Abigail was linked perceptually to the hospital routine. "If we but *execute with care and caution the details of daily routine*," the Phipps nurse read in her manual, she would find it possible to be "forever on the alert" for a suicide attempt without living in a constant state of uneasiness. A disturbing inventory of the various modes of suicide attempts "that have really been tried in this clinic" appeared in the nursing manual to reinforce the warning:

> Drowning in bathtub; cutting wrists and neck with eye glasses, watch crystals, safety razor blades, glassware from trays, window panes, glass that had been left in rooms or by the bedside; strangulation by the use of belts, pyjama or bath-robe cords, neckties, towels and bed-clothes; standing on high places, such as sink in bathroom, window sills, beds, etc. and falling on head; swallowing glass; raising bed and allowing it to fall on head; use of scissors, occupation tools, knives, forks, spoons, etc.; tipping chair over backward; taking poisons from dispensary while there for an examination, drinking cleaning solutions; starvation, picking the body to obtain infections, biting and swallowing thermometers, knocking head against wall and running head through panes of glass, escaping from the ward and purchasing drugs, plunging into machinery, trying to pluck out the eyes by removing pins from [sheet] packs; jumping off wall outside Clinic; falling headlong down the stairs.[29]

This composite account of incidents that took place in the Phipps Clinic confirms that Meyer's ideal routine for imposing psychobiological balance was, in practice, always imperfectly executed and routinely ineffective with many patients.

Recreation

Recreation activities, another aspect of the hospital routine, were supposed to encourage and strengthen social interactions. Each day various patients of the same sex were corralled in the recreation hall on the top floor of the clinic for indoor amusements such as singing and dancing classes, school-yard games, lectures, and performances by local glee clubs. Like therapeutic occupations, recreational activities at the Phipps Clinic were indistinguishable in practice from those at other institutions, but it is Meyer's particular conceptualization of them as social instruments that makes them noteworthy. At the nearby Sheppard and Enoch Pratt Mental Hospital, William Dunton organized his charges into theater troupes and choirs that performed for other patients. As was common at asylums in the period, Dunton's patients also participated in organized outdoor sports leagues. Activities like these were not workable for the Phipps Clinic, because its patient population was small and more fluid. Meyer regretted not being able to give his patients the benefits of these cooperative endeavors and also of farming, which under different circumstances he would have preferred to incorporate.[30]

Meyer saw the recreation hall, like all the spaces in the Phipps Clinic, as an experimental field in which a patient's attempt at social adaptation might be isolated and studied microscopically. The room was, quite literally, a site of re-creation that served as a stage for consciously enacting scenarios drawn from the script of the common sense consensus. Each Friday afternoon, for example, the female patients helped to plan and then attended a party on the clinic's top floor where they served tea, played cards, and danced with each other. A billiards room nearby provided a space where the male patients could socialize without female nurses or attendants. These special activities were devised to recreate social spaces and interactions of the so-called real world—in this period, of course, highly gendered and classed—in which, theoretically, staff modeled and patients practiced utilizing normative behavior that Meyer considered an essential adaptive asset.

The cultivation of meaningful social relationships, so integral to the therapeutic experiment, was contrived and deliberate. Nurses, attendants, and orderlies (and often psychiatrists and medical students) attended these social gatherings and were expected to interact with patients in a genuine manner, to collect valuable clinical data, and to serve the exercise of habit-training. One psychiatrist noticed that, although an older private patient "mingles fairly freely with the other patients and enjoys playing pool with them, there is toward his nurse and the physicians an attitude of aloofness, which at times becomes rather surly and overbearing." Another patient, Lottie, who struggled with psychotic illness for over a year in the clinic, enjoyed an organ recital in the auditorium. "I wore my violet chiffon dress," she wrote to her mother, "and one of the young doctors sat beside me." Her comment was recorded and regarded as an encouraging sign of participation in the social consensus. Nurses were warned that, despite appearances, activities were not entertainment but scientific and therapeutic endeavors. It was important to be gracious and interested but to remember that they were patients and not ordinary social contacts.[31]

The planning and supervision of social gatherings was labor intensive for the Phipps staff, and the unexpected was always to be expected. Dolly was "dressed in her own clothes and taken to the Friday party" along with the other women on East-2 and East-3. The next morning, Meyer asked Dolly to appear at his weekly teaching clinic as a "demonstration" case, which she did, and he asked her to tell the medical students what had happened while she was at the party. She reported that she did not remember. "Yes, you remember very well," he assured her, after which she recalled, "that's where I attempted to burn my dress." Meyer relayed to the audience the details of the incident observed and recorded by the staff in attendance: "[T]he patient approached a gas log and attempted to set herself on fire." She had complained at the party, too, that the candy was poisoned and that she was humiliated because she acted and dressed inappropriately. The engineered social gathering had provided an opportunity to observe Dolly's impulses and reactions. Subsequently, they had been recorded and analyzed collectively. Now, Meyer proceeded to the next phase of the ongoing therapeutic experiment. In this instance, in conjunction with his clinical teaching.[32]

In his interview with Dolly, Meyer applied extant clinical data with the goal of advancing his inquiry and modifying the patient's reactions. He asked Dolly to describe the situation in which she had tried to set her skirts afire. "Oh young ladies were playing cards and different things, and dancing," she explained. "I danced a little—that was the time I attempted to burn my dress." In an effort to identify some of the causal links in this ongoing experiment in nature (for himself, his students, and for Dolly), he asked, "[H]ow did that thought come?" Dolly reasoned as she had many times in conversations with her ward psychiatrist: "I am a bad woman and I am all alone in the world." The next thing Meyer wanted to know was how she

spent her time in the hospital. "Doing nothing at all," she replied. He continued to interview Dolly in front of the class for another ten minutes. "Do
you occupy yourself now?" he asked and Dolly shook her head. "Won't you
make an effort?" he cajoled, but got no response. "You want to do
something—won't you try?" Finally, Dolly mumbled her consent. But, he
was not done, he wanted to know about other factors impinging upon the
natural experiment. He asked about her sleep and appetite, both of which
were "very good" she assured him. "Your bowels?" he inquired. When Dolly
did not reply, he asked again. "Your bowels?" to which Dolly said "(testily),
my bowels are alright." Then, with one last prescription, the interview was
over: "Occupy yourself a little and get away from your thoughts." As she was
escorted back to the ward by a nurse, she may have overheard Meyer begin
to discuss her case with his students. He felt she had improved considerably
during her time in the clinic, which signaled the possibility of modifying
her psychobiological reactions. "She will probably be able to begin to pay a
little attention to occupations," he predicted, "and then step by step she will
no doubt regain her poise." If his staff continued to engage her instinctual
social responses, Meyer thought it possible that she could recover.[33]

Not all socialization inside the Phipps Clinic was planned formally. Patients who appeared to present fewer risks for mischief or self-harm were
granted various degrees of freedom and privileges. Some patients were allowed to leave the ward or clinic with an attendant. Others were permitted
to circulate around the hospital unaccompanied. The locked ward door was
opened for these patients, who might sit in a rocking chair under the por-
tico, stroll in the rear garden, or work on projects in the activity rooms. Informal groups could even make use of the recreation hall for singing, dancing, or games. All patients required a nurse's supervision, however, if they
wished to play the grand piano or the pipe organ. Neither was to be used "by
anyone who is not an accomplished musician nor by anyone who is too ill to
take proper care of it."[34]

After a morning ideally full of activity, patients ate lunch. As was the case
for occupation classes, patients from the semi-quiet Ward Two and the quiet
Ward Three formed a convoy to the dining room on the third floor, where
they took lunch and supper together. Some nurses and ward psychiatrists
dined with these groups. A patient that was incapacitated in some way ate
on the ward. During a separate lunch hour, the "less disturbed" of the patients from the first floor were also taken to the dining room for meals.

Another dining room on the fourth floor served afternoon and evening meals to private patients who wished, or could be persuaded, to dine with fellow patients rather than taking meals in their rooms.[35]

After lunch, patients were encouraged to rest or engage in some agreeable quiet activity on the ward, such as letter writing, reading, card playing, listening to music on the Victrola phonograph, knitting, domestic tasks, or handicrafts. Working in the ward was also viewed as therapeutic. Clearing meal trays and returning them to the pantry, for instance, was a communal task encouraged by nurses and attendants, as was sweeping, bedmaking, and polishing brass fixtures. Unlike most American state hospitals in this period, the routine maintenance of the wards at Phipps did not depend on patients' labors. In addition to nurses, attendants, and orderlies, eighteen female and four male servants (as they were called) formed a detachment responsible for housekeeping duties throughout the clinic. Meyer maintained that when a patient helped around the ward, it differed from the obligatory labor he had witnessed patients performing in the industrial shops and laundries of large asylums, if the work was "free and pleasant and profitable" and produced "helpful enjoyment." Work performed by patients—whether a benign distraction or an exploitive obligation—had long been commonplace in mental asylums throughout the world. Again, it is the conceptualization of work as an instrument of biological adaptation that endowed its use at the Phipps Clinic with new significance.[36]

In the prewar period, the ability to perform work was equated with health and well-being more generally. Meyer asked one patient, a cabinetmaker, if there was anything wrong with his mind. The patient replied in a thick Italian accent (which the transcriber tried to reflect in the case record): "I make good work—I make whole tables—there's nothing wrong with me—my wife she says you are crazy, crazy—do you think that I could work if I was crazy?" Work was also seen by many doctors as the antidote to malingering and hypochondria. Lawrence, a forty-four-year-old farmer, regurgitated large amounts of food and complained of a whizzing sound in his ear and constant back pain. He told ward psychiatrist Charles Thompson that one of the doctors at another city hospital had abruptly discharged him, implying that Lawrence was a loafer and instructing him "to go out at hard work." Lawrence told Thompson that he had left that hospital and collapsed on the street. Agnes, who had attempted to hang herself, declared her future hopeless because "nobody wants anybody who can't work." Throughout history, the capacity to perform one's vocational function—be it Dauphin or dressmaker—has remained an essential criterion by which a social group evaluates and rules on the sanity of its members.[37]

Just as handiwork produced out of deference to superiors or fear of reprimand did not serve Meyer's therapeutic strategy, the mere execution of work was not in itself a sign of psychobiological health. After Percy had lingered in an unresponsive state for weeks, the staff was pleased to see that he was "up walking around [and] seems more interested." His condition was described as a schizophrenic reaction and his catatonic state was considered typical of the disorder. It was viewed as a positive sign when he polished the floor and helped make the beds (not to mention that he "worked a while on a basket this afternoon"). Eventually, however, it became evident that these were not necessarily signs of improvement. He had merely transitioned from a stuporous state to one of extreme suggestibility, also emblematic of the condition. "Will do anything he is told," the psychiatrist observed and recorded, "no matter how absurd." Since Percy was unable to distinguish reasonably between, for instance, a command to polish the floors with molasses or to do so with floor wax, neither his capacity nor his willingness to work signaled improvement.[38]

In addition to its therapeutic value, recreational activity as framed by Meyer was a kind of experimental control that he imposed upon each unique experiment in nature. One female patient, who had been in the clinic for many months, was scheduled to go for a drive in Meyer's automobile (it was often used by the staff for this purpose). The nurse who was to accompany her noted that she "was to go out in Dr. Meyer's machine this morning but refused" because the seasons had changed since her admission and the patient insisted she did not have appropriate clothing. Psychiatrists noted that another female patient, who believed she was imprisoned at the Phipps to participate in a sexual experiment underwritten by the American government, "refuses to take a ride in the automobile because she objects against her mind being public property." For Meyer, a patient's unwillingness or inability to participate in activities, in this case a car ride, was an important experimental result akin to that obtained by a chemist or physicist after introducing a new variable into a laboratory experiment. In response to the same variable, the first patient's reaction resonated with the common sense consensus, whereas the response of the latter did not. The experimental conditions imposed by the Phipps environment, Meyer reasoned, rendered these important differentials apparent.[39]

The stratified ward structure was a feature of the Phipps environment that Meyer and his psychiatrists exploited as a therapeutic or investigational tool. A few weeks after she attempted to set her dress on fire at the Friday party, Dolly was transferred to the disturbed ward in a psychotic state. She spent over a week pacing alone in her room on East-1 "appearing very much worried and talking almost continually." Dolly had been admitted to the clinic with delusions of her own moral and legal transgressions and repeatedly assured her psychiatrist, Tedrow Keyser, that the Phipps Clinic was a penitentiary where she was locked up because she was a "bad woman." One day, however, when Keyser heard the usual sentiment rephrased slightly— "sometimes I think it is a prison [but] they say it is Johns Hopkins Hospital"—he pressed Dolly for a decisive ruling: "Is it?" He hoped her answer would align with the "they" who said it was a hospital (the common sense consensus) rather than the "I" who thought it a prison (an inefficient psychobiological reaction). Her answer was encouraging, but not unequivocal. She hoped it was Johns Hopkins Hospital, she told him, "so I can get out."

Over the next few days, more logical utterances followed, prompting Keyser to ask Dolly to move to a room on East-2, the semi-quiet ward. "No, what would I go upstairs for?" she wanted to know. It was quieter up there, he reminded her, asking, "won't you go with me?" Dolly continued to protest: "No. I was upstairs and I didn't behave myself. I get mixed up. I've done a great deal of damage to the hospital" (notably, she termed it a hospital). With a great deal of persuasion, Keyser recorded in the case history, Dolly was transferred to East-2. Once there she was heard to say, "I don't think I'm fit to be here." Nurses on the second floor tried to interest Dolly in games, but she refused. She was once again convinced that the clinic was a prison and she begged incessantly to be taken instead to an asylum. "When told this morning that she was to be transferred to another place," read the discharge note, "she would not believe it, saying we intended to put her downstairs because she was so bad." As was the case with every space, activity, or interpersonal interaction within the Phipps environment, the ward structure was endowed with social meaning and often implicated in the therapeutic experiment.[40]

In addition to valuing occupational therapy and social recreation, Meyer put great stock in physical exercise. He insisted that both physical exertion and playfulness were essential to maintaining psychobiological balance. Patients played games with balls, beanbags, or ropes in the recreation hall or ward, on the hospital lawn, or in the airy pavilion at the rear of the Phipps garden. In the gymnasium, male and female patients (at separate times) used machines rigged with pulleys and weights, turned a tensioned ship's wheel affixed to the wall, and did calisthenics or yoga with the gymnasium instructor. Both sexes participated in basketball classes. Meyer asked one patient if he enjoyed the exercise. "I never felt so tired in my life as I do now," the man replied, "playing ball or work in the gymnasium makes my arms and legs sore." One of Meyer's most consistent prescriptions—to private patients, advice-seekers, friends, and colleagues—was to purchase and adhere to a popular book by J. P. Müller called *My System: 15 Minutes' Work a Day for Health's Sake!* It was a system of daily stretching, home gymnastics, breathing exercises, and vigorous friction rubs.[41]

Meyer agreed with William James, who had asserted that performing some "gratuitous" physical exercise each day was vital to "the faculty of effort" and its continued resilience. In this case, anecdotal evidence about Meyer himself illuminates his views best. As the first exhausting year of the clinic's operation drew to a close, he contemplated in a letter to his brother his plans for a summer holiday (customarily, he traveled to Europe or California with intentions of sight-seeing and resting). Purposeful exercise, he remarked, would strengthen his mental abilities and help displace inefficient habits. "A summer with American methods of life, with riding and tennis and swimming would probably prepare me better for the fall and winter than going away and coming back to the old rut."[42] Another lighthearted anecdote from a later period is irresistible: Howard Jones, who received his medical degree from Johns Hopkins in the 1930s, reminisced that, as medical students, he and his young wife decided to use a rare day off in the middle of the week to go ice skating on Lake Roland a few miles from the hospital. When they arrived, they encountered a lone ice skater—the sixty-seven-year-old chief of psychiatry (Meyer retired at age 75).[43] In 1912 Meyer appointed a young psychiatrist named Edward Kempf to the inaugural staff of the Phipps Clinic. When Kempf reported for duty, his ardent idealism regarding the need for reforms in asylums worried Meyer. He soon cautioned Kempf that psychiatry was "exciting work" and therefore he must spend two days per week "in the recreative company of men who will help you to take your mind off this problem." These glimpses into Meyer's personal life shed light on his perception of physical exercise for its own sake as necessary to psychobiological equilibrium.[44]

Hydrotherapy

Exercise was typically followed by hydrotherapy, a significant component of the hospital routine. The new practice of hydriatics, as it was called, represented a distinctly twentieth-century iteration of water cures, a remedy dating to antiquity. The case histories reveal that most patients at the Phipps Clinic between 1913 and 1917 assumed that these impressive new procedures were the scientific medical treatment they expected to receive at a modern hospital like Johns Hopkins. The documentary evidence also shows that Meyer and his staff utilized hydrotherapeutic techniques as therapies, sedatives, and as forms of physical restraint. Each is examined in turn here, after a description of the facilities and techniques and their introduction into psychiatry at the end of the nineteenth century.

Hydriatics consisted of diverse techniques for applying water to the exterior of the body using sophisticated equipment to regulate water pressure, temperature, and intervals. The shower room at the Phipps Clinic was typical. It was made entirely of marble and equipped with curious configurations of chrome fixtures in each corner. In the middle stood a great slab of white marble, its top covered in nickel handles, gauges, dials and indicators, and its plumbing connected to boilers and water tanks in a separate room (it was this heavy fixture that symbolized the iconoclastic climax of *One Flew Over the Cuckoo's Nest*). From behind this console a specially trained hydrotherapist adjusted the temperature, pressure, and duration of various showers, called douches. In the Scotch douche, a popular technique imported from European spas, alternating streams of high-pressured water, one hot and one cold, were administered to the patient's body in timed bursts through two hoses held by an attendant from a distance of twelve feet. There was also the rain bath, a full-body shower administered while the patient reclined in a shallow tub, and the cold needle shower, which stung the skin with short bursts of tiny, high-pressured streams of water. In theory, such "tonics" stimulated the nerves. Hydrotherapy enthusiasts maintained that each technique elicited a different physiological effect, through this manipulation of the temperature, duration, and intensity of the water flow. Administered using this modern technology, baths and showers certainly looked like promising new therapies to many physicians and most patients.[45]

Beyond the shower facility was a room containing six narrow beds. Here, small groups of patients were pinned into the close confines of something called a wet sheet pack. This technique of swaddling the torso and the limbs individually in cold, wet bed sheets produced a chill that was soon succeeded by an intense flush as the body attempted to reheat itself. This commonplace physiological reaction had obvious sedative effects. The copy of *Black's Medical Dictionary* that resided in the staff library at Johns Hopkins Hospital described the wet pack as "a treatment much in vogue" but warned that "the patient, enveloped in this pack, lies absolutely helpless and should on no account be left by the attendant till the pack is removed." The new facility in the Phipps Clinic was designed so that six or seven patients in packs could be attended by a single person. This procedure was also regularly administered on the ward where, as discussed below, it was also used to forcefully immobilize patients.

Like the wet pack, the so-called continuous or prolonged bath typically produced powerful sedative effects. The technique consisted of a deep tub fitted with outlets and drains that kept water circulating continuously. An automatic control was supposed to maintain a constant temperature of 98 degrees. In some models, the bather's legs and torso could be fastened into a hammock affixed to the tub to prevent both floating and escape. Tubs at the Phipps Clinic were fitted instead with a canvas cover that had a hole for the patient's head. Patients were encouraged to drowse and sleep or to read or even work at an occupation like basket-weaving while in the bath. Before the bath, the patient's skin was coated with a thick layer of lanolin cream to prevent sores or infection. Each ward in the Phipps Clinic was equipped with one or two continuous bath tubs.[46]

The soporific effects of baths, showers, and sweating were reconceived during the last half of the nineteenth century as a humane way to manage violent or raving individuals. Encouraged by results achieved at posh health spas, alienists and neurologists began to capitalize on the sleep-inducing effects of water treatments. In 1903, while Meyer was still director of the Pathological Institute on Ward's Island, a state-of-the-art hydrotherapy facility was installed at the Manhattan State Hospital. New York authorities eagerly showed it off to the public and the *New York Herald* described hydriatics as revolutionary. The theory behind this "novel method in the treatment of lunatics," the journalist explained, was simple: "instead of the padded cell, the bathtub; instead of the straightjacket, the Scotch douche." This conflation of therapy and discipline, however, was not new and had engendered medical treatments for the mentally disordered since the eighteenth century. By 1913, modern hydriatics were popularly conceptualized as a compassionate alternative to long-demonized methods of physical restraint and as an effective means of neutralizing violent behavior in patients.[47]

About medical treatment generally in this period, the public imagination was stirred by reports of extreme health regimes and fantastical water or electrical treatments offered at health spas such as John Harvey Kellogg's in Battle Creek, Michigan. There was widespread interest in health benefits derived by accosting the body with extremes. The popular exercise book that Meyer often recommended to patients and advice seekers—J. P. Müller's daily exercises—included energetic body rubs and photographs of Müller performing his exercises in the snow while nude. The practice of dousing patients with alternating bursts of hot and cold water in the shower room at the Phipps Clinic must be interpreted within the context of these newly popular extreme health practices. Conversely, the clammy oppression of the wet sheet pack was considered a familiar, even old-fashioned, treatment that might have been used by a patient's grandmother as a remedy for stomach aches or insomnia. Importantly, then, Phipps patients and their families embraced or tolerated curative measures that today might appear unpleasant, severe, or cruel.[48]

Within medicine and psychiatry at this time, views on hydrotherapy were mixed. Many alienists insisted that hydrotherapy was far more than just an alternative form of physical restraint and that it represented real therapeutic promise with a scientific basis. Like New York, other states began to allocate funds for the installation of hydrotherapy facilities in asylums. Eventually, textbooks and scientific articles surfaced with weighty explanations of hydrotherapy's physiological effects. Yet, its theoretical foundations remained weak, which allowed psychiatrists with irreconcilable views on the essential nature of psychopathology—organic versus psychogenic, for example—to freely advocate its use.[49]

Meyer's own views on hydrotherapy are difficult to parse. Initially, he openly disparaged it, albeit within the context of his chronic frustration over the administrative or political basis of decision making in state asylums. In an address to New York alienists in 1902—just as expensive new hydriatics facilities were installed at Ward's Island—he grumbled about lack of funds for his Pathological Institute and the sizeable sums spent on "special methods of treatment" that he dismissed as mere fads. Such expenditures were public relations maneuvers, he insinuated, intended to advertise the humane and scientific character of the asylum. "Nobody will be unkind enough to ask just how much is actually being achieved," he concluded fatalistically. This criticism reflected his priorities at the time, which were to increase clinical work and research in mental institutions. During his first tour of American asylums, in 1894, Meyer himself was "turned over to the bath master" at the Battle Creek Sanitarium for many hours of elaborate treatments, but he did not remark further upon the experience in his detailed account of the visit.[50]

Given the allocation of space and funds for hydriatics facilities in the new Phipps Clinic, he evidently overcame his reservations, either by choice or by concession. Curiously, after his appointment to Johns Hopkins in 1908, references to hydrotherapy are noticeably absent from his published papers and the buoyant rhetoric he spouted for public consumption. In 1913, he passed on the opportunity to hire a physician who had devoted the previous six years to studying and developing hydrotherapeutics for the treatment of insanity in state asylums. "The main issue is efficiency of what is being done," he explained when he declined her application for a position at the Phipps Clinic, "not of mere success, but of the actual practical results of an institute." The agreeable effects of hydrotherapy were not in doubt, he suggested, as the experiences of thousands of spa-goers had demonstrated. The pathologist Meyer wanted scientific studies of causal relationships between specific techniques and their effects—the clinical research that required controlled methods and conditions and that only an academic institute (like his own) provided. Throughout the planning and building of the new clinic at Johns Hopkins, he publicized his intentions to study, teach, and develop occupational therapy and psychotherapy at the Phipps Clinic in precisely this way (in addition to evaluating the accuracy of laboratory methods of diagnosis). Yet, he appeared uninterested in hiring a qualified investigator who could make hydrotherapy a meaningful part of this research program.[51]

A published report by someone he did hire to administer hydrotherapy at the Phipps Clinic, and notations in the case histories, shed more light on his perception of the treatment's role in the therapeutic experiment. Meyer and his psychiatrists prescribed specific hydrotherapy treatments, in consultation with the hydriatics instructor and based on the predicted effects of various techniques. (The designation "instructor" at the Phipps Clinic is another indication of Meyer's belief that his patients *achieved* rather than *received* their individual treatments.) As was the case with recreation and meals, most patients descended en masse to the hydriatics facilities next to the gym in the basement of the clinic. An attendant fetched private patients at a preappointed time. In a single dressing room, alternating groups of female and male patients donned bathrobes (also called wrappers). While waiting to receive their individualized treatments, patients congregated in the adjoining "reclining room." The hydriatics instructors had a variety of techniques from which to choose. In addition to intense douches and gentle showers, there was the salt glow and mitt friction (abrasion massages thought to increase blood circulation and release bodily toxins through the skin); vapor, hot air, and sitz baths; or shampoos and fomentations (heat compounds) applied to the scalp. Attendants assisted with these procedures, while nurses placed other patients into wet sheet packs.

Lloyd Felton, the first hydriatics instructor for the male patients at the Phipps Clinic, presented a paper in 1915 to the Maryland Psychiatric Society explaining how hydrotherapy was used in Meyer's clinic. Felton's statements echoed familiar Meyerian sentiments. For example, it was critical, he said, to utilize hydrotherapeutics to tend to the patient's individual complaints and symptoms, because he or she became "easier to handle, more satisfied with the institutional life and in general able to derive much more real enjoyment out of his existence." He also emphasized that no particular water treatment was indicated for a specific disorder (hardly surprising, since Meyer eschewed diagnosis, deferring it until the patient's discharge). "The choice of the hydrotherapeutic measure," Felton explained, was inseparable from "the patient's reaction from day to day." Felton concluded with a statement that epitomized Meyer's rhetorical strategy when confronted with claims he considered specious: "We can not cure but we dare not throw any help away." The reverberations of Meyerian sentiments in this report, and the likelihood that his boss sat in the audience with other Maryland alienists, suggests that Felton's perspective on the therapeutic uses of hydriatics echoed Meyer's own.[52]

Regardless of what Meyer thought personally of hydrotherapy for treating psychiatric disorders, he certainly had no choice but to offer these techniques at the Phipps Clinic. The proliferation of technology-based diagnostics and therapies in American hospitals meant that his patients and their families expected modern treatment like hydriatics—and they expected results. Four days after her admission to the private ward, for example, a patient took a treatment in the continuous bath tub. The nurse reported that it was "rather painful because of arthritis (but she) wanted to attempt it because she wants to get out on the roof garden and see everything." This patient expected that the bath would yield curative results. Another woman, admitted in 1913 and contacted for a follow-up study ten years later, was glad to report that she had no mental troubles. Then she asked if she could be allowed to return to the Phipps, nevertheless, for the "baths and treatment." And a few lines from Lottie to her mother are also illustrative: "I am taking about one salt rub with an arm shower after it a week. I don't mind it and the nurse who does it rubs my back very nicely so it may take that tired feeling away. I asked for it myself. It certainly has ruined my life keeping me here." Lottie appeared to appreciate the medical or sensory benefits of hydrotherapeutic treatments and simultaneously to resent her confinement in the clinic. Meyer's patients anticipated relief from their symptoms by means of hydrotherapy; the case histories show that this was true of free, subsidized, and private patients.[53]

The documentation in the case histories certainly shows that Meyer considered hydrotherapy yet another handmaiden to the therapeutic experiment. One male patient demanded to know why he could not simply take a regular bath in his ward for relaxation. "To stand around in a wrapper for an hour or more is not good for me," he complained about waiting his turn for hydriatics, "and I hope there will be a change in procedure." His case history indicates that he continued to go to hydriatics after lodging this grievance. The cryptic comments of a female patient were recorded by her ward psychiatrist: "patient went down to the hydro-therapeutic room but on returning said that she had not enjoyed it at all" because "men don't play any part in my life any more." The psychiatrist interpreted this reaction as corroboration of what he suspected was a connection between her marital relations and her psychosis. She then added that her husband was a "bad man and has no right to little children any more." The enmeshment of what she was doing and what she was thinking—in this case during hydriatics— yielded useful data for the psychiatrist. It was recorded, interpreted, and incorporated into subsequent interactions with the patient. Like occupation classes and recreational activities, hydrotherapy became a variable in the therapeutic experiment.[54]

Restraint, Sedation, and Isolation

Meyer also relied on the predictable effects of continuous baths and sheet packs to calm or immobilize patients who were agitated, disruptive, or a potential danger to themselves or others. The Phipps nursing manual stated, "[T]he continuous bath and cold wet pack are given for sedative purposes."

The innovation that transformed an ordinary bath into a means of sedation—extending the submersion period to days or weeks—was pioneered in German psychiatric clinics in the 1880s. A dispatch from one of Meyer's Worcester assistants—who had been sent by Meyer to work in Emil Kraepelin's clinic in 1906—advised American physicians that the revered German psychiatrist rarely used drugs to sedate agitated patients. Instead, Kraepelin had eighteen continuous bathtubs available for this purpose. He also proclaimed the effectiveness of the continuous bath in his popular textbook. At the Phipps Clinic, the duration of the bath was prescribed by the ward psychiatrist in consultation with the hydriatics instructor. Felton told the Maryland Psychiatric Society that submersion for longer than six hours was unnecessary and potentially harmful. William Alanson White, an influential psychiatrist and the director of St. Elizabeth's Hospital for the Insane in Washington D.C., implored his peers in 1916 to ban the use of a harness that secured the patient in the tub, after a woman was scalded because the automatic temperature rose during a nurse's absence. "Restraint of any kind during this treatment is not allowed" was the official policy at the Phipps Clinic, as specified in the nursing manual.[55]

Phipps nurses were instructed to ensure that the patient received an explanation from the ward psychiatrist as to why a bath or a pack had been prescribed, in order to "gain the patient's cooperation and secure the proper mental attitude." In practice, however, these immobilization techniques were often used to manage erratic or violent behavior in patients whose mental states prevented them from appreciating this preamble or consenting to the procedure. One evening Richard became excited, shouting cryptic vitriol at the nurse: "Get your goddamned mother and a few more dead bodies, honey! Bring along 33—number 33, that is my mother." He grew increasingly noisy and threw his bedclothes about the ward. When he could not be persuaded to calm down, nurses and orderlies forcibly placed him in the continuous bath where "he became comparatively quiet." The official policy of not harnessing patients into the tub was evidenced by one patient's prolonged episode of panic: "he would jump from the bath and run about on the ward highly excited, even screaming." According to the case history, the patient was returned to the tub nine times. In the end, the technique produced the expected effect in that patient and yielded a noteworthy psychobiological reaction: "Tonight," the night nurse observed, "he is considerably more calm, and has apologized for his behavior." Nurses were reminded,

"[I]f the patient gets out of the tub, [the] attempt to give the treatment must not be abandoned immediately." Though "splashing is disagreeable," the manual acknowledged, it was to be expected and tolerated.[56]

The continuous bath presented other more serious challenges, because it provided opportunities for self-harm and unwanted behavior. Nurses learned that if a suicidal patient succeeded in submerging his or her head under the canvas cover, they were to "pull open the drain with the lever, shut off the water, loosen the strap of the cover and throw it back." The tub was designed to empty quickly and if this procedure was followed, the nursing manual promised, the patient would not be harmed. Other equipment used in conjunction with the bath could also be problematic. Under its canvas cover, for example, a patient might work on picking open an artery or give way to compulsive habits. Dolly, for instance, injured herself while in the continuous bath due to her uncontrollable urge to masturbate: "Yesterday the patient attempted to insert a glass bath thermometer, cutting herself." Despite instructions to the contrary, attendants and nurses evidently were tempted to divide their attentions when supervising a continuous bath—perhaps preoccupied by other tasks, or beleaguered by the responsibility of supervising patients with demanding needs and creative and unpredictable behavior.[57]

The wet sheet pack was by its nature restraining, and it typically produced manifest soporific effects. Under Meyer's direction, the Phipps staff routinely used wet or dry sheet packs to immobilize patients who exhibited destructive, injurious, frenzied, or compulsive sexual behavior. Large pins affixed the sheets to the underside of the bed to ensure that the wrapping did not yield to a patient's movements. A nurse was to be present at all times, to monitor pulse and respiration, record utterances, provide drinking water, and watch for signs such as labored breathing or changes in skin color. According to the nursing manual, acutely depressed or manic patients were to be told that "they will be removed if the pack does not prove helpful in about twenty minutes." If the continuous bath or the sheet pack was used as punishment at the Phipps Clinic in this period—or to alleviate demands on the staff by preventing mischief—there is no evidence of it in extant sources. [58]

The continuous bath or wet pack did not always calm agitated patients, and hydrotherapy was but one step in a progression of attempts to adequately manage an agitated patient. Chemical sleep agents, or "hypnotics," such as bromides and barbital—as well as more powerful soporifics like morphine—were also utilized in the Phipps Clinic to sedate patients when necessary. Like Kraepelin, whose early involvement in Meyer's development was profoundly influential, Meyer preferred to minimize the use of drugs to sedate patients. Beginning in the 1860s, bromide salt compounds were widely used in asylums to manage periods of violent or self-destructive behavior, insomnia, and the delirium associated with epilepsy, manic and depressive states, and alcoholism. They were also increasingly prescribed by general practitioners for the relief of restlessness, worry, and insomnia. The therapeutic effectiveness of bromides, however, was coupled with the high rates of chemical toxicity and dependency that accompanied long-term use. Bromide remedies could be purchased not only from the druggist, but also through the mail, from itinerant peddlers, and from the grocer. A significant number of Phipps patients had used bromides to assuage their symptoms long before they were hospitalized.[59]

By 1913, both bromide and morphine dependence were prevalent social ills, and their medical use by Meyer and his staff was tempered by a degree of moral aversion. One evening, a patient was disturbed, restless, and combative with nurses. She was given Veronal, a barbiturate, but a nurse observed that the drug "did not have any special effect." Around two o'clock in the morning, the ward psychiatrist "was awakened by a tremendous pounding on the ward door." When he entered the room, he found the patient lying on the floor beating the door with her feet. She demanded stronger medicine "for her nerves." He tried for "some time" to convince her to return to bed and the woman refused. Eventually, he prescribed one quarter of a gram of morphine by hypodermic needle. According to the case history, following this "she settled down and slept for a few hours." In this case, the psychiatrist preferred that she return to bed without further sedation, but eventually granted the patient's demand.[60]

If social interaction, autonomy, occupation, recreational activities, and baths and sheet packs did not adequately redirect behavior, isolation was the management technique of final resort. Percy, for example, was transferred from the quiet ward to the semi-quiet ward when he became overexcited. When he arrived on West-2, he "grasped the hand of each patient and shook it hard" telling each one earnestly "I believe in Jesus Christ." For several days afterward, his only utterances were biblical phrases and apocalyptic warnings. He refused nourishment and preached energetically, working himself into a frenzy. Attendants installed Percy in the continuous bath on the ward. As he soaked, he intoned the psalm "Blessed is He that cometh in the name of the Lord" over and over again and intermittently jumped out of the tub to be returned by force several times. According to his case history, he was in the tub from three to nine in the evening. At nine o'clock, he was still resistive and threatening, still loudly chanting, which prevented his fellow patients from sleeping. Because neither the ward reassignment nor the continuous bath had sufficiently quelled this state of agitation, he was transferred to an isolation room on West-1, the disturbed ward for men.[61]

How Meyer approached the unavoidable task of managing patients in elated or antagonistic states was shaped by his experience working in large public asylums. As a newcomer to the American asylum in the 1890s, he had been pitched into a longstanding and highly polarized debate about the use of physical restraint in asylums. His stance at that time was characteristic: the decision about whether to employ bed cuffs or a straightjacket should be a medical one and not based on dogmatic principles or the needs of the staff—the verdict depended upon the particular circumstance of each case. During his postgraduate training in Scotland, he had observed management systems based on nonrestraint and unlocked doors. Over the next twenty years, he internalized the profound association between the public's trepidation about mental institutions and the imagined horrors of physical restraint. The only way to overcome popular fears of the "so-called excited wards of the overcrowded asylums," he told colleagues in 1913, was "to make these provisions . . . much superior to what is at present available." When he had the opportunity to design his own clinic, he called for means to safely segregate agitated patients. The isolation rooms on the ground-floor wards were constructed of double-thick walls separated by air pockets for soundproofing. Whether or not the walls were lined with protective padding is unclear. They were fitted with individual ventilation systems and interior windows for constant observation. When isolated, patients were encouraged to work off excess energy by kneading Plasticine or tearing up old magazines. No documentary evidence has emerged to suggest that straightjackets or bed cuffs were used at the Phipps Clinic between 1913 and 1917.[62]

Ultimately, Meyer's institutional circumstances permitted him the luxury of marginalizing physical restraint. First, the ratio of staff members to patients was very different from that in the state asylum. When he reported on the work of his small clinic at the 1913 International Congress of Medicine in London, his European colleagues were astonished to learn that ten Phipps psychiatrists were responsible for ninety patients. Thomas Clouston, director of the Royal Edinburgh Asylum, acknowledged contritely that "our paymasters and our committees will need a great deal of education before they will provide the funds for carrying out such beneficent schemes as that!" As a private teaching hospital in the United States, Johns Hopkins was not constrained by state funding and policies or by dominant religious or national traditions. This gave Meyer an opportunity to implement an approach that, even had they wanted to, most established alienists in the Old or New World could not.[63]

Second, Meyer exercised full control over which cases he admitted and discharged from his clinic. This contrasted with the superintendent of a state asylum and, to some extent, the proprietor of the commercial sanitarium, whose livelihood depended upon an ability to manage difficult patients deftly. Meyer did not avoid admitting cases of severe psychosis or mania, but he retained them only as long as the integrity of the therapeutic experiment remained intact and the safety of other patients was maintained. One of Meyer's private patients, the wife of a civil servant, provides an example. After several months in an upscale private asylum, thirty-year-old Emily was brought to the Phipps Clinic by her father, a prominent American physician. Her local specialist wrote to Meyer that she had been in "a state of acute auditory hallucinatory mania with constant motor activity" and was "violently maniacal." Phipps nurses on the private ward observed that she enjoyed hydrotherapeutic treatments, the gymnasium, and occupation classes. She was also boisterous, easily angered, and unpredictable. Emily maintained that she was the subject of a government experiment that, if successful, would "solve the social problem" by transforming her into a hermaphrodite. It was Emily who refused to take a ride in Meyer's automobile because she objected to her mind being public property. One evening she suddenly screamed out "I can't stand this, I can't stand this!" She jumped onto the window sill, beat the panes of glass with her fists, and kicked nurses who approached. The attack lasted for one minute and afterward Emily apologized to the staff. When Chief Resident D. K. Henderson asked her to explain the outburst, she sobbed, "I am tired of being everybody else—I have been for six months—I want to be myself." A few days after this episode, she confided to her physician at home that "the Phipps Clinic is really a marvel and were it possible to import some of my truly dear friends it would *almost* be perfect." She also insisted that whenever someone entered her room it was to irritate her genital organs by "indovination," a word she had coined but could not define for herself or the psychiatrist.

Because she continued to alternate between lucid and delusional (often destructive) behavior, Meyer remained hopeful that Emily might respond favorably to the hospital regime. For example, she wrote two letters to Henry Phipps, whose name adorned both the building and the bed sheets. First, an inquiry: "As prostitute, soul mate and psychological mother of the United States I wish to ask from whom I may expect remuneration for my services? I am at present representing international relations and prostitution in your House of Life—is it your wish that I should occupy this position permanently?" In her next dispatch to the philanthropist, she wished to "express my appreciation of your truly wonderful clinic. I have enjoyed the baths and the gymnasium ever so much." She told her father that "there is really nothing definitely to write about except that things are going rather satisfactorily." The following day, she informed him that "between being masturbated by the bubble and clitoris local system, and dreaming of being saved from destruction by Wm. James Bryant, it seems to me about time to return [home]."[64] Unfortunately, Emily's delusions deepened. She attacked a nurse, "pulling out some of her hair and creating such a state of affairs as to make it necessary to assign her a private room on the first floor." Once in the isolation room, she broke the glass of the fire alarm box and "scratched her skin enough to be able to write several words on the wall" (in blood, presumably). Meyer decided he could no longer compromise the therapeutic environment or put others at risk. "The conditions here would require more restriction than the patient will put up with," Meyer explained to her father, "and as I don't like to have coercive measures, and yet have to protect my other patients against outbursts, a transfer seems to me highly indicated." Emily was discharged soon afterward and reinstalled in the private asylum near her family home.[65]

Eating, Evacuating, and Sleeping

Within the therapeutic experiment, no reaction or interaction was insignificant. This included eating, eliminating bodily waste, and sleeping. Two weeks into his admission, the clinic's first patient (the man who had lost consciousness on the streetcar) was still taking his meals in bed. He was "decidedly more confused" and required assistance with most tasks because of speech and memory loss. A nurse noted as a positive sign, however, that he "ate the pulp of a lemon from a slice with relish and of his own volition discarded the rind." She thought it worth recording that in his confused state he knew to eat the pulp and discard the rind since it signaled at least some of his higher mental function was intact. Unfortunately, he continued to deteriorate. He wept when he was told that he must leave the familiar, kind nurses and go with a strange woman that the doctor assured him was his wife. The man died at home ten days later.[66]

Patients often endowed their food with symbolic meaning, making a meal another prospective variable in the therapeutic experiment. In order to test his memory, Richard was asked to itemize his breakfast from the day before, which he did: four slices of bread, a piece of butter and two eggs. But, he added adamantly, "I ate the brown egg but did not eat the white egg." The psychiatrist was unable to elicit from him why this was significant, but recorded the data faithfully. A patient's perception of food might present unforeseen hazards, too. During her midnight rounds, the night nurse observed that Percy's cheeks and neck were blistered. When she queried him, he said only that God had instructed him: "Deny yourself. Take up your cross and follow me." He seemed quite content and fell asleep. The following morning, the ward psychiatrist questioned the orderly who was on duty the night before. After bringing Percy his supper, the orderly testified, he distributed other patients' meals. He left Percy alone for only a few minutes and when he returned, he found Percy's tea spilled. He insisted that the patient had not shown "the slightest sign of pain or discomfort." The incident appeared mysteriously meaningful to Percy who refused to discuss it. A dermatologist from the general hospital examined and treated the damaged skin around and, notably, under the patient's chin. His findings confirmed what was already suspected: "burn from chemical or very hot liquid." In following God's command to "take up your cross," it appeared likely that Percy had raised the tea pot containing boiled water ceremoniously above his head, looked heavenward, and emptied its contents onto his face.[67]

Phipps psychiatrists also used food as an experimental control to test a patient's reactions. Kasper, the young craftsman from Switzerland admitted with delusional ideas and excitation, eventually transitioned from an excited state to an unresponsive state, a common occurrence associated with a schizophrenic reaction. He did not acknowledge anyone or anything around him for several days. He continued to eat heartily, however. In an attempt to test and override the reaction, a particular food was omitted from Kasper's meal. Knowing that eggs played an important symbolic role in his systematized delusions, an experiment was executed: "Breakfast tray was served without eggs; when patient saw the tray he got out of bed, put on a robe and slippers, came to the kitchen and said *Where are my eggs?*" It was the first time he had left his bed, dressed, or spoken in three days. As were all aspects of the hospital routine, food and eating were implicated in the therapeutic experiment.[68]

The following interaction, involving food, exemplifies the simultaneity of clinical investigation and behavior modification in Meyerian psychiatry. During an interview with ward psychiatrist Tedrow Keyser one evening, Dolly—in one of her periods of believing that the clinic was a penitentiary for immoral women—assured him, as he recorded in her history, that "the doctors do not want a bad woman in the hospital, that her room is too good for her, that the bills are not paid, that she should be put on the street." She continued to fret, telling him that she did not have any "dark blood" (one of Dolly's many preoccupations). In response to her statement that she tried her best to "do right," Keyser offered affirmation: "And you succeed." At that moment, Dolly's supper tray was brought into her room. "Please don't give me that," she begged the nurse; "please give me something clean." Encouraging her to compare her perceptions about her supper with those of the common sense consensus, Keyser asked if she found it curious that, even though he would happily eat the food on the tray, she supposed it harmful. "Please take it away," she insisted, adding "I'm not a Chinese Mason." The nurse and Keyser departed without the tray. "After being left alone, the patient ate quite a good dinner," the nurse later observed and recorded in the case history, "saying she did it because the doctor asked her to." In this encounter, it is difficult to untangle when Keyser the psychopathologist was investigating Dolly's psychobiological reactions and when Keyser the psychotherapist was attempting to modify them.[69]

In fact, most forms of psychopathology interfered with nourishment instincts in various ways. Maintaining adequate nutrition was a constant concern for the nurses, who were expected to employ diligence, creativity, persuasion, and, if necessary, coercion to overcome symptomatic interferences. Food often had no allure for a patient in a depressive state, for example, and those experiencing withdrawal from alcohol or drugs were frequently severely nauseated. Like Dolly, many patients claimed that the food was poisoned. Individuals influenced by the euphoria of a manic state were commonly oblivious to hunger and unconcerned with eating. Delusional or psychotic patients might respond violently when compelled to eat. "I attacked one of the nurses when she pushed me up to the table when I didn't want to eat," Dolly explained to Keyser one day; "I tried to choke her—I don't want to be shoved around." She apologized immediately afterward, "saying she did it impulsively." In her defense, she added cryptically, "I spoke to Mrs. Meyer about the emperor" (by which she meant Meyer) to say that she regretted calling Miss Slagle, the occupation instructor, a Jewess.[70]

When necessary, liquid nutrition was introduced into the body by means of a rubber tube. Percy repeatedly resisted food and water, "closing his mouth against it and turning (his) head away." Nurses noted that he looked tired, moved his legs constantly as a result of catatonic excitement, and that his pulse was fast and weak. Eventually, with the aid of several attendants to immobilize Percy, the ward psychiatrist inserted a feeding tube into his mouth and down his throat. Through this tube he received a pint of milk, three eggs, two ounces of castor oil, and water. Generally, two persons were required to administer the procedure, one to hold the tube and funnel, the other to slowly pour the mixture. The head nurse was satisfied that the intervention was beneficial and observed that Percy "looked stronger afterwards, less perspiration, pulse stronger." According to the nursing manual, staff employed this technique when a patient refused food and water for more than twenty-four hours.[71]

Various unresponsive states were common to severe mental disorders and presented risks to bodily health. Limited unresponsiveness appeared with depressive episodes, while more stuporous states took different forms. In a catatonic state, for example, a patient might develop muscular rigidity that gave the limbs a "waxy flexibility" so that he or she could be posed, even awkwardly, and the position would be maintained for hours. Alternatively, catatonia was expressed as "negativism," a state in which the patient might remain motionless in bed for days or weeks refusing to speak, eat, empty the bladder or move the bowels. Meyer described negativism, characteristic of the schizophrenic reaction, as an automatic resistance against natural impulses.[72] This opposition was irrepressible and involuntary, a resistance to exhortations from nurses to move or speak, but also a resistance to internal commands from the body to eat, swallow, sleep, or eliminate waste. A catatonic excitement, conversely, was characterized by repetitive uncontrolled movements, or by an impulsive compliance to an inner command that was incomprehensible to others. "He may suddenly get up and jump out of the window, or smash something, or hit someone," one Phipps psychiatrist explained. A young bank teller named Saul, for example, suddenly began to scream that he had been poisoned. He trembled violently and looked at his hands and feet and exclaimed that they were dying. He rubbed his extremities and gasped for air. His catatonic state lasted for twenty-four hours. "Tonight he is considerably more calm," a nurse reported the following evening, "and has apologized for his behavior." Saul was unable to explain what had triggered his panic.[73]

Phipps nurses regularly confronted the problem of unmoved bowels due to negativism. Patients who did not produce "an action" for an extended period would be given powerful drugs called "cathartics." These forceful purgatives were not prescribed to regulate the bowels or relieve constipation, for which purpose most patients willingly consumed a moderate daily dose of castor oil, a common domestic practice. It was not uncommon for Kasper to go without an action for over one week, and recurrent efforts to induce a bowel movement were detailed in the case history. "Yesterday, his bowels not having moved for 6 days, an attempt was made to give him (2 drops) croton oil—he took this only when threatened by force." The nurse reported that it was "very effectual" as Kasper eventually passed a "large constipated stool." Reverence for the therapeutic atmosphere was weighed against the need to intervene. When his bowels did not move again for another five days, Kasper fought violently against receiving cathartics, and "the attempt was not carried farther because of the proximity to the other patients." Evidently the interaction was not without value, since the ward psychiatrist reported favorably that "in the course of the physical struggle he spoke for the first time in several days, crying out in stentorian tones (in German) *God doesn't want me to take medicine—He instructed me not to do so.*" Nine days later, for a total of two weeks, Kasper still had not moved his bowels and continued to resist medication. In consultation with the senior staff and head nurse, the ward psychiatrist decided to use physical force to administer the drug. Attendants overpowered and confined Kasper in a dry sheet pack. The account of the event, as rendered by the ward psychiatrist, is instructive.

He was this morning put into a dry pack and given castor oil and croton oil per nasal tube. He was at first very combative, glaring wildly at the attendants, attempting to bite, etc., but soon became quiet and remained so. On being taken out of the pack he talked freely for about half an hour. He said that the reason he had not talked for the last few days was that God had forbidden him to do so. God had also forbidden him to take any medicine, as his body was different from all the other people in the world; therefore he did not need the same treatment. However, God had commanded him that he had a certain time to live upon this earth, and while he was in the pack, had told him that he had better try to get along with the people here as well as possible. For that reason he said he would take any medicine in the future, and also promised to take a bath, which he did. Immediately afterwards, however, he lapsed into his state of unresponsiveness.

First, this narrative describes cinematically a set of circumstances in which the patient's autonomy and spontaneous reactions were deemed secondary to inducing more normal physiological reactions using medications. Second, it chronicles the unstoppable march of the therapeutic experiment. Psychiatrist and nurse diligently observed and recorded Kasper's psychobiological reactions to his changing circumstances—from glaring wildly, to talking freely, to promising to take a bath. Meyer undoubtedly preferred to avoid physical restraint or coercive measures whenever possible. However, once force was employed, it was not necessarily minimized or omitted from the case history. Rather, as these examples show, it constituted another variable in the natural experiment. That evening, a nurse recorded the effects of the powerful drug. Kasper produced "a tremendously large constipated stool in bed." Deliberate untidiness with urine and feces was also a common manifestation of negativism, and incontinence was a side effect of physical exhaustion or neurological deterioration.[74]

Nurses' notations reveal how quickly a patient's behavior could change from calm to odd to unruly. Whenever Percy heard running water, he immediately ran to it and put his face under it. When a fellow patient refused to take his medicine, Kasper strode across the ward, "took the glass from the nurse's hand, and made the patient take it [saying] *It's God's wish that you should take the medicine.*" Another nurse heard Kasper "talking, laughing and whistling" to himself at 5:30 in the morning. He was "profane and irritable" when she told him to hush lest he wake the other patients. She immediately realized that Kasper was displeased with her. A few hours earlier, she had forced him out of bed to use the toilet. Now, he "suddenly jumped out of bed [and] tried to attack the nurse." When she retreated to the ward office, he followed her down the hall, yelling at her. A female attendant came to the scene and "the patient threw her to the floor." When more attendants arrived, "he was carried back to bed," by which time everyone on the ward was awake.[75]

Life in the clinic was not constantly characterized by commotion, and the Phipps environment could take on a domestic tone when patients and night staff settled into the ward for the evening. After occupation classes and recreation, after exercise outdoors or in the gymnasium, after the stimulation of a Scotch douche or the relaxation of a continuous bath, and after supper in the dining room, patients were encouraged to pursue their own interests before bed. Together in the parlor, some read while others weaved, whittled, or wrote poetry. When the weather was pleasant, patients spent time on the verandas, which overlooked the city. Outbursts and crises were not uncommon, certainly, but as one patient confirmed during a word association test, the ward was perceived by some as a peaceful place. Presented with the test word "quiet," the patient promptly responded "the ward here." Meyer believed that the communal setting of a small ward nudged the patient's thinking in the direction of the common sense consensus. "The patient ceases to think that he or she is the only mortal that can be so afflicted and that, therefore, his or her own notions alone can be correct and the physician's must be wrong." He also observed that patients inspired each other to behave responsibly and strive for health.[76]

Because ward psychiatrists worked night shifts and resident psychiatrists lived in the clinic, they exploited quiet opportunities for informal social interaction with the patients in the evenings. When Kasper unexpectedly emerged from a catatonic state late one evening, psychiatrist Charles Thompson was available to take advantage of the change. Thompson's record of their conversation, three typed pages, documented Kasper's lucid remarks about his apprenticeship in Switzerland, his parents, and when he first began to believe his eyes had the power to influence others. He confided that, since entering the hospital one month earlier, he presently believed that "his eyes are quite as they should be." Thompson's notes reveal that Kasper the young artisan had reemerged, replacing the unpredictable wild man that staff had placed forcibly into a sheet pack and who, in return, had kicked and glared and defecated in his bed. No, *Kasper* had appeared unpredictably and sat calmly talking with Thompson. New opportunities for both clinical inquiry and therapeutic efforts emerged from such institutional circumstances, novel in the United States at the beginning of the twentieth century.[77]

Like any other event in the clinic, lights-out on the ward engendered observable reactions that served the therapeutic experiment. When Lawrence was admitted to the Phipps Clinic, he was complaining of chronic belching and gas, constipation, excessive sweating, difficulty breathing and a rapid, irregular heartbeat. Sitting in the parlor one evening, he complained of dizziness. After the nurse helped him into bed, he began to gag and vomit. "He made a great fuss over it," she recorded in the case history, "and caught hold of whatever came near to help him." Thompson, the ward psychiatrist, was summoned and pulled up a chair at Lawrence's bedside. "For some time," he reported, he and Lawrence discussed "how, where, and when the vomiting had begun." Lawrence did not remember exactly. Thompson noted that during their discussion the gagging ceased. Thompson stood to leave just as the lights on the ward were put out for the night. He said good night and walked away. "I hardly know where I'm at," Thompson heard Lawrence say to no one in particular; "are they all going away and leave me?" A few minutes later the nurse heard Lawrence begin to vomit again.[78]

Not surprisingly, just because patients were asleep did not mean the therapeutic experiment was on hold. Two nurses and two attendants were on duty during the night. They helped prepare each patient for bed and patrolled the open public wards and private rooms alike, recording observations that might prove important (the locking of patients' doors, by patients or by staff, was strictly prohibited). The nursing manual stated clearly that "the keeping of an accurate sleep chart is most desirable" and that "the nurse must make frequent careful rounds herself, and not leave too much to the attendants." In one patient's record, an observation made by the nurse in the middle of the night was applied to a psychotherapeutic interview the next day. The psychiatrist recorded the patient's description of a happy dream, adding "patient described by night nurse as smiling and happy in her sleep." Supported by a centralized record-keeping system, these notations came together on the typed page of the case history to provide a comprehensive rendering of a psychobiological reaction performed as the patient slumbered.[79]

The case histories frequently contain letters from patients who wrote to Adolf Meyer, psychiatrists, and nurses after leaving the Phipps Clinic. Patients who viewed their hospitalization in a positive light sent more letters than those who were dissatisfied or whose condition worsened. Of those who did communicate with Phipps staff after their discharge, many inquired about fellow patients and expressed regret over no longer being together. Perhaps this is not surprising, given the deliberately communal and domestic setting of the Phipps environment. One patient sent Meyer and his fellow psychiatrists several updates and wrote to one of the nurses:

> I remember you most gratefully. My mind constantly reverts to the strangeness of life at the Phipps last year but always with thoughts of gratitude to the good kind nurses and yours was long and lasting. The old "gang" of patients have all gone I suppose but you have, of course, only swapped the devil for a witch. Yet could any be more perplexing than the flock of which I was a member. Poor black sheep! How the Drs. never would admit strain or stress or inheritance or illness or try to make us feel blameworthy for our awful fate.

Despite his quarrels with them, and there were several disputes, this patient evidently maintained a fond impression of his time with fellow patients on West-3, where he spent five months.[80]

Meyer likely would have agreed with this comparison of psychiatric patients to black sheep—individuals whose thinking and behavior (at least at the time) did not accord with those of the flock, the common sense consensus. He was confident that habit-training would help some of these patients adjust more successfully. In 1913, he declared proudly that "Mr. Phipps has given us a wonderful center of mental orthopedics." The contrived social milieu of the new Phipps Clinic, however, bore a strong resemblance to the moral therapies of the eighteenth century. How did Meyer justify promoting this therapeutic approach as innovative? He stated that traditional moral treatment employed a "moralizing psychology" to alter behavior. "Whereas today we have an adequate knowledge of the role of substitutive reactions and automatisms," he explained, "and a more adequate sizing up not only of the odds, but also of the constructive possibilities required for an adequate reaction." For Meyer, the principles of biology and his efforts to manipulate the dynamic interaction between the human organism and its environment distinguished his psychobiological approach from those of his predecessors in the centuries before.[81]

Indeed, what marked his practices as new was the attention and significance he accorded symptoms and social interactions—in the form of observing, recording, analyzing, and applying the data of psychobiological reactions—as biological events. Resonating within the therapeutic experiment were Meyer's instrumentalist renderings of manual occupation and social interaction as methods of habit-training, and his ambivalent views on the uses of hydrotherapy. Every happening within the Phipps environment—contrived or unforeseen, commonplace or bizarre, voluntary or coerced—represented a variable in the natural experiment and a potential clue to understanding and ameliorating the patient's maladaptation. Meyer's therapeutic experiment emerged within particular institutional circumstances that afforded him the opportunity to implement a novel approach to psychiatric practice. The conflation of medical inquiry and therapeutics represented a twentieth-century and distinctly Meyerian twist on the moral therapies of earlier centuries. The other principal element in the therapeutic experiment, individual psychotherapy, is explored in the next chapter.

CHAPTER 6

Subconscious Adaptation

Psychotherapy and Psychoanalysis
in Meyerian Psychiatry

"Psychotherapy is in the air and wildly exploited in the book-market and in magazines," Adolf Meyer editorialized in 1909. Every new book on psychotherapy, he continued, was devoured eagerly by a heterogeneous group of readers prompted by curiosity or a desire to bolster their own theories. Indeed, psychotherapy—the practice of using the mind as a curative force— seemed to be everywhere in the United States in 1909. The phenomenon was the result of a widespread cultural uptake of talking cures, advocated by secular and religious groups, combined with a growing interest among physicians in the psychological causes and treatment of nervous disorders. The American Therapeutic Association held a symposium on psychotherapy in New Haven that year, and the International Congress of Experimental Psychology in Geneva designated a related topic, the subconscious, one of its special themes. August Forel, Meyer's prominent advisor in Zurich, founded the International Society for Medical Psychology and Psychotherapy in 1909 to help legitimize a science of psychotherapy, and Harvard psychologist Hugo Münsterberg published his well-received book entitled *Psychotherapy.*[1]

In September of 1909, each of the departments of the young Clark University in Worcester, Massachusetts, invited researchers from around the world to receive honorary degrees in celebration of the institution's twentieth anniversary. Stanley Hall, a psychologist and the university's president, was eager to meet the Viennese neurologist Sigmund Freud, in whose work he had a special interest. In addition to Freud, Hall invited seven other specialist researchers to receive honors, including psychologist Edward Titchener, anthropologist Franz Boas, and Swiss psychiatrists Carl Jung and Adolf Meyer. The event is memorialized today as Freud's only visit to the United States. In 1909, Freud's name and his new theory of psychoanalysis were known to only a small group of American specialists. Indeed, he was surprised to hear that Meyer and Hall had incorporated psychoanalysis into their lectures at Clark, Harvard, and Columbia universities years earlier. English-language publications by Carl Jung and Meyer's recent appointment to Johns Hopkins had made the names of these two psychiatrists familiar to some American readers, but Hall arranged press coverage of Freud's appearance in advance, for fear it would be overlooked. Importantly, psychoanalysis was one of many forms of psychotherapy under serious discussion in 1909. Meyer's characterization about the diversity of psychotherapy enthusiasts was apt, as was his observation that many used it to strengthen their own theoretical and therapeutic claims. That is precisely what he did, too.[2]

As psychiatrist-in-chief of the country's preeminent teaching hospital from 1908 to 1941, Meyer dominated American psychiatry during the first half of the twentieth century. For much of the second half of the century it was dominated by Freud—or, at least, by his ideas and his followers. Historians of psychoanalysis identify Meyer as a principal importer of psychoanalytic concepts to the New World and as an influential teacher of the first generation of Anglo-American psychoanalysts. In the years leading up to his appointment to Johns Hopkins, the New York Pathological Institute, under his direction, became the tiny epicenter of psychoanalysis in America, but by the 1930s, he was a vociferous critic of Freudian doctrine. Elucidating Meyer's particular brand of psychotherapy helps to clarify his role in the early history of American psychoanalysis and to shed light on his criticisms after World War I.

Focusing on Meyer's use of psychotherapy, this chapter deals with an amorphous group of medical conditions encapsulated by him and his contemporaries as "functional disorders." By the 1890s, adjectives such as *functional, neurotic,* and *psychogenic* had emerged within academic medicine to describe several distinctive disorders with overlapping symptoms and a common feature: a bodily or psychological disturbance with no discernible organic pathology. Sufferers reported symptoms such as chronic headaches or pain, digestive trouble, respiratory distress, or vision distortions; or they exhibited choking, paralysis, uncontrollable tics or motor movements, amnesia, or convulsions. Additionally or alternatively, they developed irrational fears, obsessions that became the basis of strange rituals, incapacitating anxiety, and, in some cases, hallucinations or delirium. These conditions could also be characterized by extreme sadness, excitement, irritability, or meekness. Symptoms and behaviors could not be linked to a known disease process using laboratory or diagnostic tests in the clinic (though many diseases could be ruled out using these means). Depending upon the dominant symptoms, sufferers might be diagnosed with nervousness, exhaustion, hysteria, neurasthenia, psychasthenia, neurosis, hypochondria, or invalidism. Experiences of physicians and patients confronting these debilitating, sometimes life-threatening, conditions varied immensely, as did late-nineteenth-century discourses about their causation and treatment. Many physicians remained convinced that (as yet undiscovered) organic processes were responsible. Others dismissed sufferers as hypochondriacs and malingerers. Much ink was spilled in medical journals on both sides of the Atlantic Ocean about how to differentiate, diagnose, and classify functional disorders.

A small number of investigators speculated that functional disorders were psychogenic—mental, rather than bodily, in origin. Meyer belonged to this minority, but he rejected any dualistic distinction. "Conduct and behavior constitute the mental life of the individual," he explained in 1911. He categorized a psychogenic disorder as a pathological process initiated or sustained (or both) predominantly by life experiences. "The essence of these disorders, where there is no poison or infection demonstrable, is that the patient, through circumstances or through fundamental peculiar make up, comes to react to the difficulties of life in ways which are bound to vitiate the life of the brain."[3] In addition to traditional functional disorders—what he called hysterical, neurasthenic, and psychasthenic reactions—he framed affective reactions (depression, paranoia) and schizophrenic reactions (related to dementia praecox) as psychogenic. He viewed them all as nebulous variations of "an insufficient or protective or evasive or mutilated attempt at adjustment." This unsuccessful attempt at adaptation, he reasoned, resulted from conflicts among unregulated instinctual drives. Left uncorrected, the maladjustment became progressively habitual and pathological. Meyer defined a mental habit (normal or pathological) as the functional expression of an "unconscious-automatic" sensory-motor response performed without conscious awareness. After 1906, he increasingly referred to this group of conditions as habit disorders. In daily practice, he marginalized the differences among functional disorders and pursued them collectively with relatively consistent investigative and therapeutic aims.[4] Accordingly, I set aside questions about how these individual disorders were culturally and clinically construed in the opening decades of the twentieth century—lines of inquiry that have been explored by other historians.[5]

His use of psychotherapy at the Phipps Clinic between 1913 and 1917 was anchored by his concept of psychobiology and was consistent with his goal of establishing a clinical science of psychiatry. Psychotherapeutics helped him to pursue what he considered the three tasks of pathology: to identify the site of a pathological process, to investigate the factors that shaped its development, and to modify those factors. To pursue this methodological strategy, he framed each patient's illness as a discrete experiment in nature—a natural phenomenon with a comprehensible chain of development. Meyerian psychotherapy served the four phases of the therapeutic experiment (examined in the previous chapter) by combining medical inquiry and treatment. First, Meyer endeavored to facilitate the patient's own recollection of the experiences that had yielded the faulty reaction and perpetuated pathological habits. Typically, he expected to accomplish this by collaborating with the patient. Using a form of psychotherapy he referred to as "ventilation," the Meyerian psychiatrist combined his or her analysis of the biographical material with the patient's own introspective insights to discern relationships between causal events. In this investigative phase, Meyer employed techniques developed by the psychoanalysts Carl Jung and Sigmund Freud to discern causal factors operating at a subconscious level, an idea formulated for psychiatry by the French clinician Pierre Janet. He reconfigured psychoanalytic methods as necessary to conform to psychobiology and his hypothesis of habit disorders. To this end, he recruited young psychiatrists trained in psychoanalysis, even as he kept committed Freudians at arms length from the Phipps Clinic. Second, Meyer attempted to facilitate the patient's own reinterpretation of experiences related to the faulty reaction and to inculcate more effective adaptive reactions—ideally to the point where they became automatic. According to the logic of psychobiology, to modify behavioral and neural patterns was a single modification—to automate behavioral and neural activity, a single act of automation. Meyer used psychotherapy to facilitate these therapeutic goals, an approach he called habit-training (as discussed in Chapter 5).

Before exploring Meyer's use of psychotherapy to realize these objectives, I describe briefly the ideas and professional activity of individuals and groups that shaped his thinking about its use and the disorders to which he applied it. For comparative purposes, I summarize Sigmund Freud's formulation of psychoanalysis as Freud presented it to American audiences in 1909. I use Meyer's scientific papers, the case histories of Phipps patients, and his correspondence with patients, their families, and their doctors to elucidate the conceptual and methodological underpinnings of psychotherapy as practiced by Meyer and his staff of psychiatrists in the prewar period. Finally, I examine closely the cases of two Phipps patients to show how these various elements combined in daily practice. Throughout, I suggest ways in which the Meyerian approach articulated with and differed from that of the Freudian.

Subconscious Ideas, Medicalized Psychotherapy, and Psychoanalysis

At the end of the nineteenth century, the work of the French psychiatrist Pierre Janet was foundational to new formulations about the causes and treatment of so-called functional disorders. Janet was a student of the famous Paris clinician Jean-Martin Charcot, who had catapulted hysteria and hypnotism into the realm of serious scientific study in the 1880s. Janet used hypnosis as an investigational tool for exploring pathological processes responsible for symptoms associated particularly with hysteria, such as inexplicable attacks of paralysis, amnesia, choking, respiratory distress, or convulsions that mimicked epileptic seizures. Janet extended his investigations to a disorder characterized by unusual obsessive and compulsive behaviors, a condition he termed psychasthenia. He proposed that hysteria and psychasthenia were rooted in "the exaggerated development of an idea, of a feeling, of a psychological state" that occurred "outside the memory and the normal consciousness." A traumatic memory often precipitated this "dissociation," as Janet described it in English at Harvard University in 1906. Emancipated from personal awareness and control, the dissociated, or, "fixed" idea developed autonomously at a subconscious level and influenced thinking and behavior. This concept had emerged in European philosophy in the early nineteenth century. It gained traction in academic medicine within broadening discourses of nonconscious mental activity that took place, for example, during hypnosis, distraction, sleep, habitual behavior, and forgetting. According to Janet, an irrational obsession or hysterical paralysis was expressed at the behest of a fixed idea, a subconscious process of which the wakeful patient was unaware. Janet related this pathological process to a "narrowing of the field of consciousness" that diminished an individual's ability to synthesize diverse psychological phenomena, such as memories, perceptions of reality, and imaginations. Meyer the pragmatist eventually adapted Janet's idea to psychobiology by speaking of a patient's inability to integrate past, present, and prospective experiences. Psychological synthesis was undermined, Janet reasoned, by defective inheritance.[6]

By the 1890s, there had accumulated a body of legitimate knowledge about nonconscious mental activity and psychotherapy, much of it drawing on Janet's work. Clinicians throughout Europe pursued research with patients on hypnotic suggestion, dream reporting, sleepwalking, the suppression of traumatic memories, subconscious ideas, automatic writing, and the effects of exhaustion and intoxication on mental performance. The findings produced by researchers working in the French city of Nancy and at the Burghölzli Mental Hospital at the University of Zurich were particularly important. Members of the Nancy and Zurich schools expanded upon, tested, and disputed not only Janet's assertions but those of each other. The influence of both schools and Janet pervaded Meyer's ideas and practices (especially August Forel's use of therapeutic hypnosis) and intermingled with his pragmatic orientation. Both psychobiology and psychoanalysis emerged from this age of exploration of normal and abnormal nonconscious mental processes.[7]

Within this context, a group of elite Boston neurologists and psychologists played a decisive role in cultivating new ideas about medicalized psychotherapy in the United States. At the core of this pioneering collective were Morton Prince, William James, Stanley Hall, James Jackson Putnam, Boris Sidis, and Hugo Münsterberg. Meyer and others also regularly attended the weekly meetings. No single psychotherapeutic approach dominated these discussions, and each member published sophisticated original research. Collectively, they marginalized hereditary factors and emphasized the reeducation of the mind rather than the restoration of the body. These academics, in competition for patients with popular mind-cure movements such as New Thought and Christian Science, promoted, within their professions and in public forums, a medical psychotherapy based on scientific principles. Morton Prince, the group's founder, launched the *Journal of Abnormal Psychology* in 1906 to encourage and publish scholarship connected to psychotherapy. Their collective efforts undermined the prevalent belief in American medicine and culture that a nervous breakdown was caused by depleted stores of nervous energy (sapped, for example, by overwork, anxiety, or masturbation). This was a theory that suffused the national discourse on "nervousness" during the period and sustained an epidemic of neurasthenia among the respectable middle class.[8]

It was during this larger psychotherapy movement in the United States, beginning in 1910, that Americans (including most physicians and alienists) first heard about a new talking cure called psychoanalysis. Ten years earlier Sigmund Freud had introduced his ideas in *Die Traumdeutung* (*The Interpretation of Dreams*). In 1910, Stanley Hall's journal, the *American Journal of Psychology*, published an English translation of Freud's lectures at Clark University. In contrast to Meyer's convoluted published formulations of psychobiology, this overview of the theory and methods of psychoanalysis was eminently readable, and Freud's explanations and examples were clear and compelling. The article disseminated Freud's ideas more widely within Anglo-American medicine. In his Clark lectures, Freud made clear that he had developed his method from and for the treatment of hysteria (for which he preferred the term *neurosis*). Working from Janet's principle of dissociated ideas, he suggested that neurotic symptoms were disguised surrogates of an unfulfilled longing (typically, a holdover from childhood of a sexual nature) that was irreconcilable with the patient's intellectual pretensions, what Freud called "cultural acquisitions." To protect the conscious personality from the emotional pain induced by this conflict, the traumatic memory of the objectionable wish was repressed in the unconscious mind. The repressed idea continued to exert itself, as Janet had also suggested, in the form of seemingly unrelated physical symptoms and strange behaviors. In contrast to Janet, whose explanation rested upon an inherited defect in the structural apparatus of the nervous system, Freud proposed a psychogenic etiology—a dynamic struggle between opposing mental forces. In this way, Meyer and Freud used the term *dynamic* similarly, to describe elements that were not static or fixed but responsive and adaptable. For Meyer, however, it was the apparatus of the nervous system that was dynamically responsive (to external and internal stimuli), not merely the imagined entity of an unconscious mind.[9]

Freud's central insights emerged when he confronted a phenomenon also described by Janet. Janet had reported that merely explaining to the patient the connection between a traumatic memory and his or her illness often did nothing to alleviate symptoms. Freud repeatedly observed that a patient's attempt to recall the painful memory during psychotherapy was thwarted by the appearance of "new, artificial, ephemeral surrogates for the repressed ideas." He hypothesized that there were two opposing forces at work in the patient: a conscious effort to retrieve the memory and a protective "resistance" that continued to protect the personality. Resistance, he suggested, functioned to continually repress the emotional pain associated with the traumatic memory. He developed a technique he called "free association" that provided him access to the patient's unconscious ideas without the use of hypnosis. In the quiet of the consultation room located in his Vienna home, with the patient reclining on a chaise lounge to facilitate relaxation, Freud instructed his patient to verbalize the thoughts that occurred to him or her while recounting dreams and everyday experiences. A dream, he observed, invariably fulfilled the unsatisfied wish. Like neurotic symptoms, however, its association to the repressed idea was obscured by resistance. Peremptorily, he commanded the patient to give voice to even seemingly innocuous drifts of thoughts that occurred while recounting the dream. As he listened, he grouped together utterances (and hesitations and protestations) that shared a common emotive tone. This group of associations, termed a complex, advertised to the psychoanalyst the nature of the repressed wish—even as the patient remained unaware that his or her train of thought had yielded significant information. Symptoms only ceased, Freud emphasized, when the patient confronted the "unconscious sexual excitations" and reexperienced the emotional pain that originally accompanied them. The curative effects of this catharsis were observed by Freud's early collaborator, Josef Breuer.[10]

While developing these ideas and applying them to his treatment of men and women diagnosed with hysteria, Freud identified yet another phenomenon significant to therapeutic success. Resistance was overcome only if the psychoanalyst examined "the transfer" (later "transference"), a process by which the patient's attitude toward the physician evolved according to the unconscious wish. A repressed longing—for example, for an object of desire in childhood (typically a parent or caregiver)—was expressed indirectly as disproportionately intense feelings for the psychoanalyst. Over time, Freud perceived that interpreting the free association narrative in light of the transference phenomenon consistently showed him how to lead the patient toward a cathartic experience. Subsequently, he proposed, the patient would be able to use rational judgment to manage the wish, and the symptoms would abate. After nearly twenty years of observations and trials while treating hysterical patients, Freud was confident in the validity of the theory and methods that he had outlined in German for his American audience at Clark University in 1909, and which were subsequently published in English in 1910.[11]

By 1900, psychiatrists at the Burghölzli in Zurich, including its new director, Eugen Bleuler, had incorporated some of Freud's insights into their studies of mental dysfunction. Working with several collaborators, the Burghölzli psychiatrist Carl Jung embraced psychoanalysis enthusiastically and applied its principles to his treatment of both private and hospital patients. Jung developed word association experiments that did much to validate the theoretical claims of Janet and Freud. Association tests had been introduced in the psychological laboratories of Wilhelm Wundt and his student Emil Kraepelin to study various factors influencing normal mental performance. In Jung's protocol, the research subject responded to a long list of carefully chosen words, having been instructed previously to verbalize the first thought to enter his or her mind. Both the reaction time between hearing the test word and responding, as well as attendant electrical fluctuations in the skin, were measured using precise instruments. Other researchers had established that a person's reaction time was discernibly longer when the test word was related to something the responder found unpleasant. When the delayed responses were grouped together, a common theme became apparent, and these clusters of associations were labeled a complex. Jung and his collaborators collected vast amounts of data from "normal and insane" individuals drawn from the city of Zurich and the mental wards of the Burghölzli. His results offered compelling evidence for the agency of subconscious mental activity. Jung's quantitative methods and findings were well received in the United States. His research and his advocacy of Freud's ideas were vital to the initial American response to psychoanalysis.[12]

Jung's word association experiments appealed to Meyer, especially because they produced factual data about psychopathological processes without relying on hypotheses about unobservable and untestable neural mechanisms. Beginning in 1905, Meyer published several enthusiastic reviews of Jung's experiments in American medical journals.[13] In 1907, he wrote to Jung asking for direction on conducting word association experiments at the New York Pathological Institute, and this initiated a friendly correspondence. Soon after his appointment to Johns Hopkins in the spring of 1908, Meyer and the architect of the Phipps Clinic toured the psychiatric clinics of Europe, including the Burghölzli, where Meyer had trained in neurology and pathology. Afterwards, Meyer remained in Zurich to visit his family and work with Jung. For several weeks, he and Jung performed word association experiments, discussed each other's cases, and weighed the merits of psychoanalysis and psychobiology. Jung soon reported to Freud that his guest was "very intelligent and clear-headed and entirely on our side," though he described Meyer's psychobiological views as "radical."[14]

Because of Meyer's institutional power—first in New York and then in Baltimore—his early support for psychoanalysis was exceptionally significant to the American importation of Freudian ideas. Historians trace the arrival of psychoanalysis from Vienna to the New World through Zurich, and Meyer—with his strong ties to both Zurich and Boston—was a vital port of entry. In scientific papers and public addresses, he promoted psychoanalysis as a corrective to material reductionism in American medicine, the focus of his psychiatric reform efforts in the years before World War I. He also deemed it a wake-up call to practitioners reluctant to discuss the biological role of sexual development in mental disorders (with each other and with patients). "Freud has opened the eyes of the physician to an extension of human biology," he proclaimed in 1906. These congenial reviews were some of the first formulations of Freud's views in English. In his own practice and teaching at Johns Hopkins, Meyer deemed word association tests and dream analysis ideally suited to those cases in which causal factors appeared to operate outside the patient's conscious awareness.[15]

Substitute Reactions and Habit Disorders

In the years before World War I, Meyer utilized psychotherapy as an integral component of the therapeutic experiment he instituted at the Phipps Clinic. In his daily practice, he attempted to simultaneously investigate and modify what he conceptualized as a unique experiment in nature—each patient's psychobiological reactions, past and present. "What reaction is at work? What are the determining conditions? What is their modifiability?" was his constant refrain. For cases that he deemed habit disorders, psychotherapy played an especially important role in pursuing these objectives. Like Freud and several others, Meyer speculated that many unwanted symptoms and abnormal behaviors resulted from conflicting instinctual, emotional, and intellectual impulses. His hypothesis was based on the Jamesian notion that the automation of complex sensory-motor responses was integral to the causal agency of mental activity. He suggested that conditions commonly called hysteria, neurasthenia, psychasthenia, depression, schizophrenia, and dementia praecox were, in fact, disorders of habit that developed when mentation was impaired and psychobiological responses were inadequately regulated.[16]

Also rooted in pragmatic philosophy was Meyer's view that each person was continually adjusting to an environment that was simultaneously constituted by subjective perceptions of external reality and an internal mental life, both of which were dominated by symbols and tacit social customs. Modern humans adjusted to this social and cultural habitat—what he called the common sense consensus—by means of mentation. Within the paradigm of psychobiology, common sense was a sensory modality that served mentation. It supplied a person with knowledge of those immediate environmental demands requiring interpretation based on previous experiences and perceived potential outcomes. Common sense helped a human organism to evaluate and select the most advantageous of all possible responses. Inherent in Meyer's understanding of mentation as an adaptive tool was associative and dissociative thinking, as described by William James, as well as John Dewey's concept of thought and action as a single stream of activity. Following James, Meyer theorized that, through an endless series of psychobiological reactions, mentation variously integrated and regulated a suite of adaptive resources accumulated during human evolution: reflexive physiological systems, innate abilities, instincts, emotions, learning by association, memory, perception of reality, and abstract reasoning by means of the dissociation of ideas.

When mentation no longer functioned optimally as a regulator of adaptive resources, Meyer proposed, uninhibited primitive nervous mechanisms competed with higher mental functions in a person's responses to social and symbolic stimuli. In the wake of this "dissolution," an ill-equipped primal response was more apt to win the competition and lead to a misinterpretation of experience by the person. If the interpretation was not corrected, the experience might become a recurring (but false) criterion for subsequent adjustments to aspects of the environment structured by common sense. Eventually, this could lead to the failure of individual social functioning. When the mind became sick, he explained in 1912, "the very mechanism through which we might be able to do the right thing is out of order" and "the feelings and moods are no longer adapted so as to work for the best."[17] The result was a "pathological substitutive reaction" characterized by "poorly planned and ill-adapted make-shifts." Substitute reactions undermined the development and maintenance of psychobiological equilibrium, the natural balance of primal and social instincts. If generalized to other stimuli and employed repeatedly, a substitute reaction eventually became an "unconscious-automatic act" expressed functionally as automatic behavior, what Meyer called a morbid or pathological habit.[18]

Underlying Meyer's concept of habit disorders was the notion that no mental state was inherently pathological. The emotional upheaval that accompanied disappointment, failure, or traumatic experiences, he reasoned, was a normal response. Instinctually, healthy persons assuaged mental pain with a variety of reactions, such as daydreaming, praying, seeking solitude, avoiding painful realities, temporizing, distraction by new interests, complaining, irritability, moping, and "complete or incomplete forgetting." These adaptive strategies typically led naturally to the restoration of psychobiological equilibrium, emotional satisfaction, and equanimity. Their overuse or misuse was apt to become harmful and uncontrollable. "What is at first a remedy of difficult situations can become a miscarriage of the remedial work of life," he suggested, "just as fever, from being an agent of self-defense, may become a danger and more destructive than its source." Left unregulated and cultivated exclusively, a normal adaptive response could develop into a harmful pathological habit.[19]

Rather than dissociation, as Janet proposed, or repression, as Freud suggested, Meyer conceptualized the pathological mechanism at work as dissolution. Initially, he intertwined his thinking about habit disorders with Janet's about dissociated and fixed ideas. "We may say of a hysterical patient," Meyer proposed in 1902, that "she has at one time gone through an experience which gave her a fixed idea." If the misinterpreted experience and subsequent false belief were not corrected, he continued, "everything that goes on is done with the attitude as an actual factor." He described the dissociated idea as an attitudinal adaptation that was comparable to a morphological one, concluding that "an attitude may persist without paying conscious attention to it." The work of Franz Riklin also resonated with Meyer. Riklin collaborated with Carl Jung at the Burghölzli, and his clinical research was informed by Freudian insights. In a 1905 review of Jung's and Riklin's experimental research on normal and abnormal associations, Meyer reported that, in some cases, Riklin found that a dissociated idea was converted into outwardly unconnected symptoms, "deceiving even the patient." In other cases, associations between symptoms and emotional trauma were obvious. In both instances, Meyer explained to American readers, "the dissociated complex acts as an automatism" and "the faulty connections become habitual and often difficult to eradicate." Around the same time, Meyer told August Hoch that he thought it best to avoid the term *dissociation*, because it was used prejudicially by Janet and others to imply that the dissociated idea itself was the pathology. He had already accepted the Jamesian understanding of dissociation as a normal function of mentation—the capacity to form abstract ideas of individual characteristics of situated objects in the physical world and then to apply those ideas advantageously to new and unrelated situations. Meyer suggested to Hoch that "the much more positive concept of habit disorder" should be used instead.[20]

Similarly, Meyer agreed with Freud that painful experiences were sometimes suppressed (he appears to have used *suppressed* and *repressed* interchangeably). He maintained, however, that temporary forgetting was a normal function of the personality. "For our social life we have learned to suppress many important factors of strong affective and even emotional value," he explained in 1909. If painful experiences were suppressed indefinitely, they became "matters which nevertheless determine our attitude even if we do not think of them and which pop up in a veiled form if we repress them otherwise." He considered the suppression of emotional pain and the deceptive character of its symptomatic expression—the latter of which Freud had elucidated with his hypothesis of resistance—the defining features of the hysterical reaction. At the same time, he did not recognize repression as a pathological mechanism in itself.[21]

As his concept of habit disorders came into focus between 1906 and 1910, Meyer increasingly spoke in terms of the exaggerated, excessive, morbid, or deficient mental activity that emerged in the wake of dissolution. He suggested that, typically, there was an "original psychogenic situation" (a misinterpreted experience) connected to a substitute reaction that had kindled progressively glaring miscarriages of psychobiological adjustment. In a case study published in 1912, he discussed a twenty-seven-year-old patient "with a growing obsession of incompleteness of his toilet, the brushing of the teeth and the drying after the bath." The man's worries increasingly impeded his ability to manage aspects of everyday life. "The ruminative episodes have become more and more automatic and they dominate the patient's life with undeniable dilapidation of all capacity." One Phipps patient provided "a good account of his conflicts during his childhood," which Meyer deemed the result of an "excessive reaction" to erotic dreams and nighttime seminal emissions.[22] A private patient of his told him about "the fight within myself"—a daily struggle to be spiritually pure in the face of sexual urges. To avoid the latter, the patient explained, he worked in isolation and secluded himself from society functions that were important to his job and social rank. Meyer told him that it was "unnecessary and wrong" to exaggerate the significance of natural sexual phenomena. His prescription was to reconcile the "thought conflict between the animal and spiritual natures."[23] The patient's inability to balance social ideals and sexual desires represented for Meyer the starting point of a progressive pathological process that eventually led to debilitating fears and anxiety. He regularly used metaphors of grooves, ruts, and canals to describe pathological habits. He reminded one patient that difficult situations in life were inevitable. "False management makes those things difficult to handle," he explained, and "people are bound to make grooves of those things in which they are liable to stick." The results of dissolution—instinctual conflicts and faulty reactions— represented the pathological processes Meyer sought to identify and modify in his clinic.[24]

Meyer considered word association tests and dream analysis specialized tools for collecting a particular type of data. "If we cannot get at the facts directly," he informed his peers immediately after the Clark Conference in 1909, "there are indirect ways which Freud has shown us." The association experiments of Jung and Riklin, he claimed, elucidated "points which would have remained odd and like products of chance" in the thoughts and behavior of the patient.[25] Not every case required these indirect methods, however. As the published case studies of Pierre Janet, James Jackson Putnam, and the Swiss neurologist Paul Dubois showed, "the situations which precipitated psychogenic troubles are frequently easily accessible." His own clinical work confirmed that, by examining the patient's previous life experiences, the origins of symptoms often became obvious (to himself and the patient). But, he recognized Freud's concept of resistance as useful for those cases in which such associations appeared actively obscured.[26] He conceptualized suppression and resistance in the pragmatic terms of a stream of consciousness: "The conventionality and censorship of what one chooses to express affects the assumed and open stream of thought to such an extent that the data of mental causation (or influence of one part of mental activity on the rest of the stream) are often completely hidden." Psychoanalytic techniques could furnish "excellent material for the discovery of undercurrents" when causal experiences appeared persistently distorted by resistance.[27]

In the period before World War I, Meyer viewed psychobiology and psychoanalysis as compatible, but not equal. Psychobiology studied the interrelations of all adaptive assets implicated in a balanced adjustment. With its specialized focus on sexual conflicts and resistance, psychoanalysis certainly had a useful role to play; yet Meyer took for granted that Freud's ideas would be assimilated into his own pluralistic psychobiological model, an assumption that was evident in his early accounts of psychoanalysis. In 1905, for example, he interpreted Freud's explanation of hysterical symptoms: "some emotional (frequently sexual) traumatism is suppressed, never thoroughly *reacted to* or disposed of; the topic becomes *split off*, and *replaced by* or converted into the various symptoms." Meyer instinctively expunged or transcribed some concepts used by Freud—dissociation, repression, disposal, and conversion—into psychobiological language of reactions and substitutions.[28] His ambivalence toward psychoanalysis may explain why Freud was relieved when August Hoch became director of the newly renamed New York Psychiatric Institute (at that time, Freud's most important American outpost). "Hoch is certainly a good replacement for Meyer," Freud wrote to Jung in 1910, "who is rather tricky." Perhaps Freud did not appreciate the amplification of institutional power that accompanied Meyer's move from New York to Baltimore.[29]

On the other hand, Meyer was enthusiastic about the possible applications for psychoanalysis at his new clinic, and he and Jung corresponded throughout its planning and construction. He brokered several apprenticeships with Jung on behalf of American colleagues and students. These included his ally and successor in New York, Hoch, and Macfie Campbell and Trigant Burrow, both of whom he recruited to the inaugural Phipps staff.[30] At the same time, he blocked the institutional advancement of two of Freud's earliest Anglo-American disciples. In 1907, Jung encouraged him to promote Abraham Brill in New York, but Meyer refused, because he thought Brill dogmatic and unlikable. In 1911, Meyer deliberated about a faculty position at Johns Hopkins for Ernest Jones, whom he praised in a letter to his brother as "the best Freudian." Jones, however, had a reputation for inappropriate behavior in and out of the consultation room. This history triggered Meyer's allergy to controversy, and the job offer that Jones had been anticipating never materialized. It was at this moment, historian Ruth Leys concludes, that Meyer made a choice to not associate psychiatry at Johns Hopkins explicitly with psychoanalysis. Notably, only a few months after this decision, he reminded Jung to encourage American medical students interested in psychoanalysis to train at the Phipps Clinic because he was "anxious to have a number of men of first class training in psychoanalytic methods." Meyer was not interested in staffing his new clinic with Freudian psychoanalysts. Rather, he actively recruited both men and women that he thought capable of employing psychoanalytic techniques—as modified by him to conform to psychobiology—to advance his objective of identifying and correcting faulty and habitual adaptive patterns.[31]

Meyerian Psychotherapy

The patient's life experiences constituted Meyer's primary material for investigating substitute reactions. During psychotherapeutic interviews, patients were asked to respond to psychiatrists' questions and describe their experiences, past and present. These inquiries evolved from ongoing observations, analyses of previous interviews, and observations of the patient's reactions to the other elements in the Phipps environment: occupational therapy, meals, hydrotherapy, recreational activity, and informal interactions with staff and fellow patients. Meyer expected individual psychotherapy to generate data for reconstructing causal chains in the natural experiment. As he explained to one patient, the daily interview provided "a chance to take apart the morbid tendency into its component factors."[32] Just as the traditional pathologist examined bodily tissue methodically to identify the telltale signs of a particular organic disease process, the same methodological rigor, he claimed, would reveal abnormal adaptive patterns and the conditions that instigated and sustained them.[33]

The first step of Meyer's investigative strategy was to identify the original faulty reaction that initiated and perpetuated pathological habits. He encouraged the patient to recall the first time he or she adjusted to a new situation by using the insufficient substitute reaction. "You feel as if you cannot get even with something," he asked one patient who talked of suicide; "[W]hat is this thing? What was the rub in the former situation?" Later, the same patient told his father how his doctor at Johns Hopkins had instructed him to "dig down into my worries before my breakdown." Meyer conceptualized his psychotherapeutic approach as analogous to the targeted therapies used by his peers in other specialties. A concerned husband requested advice about his wife. She experienced "depressive attacks" and was convinced that she had a physical ailment. "A careful analysis of the foundation of the first attack and the previous disposition," Meyer explained to him, "has occasionally uncovered matters which the patient could be taught to meet in a way that would not make them liable to upheavals in the form of depressions." He added: "A correction in this sphere works, I suppose, very much as the correction of some physical drawback such as your wife hopes to get discovered." If the woman could be helped to acknowledge, first, the "psychogenic experience" that had prompted the original attack, and, second, that she might be capable of consciously choosing a better way to adjust than by the faulty, a recovery was possible.[34]

Meyer viewed the investigation and modification of psychobiological reactions as a collaborative process between psychiatrist and patient. During a teaching clinic, Meyer explained to medical students that his patient Rose's symptoms were caused by "a difficulty of adjustment in the patient on which we hope to be able to get more light by finding out what her attitude was in her earlier married life." He deemed a patient's own account of his or her life story and illness to be important. "The manner of presentation often gives us a clue to the nature of the disorder," he advised in his 1918 guide to examinations, and "we would be unfair to the patient if we did not allow him to present his own story and interpretation." The clinic's assistant director, Macfie Campbell, described psychotherapy to another patient as "a thorough study of subconscious adaptation" that combined Campbell's own medical analysis with "the introspection of the patient." Meyer related how he encouraged the same patient "to review the facts of his life and to reconstruct the situation from which his symptoms arose." Campbell described clearly this first phase of Meyerian psychotherapy: to scrutinize life experiences and identify faulty adaptive habits utilizing both the introspective insights of the patient and the psychiatrist's judgment as an expert on psychopathology and also as an agent of the common sense consensus.[35]

The clinical and social structures of the clinic were designed around what Meyer viewed as the collective enterprise of psychotherapy. In its contrived social milieu, the staff and patients maintained close and frequent contact. Each of the eight wards in the Phipps Clinic, divided equally by gender, was assigned a ward psychiatrist. He or she conducted psychotherapeutic interviews at a rate consistent with the patient's capacity to collaborate. Some patients were interviewed twice each day, others as their physical and mental states permitted. Meyer or a member of his senior staff interviewed each free or subsidized patient in the communal wards twice weekly and spoke with private patients daily. Each morning, the ward psychiatrists and senior staff congregated in the library of the clinic to discuss cases and determine strategies for further interviews. In this way Meyer directed and shaped the investigative and therapeutic practices of his staff. One patient's comment hints at the interplay between Meyer, other Phipps psychiatrists, and patients: "You, Dr. Meyer, give me the big idea," she declared in a note to him during her hospitalization, while her ward psychiatrist "helps me with the practical business of nailing it down." In the Phipps Clinic, interviews often took place in the psychiatrist's office on each ward. References scattered throughout the Phipps records indicate that psychotherapy was also conducted while the patient was in bed, weaving a basket, eating, walking in the garden, sitting on the terrace or porch, or knitting or whittling in the sun room. Meyer expected his psychiatrists and nurses to be responsive to spontaneous opportunities to advance psychotherapeutic goals. Patients were encouraged to record their thoughts in between formal interviews. These texts—in the form of dream journals, daily logs of progress or complaints, letters to Meyer or the staff, family, and friends—were added to the case history and guided the psychiatrist's approach to subsequent interviews.[36]

Meyer's approach to psychotherapeutics reflected his instrumentalist reading of experience, knowledge, and common sense as adaptive tools used by both psychiatrist and patient. In his private scientific notes, he described as part of the psychiatrist's role to furnish the patient with food for thought. He cautioned that the psychiatrist's view of the patient must never crystallize. "It has to organize and grow, to expand or shrink itself," he mused, "according to demands and opportunities." Many of his contemporaries remarked upon his skill in this regard. D. K. Henderson, inaugural chief resident psychiatrist at the Phipps Clinic, admired "the delicacy with which he was always able to probe into the most acute situation" during ward rounds. Stanley Cobb, who was on the Phipps staff from 1916 to 1918, echoed this sentiment much later in an obituary for Meyer, adding that "ward visits were also fun, because the Professor had a gay quality and warm sense of humor that lightened the day's work and saved his pupils from being depressed by the tragedy that often surrounded them." It was a lesson never forgotten, Cobb attested, "to see him interview a new patient and, with few words, but manifest, quiet sympathy, reach the heart of the problem." Thomas Rennie, a later Phipps staffer, commented that the chief possessed limitless patience and great respect for the patient's own spontaneity. "Meyer had a knack for posing questions that provoked curiosity in the patient," Rennie recalled after Meyer had retired, "leading him or her to continue the search for further reasons for the behavior."[37] One patient commented, "[I]t was easy to speak to him of the best that is in me." Meyer's collaborative and instrumentalist approach to psychotherapy receives support from stenographic transcripts of his interviews with Phipps patients.[38]

Embedded as it was in the dynamic social milieu of the Phipps environment, Meyerian psychotherapy was mobile and opportunistic, as the following two cases demonstrate. Ruben was a successful public figure, married, and sixty-five years old. He was a private patient of Meyer's who experienced paranoid thinking and incapacitating fears. During psychotherapy, he rubbed his hands restlessly as he recounted his troubles. He complained about the furnishings in his private room and criticized Meyer's staff. During several weeks of daily interviews, Meyer attempted to reconstruct the development of symptoms by asking Ruben to "ventilate" about his experiences. Ruben regarded questions about his personal life as immaterial and ungentlemanly. Eventually, however, he did disclose that he had committed adultery and that his mistress blackmailed him. This had forced Ruben to confess the affair to his wife and he had been waiting many years for the inevitable outcome that she would leave him. Meyer urged him to describe the events that led to this state of affairs. "Don't punish me anymore, I am punished enough," Ruben begged Meyer, asking him to drop the matter. "Why not speak of it to a physician?" Meyer asked. "I consider it of importance for you to state in what connection you and your wife spoke of this thing—tell me of the incident." His patient became tearful and sobbed loudly. "It may be interesting as a histological fact," Ruben commented perceptively, "but I have been trying to forget." Meyer concluded the interview with a reiteration of his conviction about the automatic nature of psychobiological habits: "You make yourself believe that you forget—no one ever forgets—you must realize it is an adjustment." Ruben, nonetheless, remained unreceptive to his new physician's strange "ventilative" treatment.

D. K. Henderson discovered that, while in the hospital, Ruben missed the intellectual exchange customary within his elevated social circles. As resident psychiatrist, Henderson lived in the clinic; late-night talks became routine for him and Ruben. Their conversations were fueled by mutual interests and class backgrounds. Henderson capitalized on the opportunity, steering their discussions toward Ruben's worries and behaviors. Because of this prolonged social contact, Henderson was able to cultivate a flow of therapeutic talk that the patient had resisted previously. It was noted in the case history that Ruben also began to respond more frankly to Meyer's daily probing into his personal affairs. The problem of what Freud might have termed resistance was mediated by Meyer, using the domestic setting of the Phipps environment and a collective approach to conducting psychotherapy.[39]

The second case begins with another late-night discussion, this time between two patients on the semi-quiet ward. One of the patients, an older gentleman who was a minister, sat at the bedside of a young man named Percy. Percy had been markedly confused and withdrawn for several weeks (it was Percy who scalded himself with tea, as described in the previous chapter). Ward psychiatrist Charles Thompson, observing at a distance, watched as the minister said goodnight and kissed Percy's forehead "in quite a fatherly way." Thompson recorded in the case history that "for a few moments afterwards the patient brightened up considerably." Sensing an opportunity, Thompson initiated a conversation. "What do you think caused your trouble?" he asked directly, to which Percy stammered: "I—I—I raped a girl." Thompson's surprise was palpable in his clinical notes documenting the encounter. Percy had lingered in a catatonic state for almost a month. For three months before that, Thompson and his patient had discussed Percy's life experiences at length, analyzing together his word association tests and dreams. Never had Percy mentioned, even hinted at such a thing. Thompson pressed for details. Percy explained that just before his hospitalization at the Phipps Clinic he had accosted a woman who was employed in his mother's household. "I forced her and she yelled for help so that I never got to her," he told Thompson. "Then she balled me out to my mother and told everybody we were a bad lot." Unfortunately, Percy's lucidity did not last. Soon after, he slipped again into a catatonic state. Thompson appeared inclined to accept Percy's statement at face value. Yet, even if this experience had inspired the faulty reaction, his patient's continued deterioration indicated a poor prognosis for retraining psychobiological habits.[40]

Not surprisingly, the function of the psychotherapist emerged disparately in Meyerian and Freudian psychotherapy. The spatial, temporal, and social contexts of talk therapy in the Phipps Clinic were dramatically different from those of Freud's study tucked away in his family's apartments in the Berggasse. The one was public, communal, and perpetual; the other confidential, dyadic, and circumscribed by appointment times. In contrast to Meyer, who thought the psychotherapist catalyzed the patient's own reconstruction of causal experiences, Freud tasked the psychoanalyst with deciphering distorted associations wholly indiscernible to the patient. In 1909, Freud emphasized that the transference phenomenon was an indispensible interpretive device in this undertaking. He explained that transference was not a product of psychoanalysis but occurred in daily life unnoticed. Due to the nature of psychoanalytic treatment, he suggested, it was amplified and became a valuable tool. Meyer engineered the Phipps environment to resemble daily life because he wanted to scrutinize and restore the patient's individual social functioning which included, as Freud had stated, the occasional transfer of erstwhile feelings to a new object. This helps to explain why, in his selective uptake of psychoanalytic ideas and techniques, Freud's concept of transference appeared insignificant to Meyer. Little did he know how divisive the issue would become for Meyerians and Freudians in the decades to follow.[41]

Meyer and his psychiatrists employed psychoanalytic concepts and techniques as investigational tools in their search for causal factors in the experiment in nature. One patient was informed that his answers to word association tests and discussions about his dreams would "expose subconscious thoughts to the light of day."[42] Of another case, a Phipps psychiatrist observed that the "association test brought out particularly [the patient's] desire for children and disappointment about [her] sterility." Meyer told another patient that "dreams are valuable assets" and encouraged the man to make a note of them each day.[43] He welcomed word association experiments and dream analysis as practical aids for studying situated biological events. He remarked in 1909 that traditional medicine considered the content of dreams as epiphenomenal symptoms of brain activity. "What counts is also the situation," he insisted, "in which the body with its nervous system works." What psychobiology shared with the psychotherapeutic methods developed by Janet, Jung, and Prince, he contended, was a means to study the adaptive work performed by an organism. Meyer concluded, "It is this same method that Freud tries on the dreams; a study of the events in their settings." Once again interpreting psychoanalysis in psychobiological terms, he declared that Freud's method required that each element in the patient's narrative be described "as an event in a situation," so that the "determining factors of the development" could be analyzed and reconstructed. He remarked to Prince a few months later that psychoanalysis appealed to his "inborn need for causal and dynamic chains." Meyer appropriated psychoanalytic techniques to serve his own strategies. In doing so, he sidelined transference as a method and catharsis as a therapeutic goal. The result was that he and his psychiatrists often employed them in ways that contradicted Freud's theory.[44]

A few Phipps patients had read about psychoanalysis in newspapers and magazines, but neither Freud's theory nor his techniques were widely understood in prewar America. In 1911, the *Chicago Daily Tribune* ran an article with the headline "Psychanalysis, A New Science." The article described psychoanalysis as a "scientific reading of the innermost secrets of the human mind without that person's knowledge of the experiment." In the same year, the *Baltimore Sun* explained to readers that psychoanalysis was based on the idea that "every person is possessed of two personalities—the conscious and unconscious" and that disagreement between them was responsible for "nervousness."[45] One patient told Meyer that he sought admission to the Phipps Clinic because he had read a book about psychoanalysis. He expected that "he would sort of be hypnotized and that things in his past life which he didn't know about and which were doing him harm, would be brought to light." Stories about the unconscious personality and, especially, multiple personalities abounded in magazines and dime novels. One Phipps patient reported that she had a "double personality," having read about psychoanalysis in the *Ladies Home Journal.* Asked what that meant, she explained that she was "entirely different out of her home from in it." Sporadic news items and comments by Phipps patients illustrate that the conceptual basis of psychoanalysis was widely misconstrued during this early period of its use in the United States.[46]

Meyer and his staff of psychiatrists asked their patients to convey dreams to generate material for the collaborative work of psychotherapy. For example, Phipps psychiatrist Edward Kempf asked a patient named Charlie to recount his dreams. Charlie was a teacher from Chicago who experienced severe worry and insomnia that prevented him from working. When he began hinting about ending his own life, his wife urged him to go to Johns Hopkins. During one psychotherapeutic interview, Charlie told Kempf that he had dreamed of his young son sitting in a handmade Morris chair with a bandage wrapped around his head. Systematically, as per Freud's instructions, Kempf inquired about each element in the dream. Regarding the significance of the Morris chair, Charlie recalled that he used to sit in it every evening to relax though he had not done so for a couple of years. He also mentioned that the chair's cushions were often used to shield a draft from the window next to his son's bed. Kempf asked Charlie to talk about his son. His patient explained that the boy was a sickly child, about whom he worried. The following day, Charlie reported something else he recalled in connection to his dream. Kempf rendered the report in psychobiological terms: "It dawned on him that for years he had carried about in his reactions a feeling that *some day I would break down and not be able to do anything.*" After adopting this conviction as a young man, explained Charlie, he had taken vacations calculatingly to avoid a nervous breakdown. Since winning a work promotion two years before his hospitalization, however—about the same time, Kempf observed, he had stopped relaxing in his favorite chair—Charlie had forgotten this premonition and steadily increased his workload without taking holidays. Kempf asked Charlie directly when he first began to worry about a nervous breakdown. As a teenager, Charlie recalled, his father had accused him of doing shoddy farm work and compared him to the "town loafer." He confessed, "[I]t made a deep impression on me." In between hydrotherapy treatments, occupation class, and recreational activities, Charlie and Kempf continued to examine Charlie's attitudes toward work, relaxation, sickness, youth, and fatherhood. Kempf, combining his own interpretation with Charlie's introspective insights, had guided his patient toward self-discovery of a faulty and harmful mental habit (Charlie's longtime conviction that a nervous breakdown was inevitable) based on a misinterpreted experience (his having interpreted his father's remark as an insinuation that he was a degenerate).[47]

If faulty mental habits were ruts and grooves, Meyer spoke of healthy adaptation as flowing rivers of associations that allowed primal and social impulses to coexist under the regulating function of mentation. He thought this was accomplished by helping the patient to acknowledge his or her original faulty adjustment and the misinterpretation at its core. "Conscious ventilation takes away the automatic nature" of a subconscious fixed idea, he remarked in 1905, by assimilating the source of emotional pain into the personality. He asked one patient to recall why he had reacted to a particular situation with unfounded suspicion. "The way to find out is to go over those remembrances and to find out the problems," Meyer advised. "Try to plot the whole course of the stream and let it have free flow." He discharged another patient with the reminder that "a complete feeling of ease about the problem of the sexual life" would help her maintain the recovery she achieved at the Phipps Clinic. He equated an easiness about the inevitable push of lower drives as a sign of healthy psychobiological equilibrium, and he thought this could be achieved by ventilative psychotherapy.[48]

Conscious acknowledgement of the faulty reaction, he suggested, reduced the depth of a habitual rut. Moira, a thirty-year-old school teacher, experienced many distressing physical symptoms, such as constipation, pains, headaches, and abdominal discomfort. She explained to psychiatrist Roscoe Hall that she felt wooden and empty. Initially, Hall found it difficult to get Moira to "ventilate" because she preferred to discuss the misery she experienced due to her symptoms. "She did not admit any emotional experiences," he noted; "one can hardly get her to discuss other problems of life." In subsequent interviews, Moira recalled reading a book as a teenager about the innate brutality of all men. Further ventilative psychotherapy revealed that she was not ignorant or fearful of sexual intercourse, or ashamed of her desire for it, but she dreaded the violence that she assumed accompanied its prerequisite: marriage. Hall's preliminary analysis was that Moira's problem was "not prudishness" but "perhaps subconscious antagonism to courtship." He speculated that her symptoms were provoked by incompatible desires: to avoid this brutality and to become a wife and mother. Within the common sense consensus, which constituted Moira's environment to such an extensive degree, these goals conflicted. Hall explained to Moira that her "invalidism" acted subconsciously as a way to avoid these painful realities. He recorded in the case history that, upon hearing this explanation, she thought for a moment and then asked if this principle might apply to a matter she had not yet divulged. He encouraged her to elaborate. Six years earlier, she explained, a man had proposed to her. Even though she longed to marry, she felt this particular man was beneath her. He was still waiting for her answer. Pulling together data from their collaborative study of her experiences (while awake and asleep), Hall presented Moira with a plausible chain of events to explain her ailments. He reminded her that on the day she was admitted to the clinic he had asked for her marital status. She had replied incredulously, "Should I desire to marry when I have such a sickness?" He now suggested to her that "evading the vital decision may therefore be the basis for the elaboration of minor symptoms." Shortly afterward, Moira remarked that she had "gained something by the last two interviews." Hall had identified her uncorrected notion about the realities of marriage as the source of her faulty subconscious adaptation. For Meyer, her misinterpretation became a miscalibrated adaptive tool, employed automatically by Moira for a decade as she adjusted to new situations.[49]

The correction of misinterpreted sexual experiences was a regular aspect of psychotherapy in the Phipps Clinic. Within several academic discourses at the end of the nineteenth century—of subconscious mechanisms, dementia praecox, psychotherapy, and sexology—it was considered axiomatic that sexual functions were commonly implicated in psychopathology. Within these discourses, the sexual inhibition that had pervaded the Victorian age was increasingly portrayed as potentially harmful to physical and mental health. Bert, a young doctor from Ohio and a patient at Phipps, was plagued by bowel discomfort and chronic headaches, something he discussed at length with nurses and fellow patients in the clinic. Single and thirty-two years old, Bert had been raised in an affluent family in which propriety and sexual inhibition reigned. During psychotherapy with Macfie Campbell, he desired only to discuss his symptoms and promising diagnostic tests for his condition. Campbell assured him that to talk of medical symptoms and procedures was to speak merely of "drawing-room factors." He explained to Bert that the chronic headaches were, in fact, "defensive headaches" resulting from attempts to avoid acknowledging natural sexual thoughts and urges. "By training and early atmosphere," he told Bert, "you have ruled these things out of consideration." He reminded his fellow physician that sexual functions were normal biological phenomena. "Troubles that are shelved," Campbell explained, continued to cause problems in the form of physical symptoms. He continued to urge his patient to ventilate about his sexual experiences with only limited success.[50]

Freud's attention to matters of sexual development garnered Meyer's initial support for psychoanalysis. Neither man condoned sexual freedom or experimentation. Rather, both were interested in the causal role of sexual experiences in psychopathology. Hardly a novel line of inquiry, it was the intelligibility with which Freud correlated early sexual behavior with neurotic symptoms in his case studies that was striking. Immediately after the Clark Conference, Meyer reported to New York alienists that psychoanalysis opened up "many unsuspected paths of undisputable importance" hitherto ignored by physicians due to "exaggerated prudery." He sermonized that, just as doctors and nurses trained themselves not to flinch at the disgust of incontinence or putrid discharges, they must also learn to adjust to the "startling words" that described sexual activity. The sexual aspect of life, he explained, was a mainspring of multiple biological drives. He reasoned, as Freud did, that the social conventions of modern civilization (what he called the common sense consensus and Freud called cultural acquisitions) "disfigured" the experience of natural sexual impulses. Drawing on examples in Freud's and Jung's cases, Meyer elucidated for the alienists how childhood experiences such as thumb-sucking, reprimands against peeping, corporal punishment, or constipation could develop pathologically in the adult personality as autoeroticism, exhibitionism, masochism, or neurotic constipation and anal masturbation. He emphasized that his analyses of case histories of patients at the Worcester asylum diagnosed with dementia praecox had yielded correlations similar to those of Freud.[51]

Meyer disagreed, however, with the sexual reductionism in psychoanalysis, because it threatened the two pillars of his program for psychiatric reform in America: pathology as method and mind as a dynamic biological function. First, to seek out subconscious machinations of ungratified sexual impulses and deliberately rule out all other possibilities was to limit and jeopardize the scientific inquiry—not unlike a myopic search for anatomical lesions. Second, a deterministic mechanism of repression encroached too nearly on that vital Jamesian principle: the selection of the most advantageous adaptive response from multiple possibilities to any single stimuli— the very means that endowed mind with causal agency in biological adaptation. "The whole inquiry takes a different aspect," wrote Meyer in 1909, countering Freud's repression hypothesis, "if we make it a point to compare the faulty reaction with the most efficient reaction in the situation." Symptoms resulted from any number of competing nervous responses normally regulated and integrated by mentation, he argued, including physiological, instinctual, emotional, and intellectual processes. Moira's irreconcilable impulses to avoid physical harm and to procreate, for example, were concurrently *both* primal responses and *both* dependent on common sense (or, in Freud's rendering, cultural) notions of marriage and mothering. And Charlie's recollection of his traumatic memory—his father's censure about his farm work—was rooted in feelings of inferiority, which Meyer emphasized were frequently connected to unsatisfied impulses to be productive or altruistic, and not exclusively the result of sexual longings. Whereas Freud located a dynamic psychosexual conflict within the personality, Meyer situated multifactored (or pluralistic) conflicts within the dynamic interaction between the organism and its uniquely constituted environment. "Freud himself would be the last person to claim his information and method were the last word," Meyer informed one critic of psychoanalysis in 1908. Indeed, in 1909 Freud was far from dogmatic in these assertions (at least publicly). Only when the psychosexual reductionism of psychoanalysis became increasingly unconditional in the years after World War I, was Meyer confronted with the reality that psychobiology and psychoanalysis were not only incompatible but potentially antagonistic.[52]

The final phase of Meyer's psychotherapeutic strategy was to guide the patient to consciously formulate and practice a better adjustment in light of reinterpreted experiences, past and present. The life history of the patient, he maintained, was "the material that you ought to be able to reconstruct into a sufficiently successful adaptation." He instructed his psychiatrists to use the data generated by ventilation to help the patient reconstruct a better, more efficient adaptation based on his or her capability. "The physician puts something in," Meyer mused in his private notes about psychotherapy, "not content—but as a catalyzer which will enable the patient to rearrange the components of his own life, to become more and more focused and capable of performance." After one patient's discharge, Meyer advised the patient's family doctor to continue "a more substantial reconstruction of the residual from his early conflicts and misconception" and "a re-education into automatisms of more direct and controllable constructive activity." This was the habit-training he employed for habit disorders.[53]

Underwriting Meyer's use of psychotherapy as a therapeutic device for habit-training was the pragmatic notion that thinking and action were a single adaptive response. "Psychotherapy is a regulation of action," he told the Medico-Psychological Association in 1908, "and only complete when action is reached." He asserted that thoughts—conscious, subconscious, or habitual—were internal impulses satisfied only by a completed action in the external world. "Playing pool and various activities," he testified about one patient, "quite obviously roused some natural and helpful instincts of self-activation." Janet had also emphasized that thought was but an activation of evolutionary drives to fulfill instinctual needs. The physical and social spaces of the Phipps Clinic represented for Meyer an optimal environment in which a maladjusted individual might consciously practice more effective psychobiological reactions. "Habit training," he concluded, "is the backbone of psychotherapy." Constituted by a routine of meals, occupational therapy, hydrotherapy, recreation, social engagement, and psychotherapy, life inside the clinic was supposed to spark and reinforce successful adaptive habits. One of Meyer's private patients documented an achievement in this regard in a letter to him: "It looked as though I was about to go off again into hysteria. I tried hard for control—turned away my mind and started to play cards."[54] Manual occupation and social activities interrupted rumination, directing the patient's attention away from "temptations or attractions alluring to his morbid appetite, longings, and fancies." Meyer counseled another patient to "build up a foundation of habits and interests of activity of a conservative character which will crowd out the feelings which have become morbidly habitual." Requiring patients to talk about thoughts and feelings was, for Meyer, worthless without also insisting that they set and achieve concrete goals. "Actions and attitude and their adaptation" were at the core of his psychotherapeutic approach.[55]

Case Studies: Henrietta and Irving

Fragments mined from the clinical case histories help to elucidate Meyer's reasoning and practices regarding psychotherapy, but they cannot fully convey the dynamism of the interactions between Phipps psychiatrists and their patients. What follows are two case studies, those of Henrietta and Irving, that shed more light on Meyer's use of psychobiology as a conceptual framework for interpreting psychopathology and guiding daily psychotherapy. Even so, the interpersonal nature of a psychotherapeutic encounter, as well as its highly subjective and contextualized content, makes any representation of such interactions seem woefully inadequate. The portrayals below are meant be illustrative and instructive.

Henrietta was a twenty-three-year-old teacher. She traveled from North Carolina in 1913 to consult doctors at Johns Hopkins about frequent episodes of fainting and uncontrollable shrieking. She experienced one of these attacks at the hospital. "She suddenly began to scream, refused to speak, lay with her head drawn back and neck rigid," a physician in the general hospital noted, and she cried "Oh! that man!" and "that knife hurts." She reported that she had no memory of the experience afterward. She was described as well nourished but uneasy and worried. She was admitted to the semi-quiet ward on the second floor of the Phipps Clinic. The ward psychiatrist, Charles Lambert, interviewed Henrietta that night. His notation in the case history indicated that he did not employ psychoanalytic techniques at that time: "the patient was approached first perfectly directly." From this interview, he obtained "a description of her life and her ambitions and anxieties." She came from a successful farming family that, much to her concern, valued money more than religious devotion. She was a devout Christian, who had "made the full surrender" three years earlier at a Baptist revival. "I wanted to become a missionary," she told Lambert, "but father and mother objected." Asked to describe her feelings at present, she reported that she was melancholy because there was so much Christian work to do. Asked what she thought caused her attacks, she listed overwork and worry over her family's evident damnation.[56]

The next day, Lambert performed the extensive physical, neurological, and mental status examinations that were part of the Phipps admission routine. He found no neurological or cognitive abnormalities. He asked about menstruation and Henrietta reported that it had been irregular during the last three years and "sometimes brought on by excitement." Lambert asked for specific examples. "Unpleasant things," she told him, smiling meekly, such as fellow teachers wearing necklines that were too low. During the examinations, they talked in some detail about her worries. The following morning, nurses observed that Henrietta was restless and apprehensive. When she was discovered weeping in her bed, the nurse asked her what was wrong: "I can't tell—I can't tell," was all she would say. She pulled the bed covers over her head and refused to speak to psychiatrist Ralph Truitt, a member of the senior staff. She did begin hydrotherapy and work in the gymnasium, nevertheless, and she remained receptive to Lambert's inquiries and examinations.

Lambert's clinical notations show how he utilized dream analysis in conjunction with word association testing to discern potentially significant experiences or mental habits that were unapparent to him and his patient. Except for the attacks, Henrietta appeared physically and intellectually normal. She was eager to advance in her career as a teacher and to pursue charitable work. Indeed, she attributed her attacks to worry and exertion. "What comes under worry?" Lambert asked her. She again enumerated her concerns about the salvation of souls, adding that she was dissatisfied with work and her lack of social refinement. Lambert urged her to continue ventilating about her worries. "I've told you all I can think of," she winced, as she bit her lip and turned her face away. Freud had asserted that such a declaration signaled resistance and he instructed his patients in advance to verbalize all thoughts no matter how upsetting or seemingly unrelated. Meyer's emphasis on observing and modifying adaptive reactions instead guided Lambert to describe precisely the behavioral expressions of Henrietta's discomfort (twisting in her chair and biting her lip) and to end the interview, in order to avoid inducing an artificial defensive reaction.

In subsequent interviews, Lambert asked Henrietta to describe her dreams. He listed their subjects along with his preliminary interpretations in the case history. He recorded that she dreamed "of getting up to speak and forgetting my subject, and becoming greatly embarrassed" to which he appended: "lack of clear conscience." She dreamed of returning to her family's farm and immediately wanting to leave again," which Lambert interpreted as "not rid of her trouble." Lambert continued to interpret dream elements to gather information for the reconstruction of causal chains. Henrietta also relayed a nightmare in which strange men entered her room. Asked to elaborate, she told Lambert that she had read many books about "white slavery" (a salacious topic popular in yellow journalism and dime novels in which middle-class American women were kidnapped and sold into prostitution). "Ever had a serious fright in relation to a man?" Lambert asked. "No, but I've heard of girls who had," she explained, "and I've read things." Lambert concluded the interview and presented his findings to Meyer and the rest of the senior staff.

The following day, Henrietta and Lambert discussed something she had dreamt the night before. She recounted a white lily that she wished to pluck, but it was guarded by a snake, coiled up and ready to strike if she did. Lambert asked her what she thought the dream meant. The white flower, she surmised, represented purity. "A force, something terrible, kept me from it," she recalled emotionally. Without a further prompt from Lambert, Henrietta began explaining that she felt oppressed by a heavy burden, one so terrible she contemplated suicide. Her religious conversion was "the first ray of light, but it kept creeping on me." The serpent in her dream, she concluded, must represent the oppressive force. "A secret sin," she whispered, "the demon masturbation." Henrietta began to cry and convulse, telling Lambert that she had practiced "self-abuse"—a common euphemism for masturbation at this time—from age three until a teenager when she "read a book saying how terrible it was." Attempting to guide her toward self-discovery, Lambert asked her what she considered significant about the lily and the snake appearing in the same dream. Henrietta felt certain that the reptile represented a white slave trader.

Lambert prepared a word association test based on his interviews with Henrietta. It combined a standardized series of test words with words chosen by him for a total of 156. He instructed her to respond to each test word with the first thought that entered her mind. He timed the duration of each reaction using a stop watch and a stenographer recorded Henrietta's responses and reaction times. The entire process was repeated immediately. Lambert then analyzed the results, carefully noting recurring responses and reaction times considered longer than normal. He noted, for example, that the word *man* occurred six times, and he grouped Henrietta's responses to the following test words, because of their extended reaction times: *deceive, duty, finger, to rub, hesitation,* and *false.* He inferred that this complex of responses was associated with Henrietta's exaggerated ideas about masturbation. He also noted that the test word *minister* elicited an especially protracted reaction time of twenty-one seconds. He concluded that the word association test and dream analysis "showed chiefly her preoccupation with sexual ruminations and much reading of white-slave literature," factors that appeared to be prominent links in the causal chain leading to her symptoms. The next day, he asked Henrietta to ventilate on the subject of each atypical association on the test. Her statements confirmed for him the hypothesis about a masturbation complex. Asked about her delayed response to the test word *minister,* she readily disclosed a "secret love for my minister" that she characterized as sinful.

In a subsequent interview, Lambert revisited the dream of the white lily and the snake. He suggested to Henrietta that the snake "might represent a phallic symbol, for which she wished, but feared to boldly pluck." Perhaps, he prompted, it was "a disguised thought or fancy in relation to her secret habit?" This notion appeared to resonate with Henrietta and she talked openly about fantasizing about men. Lambert described for Henrietta how she had reacted inadequately to sexual urges based on her misinterpretation of them. "The simpler mechanism of the dream was pointed out," he wrote in the case history, "and the method in which suppressed fancy and memory might assert themselves in dreams, or in conduct, was described." The next day, he explained to her "the relation of the hysterical attacks to the conflict" and reported that his patient afterward recalled that "the masturbation struggle usually precipitated an attack." Henrietta was reportedly relieved by these interviews.

Based on data accumulated from dreams, word association tests, and ventilation, Lambert composed a "plan of reconstruction" that would serve as the foundation for Henrietta's recovery. First, there must be a correction to her "internal adjustment"; the psychiatrists would help her to reinterpret her sexual experiences in accordance with the common sense consensus. "The general biological aspects of the sexual function were discussed with the patient—the rather central and controlling influence in the life of the plant and animal series and the significance to the individual." This biological formulation clarified that sexual drives played a natural, not harmful, role in a healthy adjustment. Meyer described the goal of Henrietta's treatment as "an attempt to readjust the sex sensitiveness to a more spontaneous expression with freedom in associations." Second, Henrietta was instructed to follow "a plan of adjustment with respect to practices, baths, more outdoor life and exercise, [and] healthy, wholesome companionship and diversions." Whether this referred to her sexual practices and, if so, what instructions she received about masturbating or marital relations, is unclear. Lambert's explicit prescription was to take "a balanced approach to life," which he explained was the only antidote to "narrow-minded habits of thought." According to the clinical documentation, Henrietta's attacks ceased during her four-week hospitalization at Johns Hopkins.

The following notation in the case history of Henrietta shows the therapeutic connection that Meyer and his psychiatrists drew between habit-training, a realignment of the internal environment with the common sense consensus, and mental well-being: "She talks freely and frankly about her difficulties; her mental attitude is wholesome and normal; has made her plans to return to school; to divide her time between work and diversion more judiciously; has learned to play cards and is going to dance—heretofore, unpardonable sins—and participate altogether more fully in life." Henrietta left the clinic one month after she had arrived "apparently well and hopeful," and Lambert reckoned that a "permanent adjustment seems certain." A letter from Henrietta to Meyer in 1922 appeared to confirm the prognosis. After leaving the Phipps Clinic, she finished college and became the director of an industrial school for girls. "I believe that the instructions I received from Dr. Lambert," she professed, "made it possible for me to recover my health." She reported that she had recently married a professor and was "happy in my work and plans for the future." From a historical perspective, the case of Henrietta appears relatively straightforward and exemplifies how Meyer pursued his psychotherapeutic strategy to identify a faulty substitute reaction and replace pathological habits—employing, when necessary, the "indirect" psychoanalytic methods of Jung and Freud.

A more difficult example—one Meyer acknowledged as a failure on his part—shows his application of psychotherapy to a case he described as a "depressive reaction." Irving was a wealthy fifty-year-old bachelor who lived with his housekeeper and longtime nurse, Mrs. Shire, on a country estate in New York. He did not come from an affluent family; in fact, his father had struggled to sustain the family and had died when Irving was twenty. After college, both Irving and his older brother had found success in the business world. While Irving experienced little inspiration there, his brother had thrived and amassed a fortune. As a young man, Irving had had few friends except his brother (whom he idolized). He had spent most evenings at home with his mother who, throughout his life, had experienced regular periods of acute depression and mania. In 1903, at age thirty-six, Irving himself had suffered what he described as a year-long "depressive attack." When his brother died suddenly in 1909, Irving had inherited the fortune and stopped working. Soon afterward, his mother died. Once again, Irving had fallen into a "depressed state." He had been cared for by Mrs. Shire. The housekeeper had promised his mother and brother that she would always take care of Irving. Alternating periods of depression and remission had followed. In 1917, his longtime doctor, Josiah Schmitt, advised him to consult Adolf Meyer at Johns Hopkins.[57]

Meyer examined Irving at home and the next day accompanied Irving and Mrs. Shire to Baltimore by train. The housekeeper was installed in a nearby hotel and Irving settled into a private room on the fourth floor of the Phipps Clinic. Some wealthy patients were permitted to share their rooms with a domestic servant, but Meyer denied the request in this case. The day after his admission, Irving underwent exhaustive physical and mental examinations. He was described as a "stocky, undersized, heavily built white man." In the days that followed, Irving provided details about his family and medical history to John Oliver, the ward psychiatrist, and responded to questions about his life experiences. He found the exams and questioning taxing. He was described as "depressed and discouraged," very restless and tense, but exceedingly agreeable. The notes also record that his hands trembled and his facial muscles twitched and he was frequently tearful and unable to speak.

Meyer concluded from these initial examinations that Irving showed "symptoms of a poorly balanced emotional equilibrium" that was practically devoid of social independence or self-reliance. He hypothesized that, beginning in childhood, powerful environmental factors fostered a pathological tendency toward primal impulses of fearfulness and dependence. Meyer soon discovered that Irving's mother had been unable to care for him because of her illness. Indeed, as a child Irving was frequently tasked with following her to ensure that she did not engage in reckless or harmful behavior. Previously, at Irving's home, Meyer had learned that his new patient had no family or friends and did not pursue any interests or hobbies. His constant companion was Mrs. Shire who fretted incessantly about his nerves, diet, evacuations, and sleep. "The watchful doubt and anxiety which surrounds him now," Meyer observed, "is bound to feed the morbid trend." He described Irving's social existence as a "real starvation diet" in an environment dominated by "self-protection and dependence."

Irving's habitual impulse to retreat was profoundly ingrained, and Meyer decided that interaction with the social and physical elements of the Phipps environment was a priority. The goal of treatment, he explained to his patient, was to teach Irving to adopt an "attitude of self-confidence and a natural reaching outwards, rather than the retraction from the exterior world." Irving explained that he was incapable of effort and Meyer assured him, "I don't ask for effort—I ask for taking a few chances—some things are painful but let's work them out." Working things out involved retraining Irving to discharge his thoughts as actions. Meyer proceeded to use psychotherapy in conjunction with the other activities in the hospital routine. When Irving expressed confusion about this method of treatment, Meyer gave him a wonderfully clear explanation of his therapeutic approach: "A study of associations, of dreams, of obsessive thoughts and whatever comes up furnishes the starting point for bringing out the affective tendencies and preoccupations, conflicts, and wishes. While this is going on in the daily interviews with the physicians, certain habits and interests are developed in work and recreation and in things and events outside of oneself." He wanted Irving to collaborate in the process of identifying substitute reactions and faulty habits, and, eventually, to reinterpret his experiences and to practice a new way of adjusting.

Using the Phipps environment as a training ground, Meyer reasoned, Irving would learn to be independent and self-reliant—initially in small ways, such as walking in the garden, playing billiards with fellow patients, or going for a drive. Irving complained of a chronic headache that prevented him from resting or participating in activities. Meyer suggested that he weave a basket on the outdoor veranda: "Basketry is regular, quiet work that builds up and something that you may give your attention to." Any activity, manual or social, that interrupted Irving's unconscious-automatic psychobiological reaction of fearfulness would achieve Meyer's immediate goal. He tried to motivate Irving with a simile about a harpooned whale. When the whale attempts to escape and swim toward freedom, he suggested, the harpoon is plunged deeper into the flesh causing pain. To avoid the pain, the whale stops swimming, surrenders, and condemns himself to a fate worse than a temporary wound. "It naturally hurts somewhat," Meyer told Irving regarding behaving in new ways. He encouraged his patient to set and complete small goals consciously until doing so became less painful. Irving wanted to feel better before doing things. "You are becoming an ally of your sickness," Meyer cautioned. "Why do you want to await it? Why not attack it?" One decisive maneuver, he coaxed, would be the best medicine.

Irving's treatment at the Phipps Clinic included three interviews each day—a consequence of both his constant lucidity and his status as a private patient. Meyer conducted psychotherapy for thirty minutes at midday, noted Oliver, who himself talked with Irving every morning and evening. "The best way you can show your cooperation," said Meyer when he explained ventilation to their patient, "is to mention any specific point freely." If something occurred to him when the psychiatrist was not present, he was to write it down. Oliver and Irving spoke "in the quiet atmosphere of the roof garden," because Irving appeared more at ease there than in his room. In addition to discussing Irving's childhood, they examined his many irrational fears and delusions. Irving had a great fear of syphilis, for instance, and experienced intense trepidation at the thought of reading. "It was pointed out to him how great the gain would be to abolish these phobias," Oliver noted in the case history, "one at a time and once and for all." They would attempt to do that by identifying and reinterpreting pathological experiences at the root of Irving's glaringly inadequate adjustment to the vicissitudes of adult life.

One evening, Oliver asked Irving why he avoided reading books and newspapers. Irving replied that "a leaden depression falls upon him" when he did so. Oliver then asked him to state the association between the act of reading and desolation, and Irving talked of being terrified by the prospect of forgetting any name. He subsequently explained that reading often initiated hours or days of frantically probing his memory to recover the names of people and places from the past. This point prompted Irving to talk about his fear of syphilis. He recalled that, as a young man, he had learned that memory loss was a prominent symptom of syphilitic insanity. After that, he recollected, he had resolved never to forget anything, as this would prove that he was free of disease. Together, the psychiatrist and Irving concluded that his fears of forgetting and of syphilis were, in fact, associated. Oliver reassured Irving that he exhibited no signs of syphilis, a fact that he hoped would allow his patient to begin reinterpreting his experiences regarding both the disease and reading. Oliver felt confident that at least some of Irving's many phobias had been traced to their original faulty substitutions.

Other efforts were not as successful. Oliver prepared a word association test, but his explanation of the procedure made Irving so apprehensive it was postponed. He was able to conduct the experiment another day and recorded in the case history that "association test shows antagonistic attitude towards all laws and regulations, especially those of Clinic." It was also difficult to muster Irving's collaboration for dream analysis. Meyer persuaded him to recount several dreams, but when asked for "any association or clue," Irving dismissed Meyer by insisting, "I never can tell what leads up to my dreams." The staff continued to urge him to participate in occupational and recreational activities, and Irving finally agreed to a short walk with a nurse in the garden. Another day, Irving ventured outside with psychiatrist Stanley Cobb, and they talked about his interest in color photography and the possible causes of his illness.

"Good morning," Meyer greeted Irving the day after his walk with Cobb. His patient sat on the edge of his bed staring at the floor. "I saw you outdoors—I feel there is an issue on which we should win out." Simply by venturing outside, Meyer declared, Irving had advanced. But Irving did not recognize the victory. "I feel so—so entirely alone," he stammered as he rocked back and forth, holding his head in his hands. He wanted to go home, he confessed, but was afraid to say so directly. "You must lay a foundation for the many years ahead," Meyer told him. Irving mumbled that he did not wish to live to see those years. Meyer balked at the idea of someone whose thinking was so discordant with the common sense consensus making such a decision: "Do you think one's plan of life should be shaped when under those feelings?" Irving was silent. "I sympathize with you," Meyer assured him, "but it is not sympathy that is needed." Irving began to weep, but Meyer held his ground. Unlike the harpooned whale, Meyer reminded him, a human employed higher reasoning to adjust to painful experiences rather than surrendering to them. Irving continued to cry and begged to go home. "You know how you felt at home, you didn't have to make any decisions," Meyer continued. "Make a decision, clinch it," he implored Irving. In a report to Irving's doctor, Schmitt, Meyer admitted that it was difficult to witness such suffering. He felt that, like Irving's family and housekeeper, Schmitt had mistakenly indulged the patient's desire for sympathy. Meyer instructed his fellow physician to no longer reinforce the pathological habit of infantile dependence.

The Phipps environment was designed to rouse what Meyer considered a person's instinctual drive to select and achieve advantageous goals. After three weeks, he and Oliver felt that Irving was improving, slowly, under the regime: "After his Hydriatic treatment he seems to clear up especially and he will tell humorous stories." In these moments, Oliver observed, Irving was affectionate and "likes to put his arm around the waist of his companion" (he recorded no further comment on what appeared to be Irving's growing affection for him). Meyer was pleased to report to Schmitt that habit-training was producing results. Despite a "tremendous inclination to follow the lines of self-protection and self-pity," Irving was making a "fair adaptation where he allows himself to follow the natural instinct of activity." The Phipps psychiatrists were optimistic about improvement as long as Irving engaged with the physical and social reality he shared with others in the clinic.

Meyer continued to interview Irving each day. "Any dreams last night?" he began one session. "No," Irving replied, diarrhea had kept him awake the entire night. "As a young man, were you habitually diffident?" Meyer asked, redirecting his inquiry to Irving's extreme meekness. His patient nodded. When Meyer asked about childhood experiences, Irving reported having few friends except for one special chum who died when they were twenty years old. When Meyer inquired further about this friend, Irving described an incident in which the two boys were playing in the dockyards and his friend fell into the water. The playmate did not want his mother to find out about the accident from his wet clothes, so the companions retreated to Irving's house. "I took off all his clothes and while they were drying, dressed him up in mine—this gave me great pleasure." Meyer asked if Irving had any "girl friendships" to which Irving replied that he had had female acquaintances but no serious attachments. "You depended upon your chum and put away certain sex interests," Meyer hypothesized aloud. "Any conflicts here?" he asked his collaborator directly. Irving said no and Meyer concluded the interview. Irving later remarked to Oliver that he felt "clearer" after discussing his childhood friendship, and he understood better what Meyer wanted from him during interviews.

Meyer recorded his analysis of the material generated during psychotherapy in the case history. "It is evident that this friend and the patient's brother were to him outlets for his great need of affection." His mother, on the other hand, gave him an outlet for "his protective instincts." Their deaths, he surmised, left the patient "unable to fill the vacuum in his somewhat homosexual affective life." Unable to cultivate satisfying social relationships in adulthood, Irving had substituted the "satisfaction of the sweetish atmosphere of sympathy and comfort that is carefully spread about him by his devoted housekeeper." This environment, subsequently, had fostered a tendency to seek shelter from the world that had become progressively habitual. For Meyer, the pathological process at work was "essentially that of a child who does not want to emancipate." Beginning in childhood, Irving had worked himself into a very deep rut of dependence and fear. Meyer was confident that he could help Irving modify this subconscious adaptation using the Phipps environment, but only if another significant variable in the experiment in nature was altered.

Meyer wrote to Mrs. Shire at the hotel in Baltimore where she was staying. He had revoked her visiting privileges weeks earlier because the worry and pity she heaped on Irving undermined his small therapeutic successes. In a long and kind letter, he informed Mrs. Shire that it was necessary to separate Irving "from those on whom he depends too pathologically." He explained that her relationship to Irving contributed significantly to his illness by mimicking that of a mother and child. "Have you not seen in your own experience," he proposed to her, "it is of the greatest importance to separate a child from too devoted and tender parents?" Meyer went on to say that during his youth Irving's mother had absorbed too much of her boy's attention, which "ought to have been invested in some interests that would have developed him more into an independent personality." Without a thorough change in Irving's current situation, Meyer declared to Mrs. Shire, Irving would find it "hard to get out of his groove." If she wanted to see her charge well, Meyer advised her, it was imperative to separate herself from him. Mrs. Shire protested, but had little choice as long as Irving remained at the Phipps Clinic.

Irving, however, declared that he was unwilling to remain. He disliked the urban noises, the other patients, the food and, especially, the locked wards and bars on the windows. Meyer had not yet told Irving his conclusion—that the illness resulted from a failure to emancipate from parental dependence. "This is a thing which I should not like to throw at him before he has gained the necessary foundation" he had written a few days earlier. It appeared now that he had lost his chance. "So far his attitude has been all for flight," Meyer told Schmitt, and he reiterated that the patient could not return to his old environment if he wanted to get well.

Schmitt was convinced that Meyer's expertise was unsurpassed. Nevertheless, he was beholden to the wishes of his wealthy patient. He suggested that Irving rent a house in Baltimore to continue as Meyer's private patient. Money, he insinuated, was not an issue. Meyer's reply was unequivocal. To yield to Irving's pathological tendencies, his primitive desire to hide and remain helpless, was "serving the perpetuation of his disease rather than the cause of health." He was confident that he and Oliver had isolated the underlying pathology: Irving's failure to emancipate from maternal protection. The therapeutic goal was to arrest this pathological process, he protested crossly to Schmitt, not nurture it. In the case of organic pathology, medical practitioners did this by excising, amputating, or counteracting with drugs. Meyer's intervention was habit-training. The Phipps environment, he tried to explain to Schmitt, was designed not to comfort the patient but to "bring out sound instincts." He refused to treat Irving outside the clinic. "It is not willfulness or stubbornness on my part," he assured Schmitt, "but a devotion to what I consider well-founded convictions." Irving was discharged from the Phipps Clinic and from there traveled to a private sanitarium to rest. He informed Mrs. Shire, who immediately telegrammed to say she would meet him there. A few weeks later, Irving wrote to thank Meyer, who, in reply, apologized for failing Irving. "Even when a transitory hurt seems inevitable," Meyer confessed with echoes of the harpooned whale, "the experience was very painful to me because I do like to help and do not like to hurt." He wished Irving well.

An epilogue to this case foreshadowed the growing acceptance of psychoanalysis within American psychiatry and culture after World War I. In 1924, Schmitt informed Meyer that Irving had remained depressed throughout the intervening years but recently had experienced "a marvelous quick recovery with the help of psychoanalysis." Indeed, Meyer soon received a request from Irving's new psychoanalyst: would he please send written confirmation that, in 1917, he had instructed Mrs. Shire to separate herself from Irving for his own welfare? Meyer obliged and asked to be updated on Irving's condition. In reply, the psychoanalyst recounted how he had shown the documentation attesting to Meyer's original prescription to an incredulous Irving. "She had never told him of your advice," the analyst reported, and news that the Hopkins specialist had recommended an emancipation from Mrs. Shire seven years earlier reportedly inspired Irving to sever all ties to his housekeeper. Irving's psychoanalyst explained to Meyer how he intended to analyze the case. He planned to use not Freud's sexual theory but rather an explanatory principle recently put forward by the Freudian Otto Rank of Vienna. Rank speculated that "there is a mother craving antedating the sex attraction of a boy for his mother—more of a biological craving of dependence." This description of Rank's hypothesis, as embraced by this New York physician, reflected rather closely Meyer's original conclusion about the psychobiological origins of Irving's illness. The psychoanalyst praised Rank's sage revision of the Oedipal theory and asked Meyer if he was familiar with it. Because Meyer agreed to meet with Irving's new psychiatrist in person to discuss the case, the correspondence ends there. His response to this psychoanalytic recasting of his initial diagnostic formulation must be left to the imagination.[58]

A few months before the Clark Conference in 1909, Carl Jung appeared disappointed at Meyer's lack of commitment to psychoanalysis, remarking to Freud that the Americans did not understand its import. Nevertheless, Jung assured his mentor, "it will rub off on some of them and is doing so already, despite their audible silence."[59] Here, Jung specified that it was Meyer and August Hoch who had not vocalized their support loudly enough. In the decade preceding World War I, however, Meyer was an outspoken supporter of psychoanalysis. Nevertheless, he had no intention of importing it wholesale and acting as a retailer of Freudian doctrine. His support contributed to the uptake of psychoanalysis within American culture, but a lack of support from him could not have stemmed it. In 1911, for example, American newspapers featured sporadic articles about a strange new science of dual personalities and dream cures. In 1915, a *New York Tribune* headline asked readers: "Has the Psycho-Analyst Fad Struck You Yet?" The writer of the article lamented the increasing frequency with which dinner parties were spoiled by the insights of self-styled psychoanalysts.[60] Freud's increasingly doctrinaire claims were adopted enthusiastically by adherents on both sides of the Atlantic Ocean, and psychoanalysis continued to pick up speed within American psychiatry.

In the prewar era, psychobiology and psychoanalysis were comparable in many ways and diverged on only a few key points. The patient's life story and the causal role of experiences were central to each. Both Freud and Meyer, influenced by the work of John Hughlings Jackson, conceptualized emotional conflict as a dynamic struggle between primal and intellectual impulses (Freud called the latter "cultural acquisitions" whereas Meyer preferred the instrumentalism implied by "common sense consensus"). For Freud, the struggle ensued within the unconscious mind; for Meyer, as part of the interaction between organism and environment. According to Freud, the conflict was invariably traced to unfulfilled longings and sexual development in childhood. Meyer readily recognized the prevalence of sexual conflicts in mental disorders, but he insisted that any instinctual drive— parental dependency, nutrition, avoiding harm, competition, aggression, imagination, as well as sexual gratification—might challenge more-evolved social instincts when mentation was impaired. The dramatic difference between the contexts in which they practiced psychotherapy illuminates why they construed the nature of the therapeutic encounter so differently— Freud as transference and Meyer as collaboration. It was their respective uses of the term *unconscious*, however, that represented the critical point of divergence between psychoanalysis and psychobiology in the prewar period. For Freud, *unconscious* was a noun, a container of repressed ideas. For the pragmatist Meyer, it was an adjective that was synonymous with *automatic* and *habitual*. Automation, not repression, subsequently became the focus of Meyerian investigation and therapy.[61]

Meyer's use of psychotherapy at the Phipps Clinic between 1913 and 1917 was rooted in his conceptual framework of psychobiology and consistent with his goal of establishing a clinical science of psychiatry. Relying on the collaboration of psychiatrist and patient, he sought to identify and modify faulty psychobiological reactions and habits. When necessary, he selectively appropriated and adapted psychoanalytic concepts and techniques to meet those objectives. He constructed psychobiology as a pluralistic interpretive framework that integrated incongruent methods and data sets related to mental disorder without manipulating, conceptually, phenomena that were fundamentally different.

By the 1930s, Meyer was battling the pernicious threat of reductionism on a new front—the psychological reductionism of psychoanalysis. He turned to a familiar rhetorical arsenal. In a maneuver reminiscent of his 1908 indictment of physicians who blindly accepted the superstitious dogma of anatomical reductionism, he editorialized that the transference relationship made the psychoanalyst little better than a faith healer who commandeered a personality rather than restoring it to health.[62] Elsewhere, he compared Freudian essentialism to the pointless search for the Kantian *Ding an sich*.[63] Meyerian psychiatry, he countered, was a "collaboration among all those concerned—physicians, patient, nurses, ward group, and family in settings which safeguard the integrity of the personality."[64] Ultimately, Adolf Meyer rejected the reductionism and determinism in psychoanalysis for the same reasons he had rebelled against the idea of an organic machine: because universal mechanisms and preconceived notions about essential causes showed no confidence in the instrumentalism and spontaneity of what the person thought and did.

Conclusion

Looking out from the dais in the auditorium on the fifth floor of the Henry Phipps Psychiatric Clinic—quite literally standing atop his new domain—Adolf Meyer met the gazes of scientific luminaries and social elites gathered in April of 1913 to celebrate the opening of the first academic institute of psychiatry in North America. With the conclusion of the official celebrations, and anticipating the small hospital's imminent occupation by staff and patients, Meyer confided to his audience, "I feel that at last another vital step, not merely in my career, but in my hopes for psychiatry, is achieved." As a young neurologist and pathologist in 1892, he had felt that he lacked the verbal skills necessary to become a clinical psychiatrist and work with patients, and he was unsure of his abilities as a teacher. Now, he occupied what he himself had described as the most important professorship in psychiatry in the English-speaking world. His vision of a clinical science of psychiatry in the United States, shaped by his opportunities and experiences during the previous two decades, was finally realized. At Johns Hopkins, he would train promising men and women to subject the data of psychobiology to the methods of general pathology, at the bedsides of eighty-eight patients each day and in the clinic's many laboratories. He was the inaugural chief of psychiatry at the institution deemed exemplary by Abraham Flexner, author of the 1910 report that dictated the terms by which North American medicine developed in the twentieth century. Meyer's reformist influence during the decades preceding the Second World War proved analogous to Flexner's within general medicine. And, like Flexner's, Meyer's successors and his historians have viewed the merits of his reforms variously. Many accredit him with making psychiatry a part of science-based medicine. Others characterize Meyer as a mere facilitator and popularizer of otherwise inevitable developments. A few insist that his approach and his leadership proved detrimental to psychiatry.[1]

At the heart of that approach and leadership was Meyer's concept of psychobiology. It served as the basis for an all-out campaign, as far as he was concerned, to liberate psychiatry from Old World dogma that explained mental activity in reductive, dualistic, or deterministic terms. He insisted that such terms excluded some of psychiatry's most valuable sources of data: environment, experience, instinct, emotion, cognition, and behavior. He premised psychobiology on the notion that mental activity, as well as brain activity, played an active role in human evolution and everyday adjustments. For Meyer, trained as a comparative neurologist and neuropathologist, this assumption was validated by experimental research. Having also internalized views pioneered by other scientists and intellectuals in a variety of disciplines, Meyer concluded that the interaction between a human organism and its environment was dynamic, transactional, and—critically— malleable. He suggested that most mental disorders were best understood not as diseases, but as adaptive failure and potentially correctible maladjustment. He shared this optimistic view with a vanguard of like-minded American psychiatrists and intellectuals in the decades before World War I. The great challenge for Meyer was how to transform this conviction into more than a collection of fuzzy platitudes—to make psychiatry a recognizable clinical science and respected medical specialty.

He judged that the instrumentalism articulated by American pragmatists was a productive means for mediating a biological view of mind with the established methods of scientific medicine. At Johns Hopkins Hospital between 1913 and 1917, Meyer had the opportunity to embody his progressive and scientific ideals in the Phipps Clinic. It was the first academic department and university clinic in the United States devoted to integrating clinical research, medical training, and therapeutic innovation in psychiatry. There he implemented the case history as a clinical technique to reify and objectify the ephemeral data of experience and behavior. He combined clinical methods with a treatment approach resembling traditional moral therapies to investigate, teach, and apply the science of adaptive reactions. He insisted that mind and body do not occupy separate domains and that the result of their integration was an irreducible biological unit that he called a person. He taught his medical students and specialist trainees that the personality as a whole influenced the functioning of the integrated parts as the individual reacted to his or her social environment. Consequently, he widened psychiatric inquiry to include emotions, instincts, behavior, and individual social functioning, as well as the anatomy and physiology of the nervous system. Disseminating his person-centered approach through his authoritative position at Johns Hopkins, Meyer influenced the overall trajectory of the young discipline, successfully instituting new clinical methods, concepts, institutional infrastructures, and curriculum standards throughout North America.

Not long after the pomp of the opening celebrations in 1913, and the optimistic flurry of clinical and therapeutic activities in the years that followed, the Phipps Clinic began to empty. By 1917, Meyer would watch helplessly as the psychiatrists and nurses he had trained to execute the therapeutic experiment, as well as his bright medical students and the financial support that had been committed to realize his clinical vision, were all dutifully relinquished to the war effort. Lack of resources and the influenza pandemic of 1918 eventually forced him to close some wards temporarily, and he himself became preoccupied with training military psychiatrists for the War Department.

Like the progressivism in which it was embedded, Meyer's "new psychiatry"—consciously crafted as an ally of John Dewey's new psychology and related pragmatist projects in the Progressive Era—was absorbed and transformed by the cultural and social realities of postwar America. Psychiatry underwent rapid and fundamental changes after World War I, the result of several historical developments. In 1917, Julius Wagner-Jauregg successfully cured syphilitic insanity, for which he later won a Nobel Prize. He introduced malarial parasites into the body, causing a fever that acted upon the spirochete already identified as the syphilis pathogen; he then treated the malaria with quinine. For many, the discovery was psychiatry's long-awaited magic bullet, a triumphant confirmation of the somatic basis and ontological nature of mental disease that called into question the causal significance of mental and environmental factors. At the same time, the evident effects of war on mental functioning confirmed for others the causal role of environment and experience. Military hospital beds filled with soldiers and officers incapacitated not by physical injuries but by inexplicable paralysis, mutism, or sudden episodes of panic or terror. They were diagnosed with "shellshock" or "war neurosis," and their symptoms and physicians set the terms for renewed discourses of psychogenesis. The pervasiveness of shellshock among men who had been considered the nation's healthiest and brightest "stock" challenged assumptions about hereditary degeneration and altered cultural perceptions of mental illness.

In the 1920s, psychoanalytic ideas proliferated within mainstream American culture, generating novel popular and medical understandings of the personality. Psychological reductionism increasingly characterized Freudian theories. Concurrently, the new discipline of psychosomatic research, based on psychobiology, emerged with strong support from the Rockefeller Foundation. The equally reductive behaviorism of psychologist John B. Watson flourished, too, challenging the functional psychology of the pragmatists. In the psychological laboratory at the Phipps Clinic, under Meyer's supervision, Watson began conducting experiments in 1914 that became the foundation for behaviorism. Watson's involvement in a sex scandal in 1920, however, provoked Meyer finally to dismiss his protégé, whose increasingly dogmatic claim that human behavior could be analyzed independently of consciousness constituted heresy under the banner of psychobiology. Historian Jack Pressman argues that, by the 1930s, the Meyerian model had become the master paradigm of American psychiatry because it articulated a vision of a single, respectable medical specialty united by theory, education, and practice. Psychoanalysis, however, continued on the path to psychiatric hegemony that it achieved after World War II. New research findings in brain chemistry, psychopharmacology, and genetics in the 1960s generated momentum for a paradigm shift in psychiatry, redirecting the discipline's focus from the workings of the personality to those of the nervous system. In the 1970s, stable disease categories became essential for continued comparative experimental research. This catalyzed a group of self-styled "neo-Kraepelinians," psychiatrists and researchers who advertised their commitment to determining the somatic bases of mental disorders by associating themselves with theories of organic etiology that were forwarded—though not substantiated, as Meyer never tired of pointing out—by Emil Kraepelin. This materialist "biological psychiatry," as it was called, had displaced psychoanalysis as the driving force within the discipline by the end of the 1980s, and overshadowed emerging "bio-psycho-social" models that had been widely considered the progeny of Meyerian psychiatry. President George H. W. Bush declared the 1990s the Decade of the Brain, as academic psychiatry pursued the science of neurotransmitters and nerve cell growth. The significance of the patient's individual life circumstances—while still relevant to everyday practice—became marginal to neuroscientific research and training.[2]

Current developments in twenty-first-century psychiatry make the rediscovery of Meyerian principles timely. The first is the growing importance of the concept of "plasticity" in neuroscience. In general terms, *neuroplasticity* refers to the reorganization of neural pathways in response to experience. Recent research findings indicate that processes related to learning, memory, and recovery from neurological injuries involve long-term morphological and biochemical changes in the brain throughout the life of the individual.[3] Meyer's earliest conclusions about the architecture of the nervous system—his segmental-suprasegmental model that was the foundation of psychobiology—contain hints of these findings. When he introduced the model in 1898, the orthodox view was that the adult nervous system was anatomically fixed, a passive apparatus that conveyed nervous impulses much the way electrical wires conducted electricity; it was an organic machine that operated deterministically. On the basis of experimental findings—his own and those of others—Meyer hypothesized that the hierarchy of nervous mechanisms acquired throughout human evolutionary development was structurally and functionally integrated. Based on the neuron theory—which his doctoral advisor, August Forel, had helped to establish—he took for granted that this integration was achieved by the selective interaction of autonomous neurons (not their connectivity as a closed network). His model made conceptual room for both cellular and functional regeneration after a brain injury, for which he thought there was emerging evidence.[4] He was also confident that his conceptualization of the nervous system corroborated William James's instrumentalist reading of experience as a causal agent in adaptation. The adult brain was not entirely fixed or mechanistic, Meyer insisted; it was an "apparatus of biological plasticity" with adaptive capacities that responded to the experiences of the person.[5]

It is interesting to note that neuroscientists today explain plasticity by combining the very analogies that Meyer and his fellow pragmatists considered antagonistic: deterministic organic machinery and dynamic biological adaptation. One scientist describes the plastic brain as "an exquisitely adaptable machine with a person on board," while a group of neuroscience researchers offers the following explanation: "Neurons do the equivalent of growing telephone wires that allow them to communicate with and influence one another. Rather than waiting, like the passive computer, for a technician to hook them up, they physically grow their own connections to other cells."[6]

Upon Meyer's death, the neuroanatomist Louis Hausman commented that the segmental-suprasegmental concept had created a new horizon for neurological research in 1898: "Meyer lifted neuroanatomy out of the adolescent category of centers and tracts, and gave it a live dimension, that of functional development without which the nervous system cannot be understood." Meyer had predicted (accurately, it turns out) that it would prove impossible during his own lifetime to observe directly the physiological mechanisms of neurons, and this led him to conclude at the beginning of the twentieth century that clinical psychiatry should be focused on what he regarded as the observable "functional expression" of neural activity in the form of thoughts and behavior. Today, investigators use a variety of nuclear imaging technologies, collectively known as functional brain imaging, to detect and measure neural activity by analyzing changes to blood flow in the brain. Techniques such as functional magnetic resonance imaging (fMRI) and single-photon emission computerized tomography (SPECT) facilitate experimental and clinical research aimed at identifying specific relationships between changes in neural patterning and aspects of experience involving, for example, perception, learning, emotions, decision making, language impairments, epileptic seizures, or psychosis. These technological innovations have provided neuroscientists with ways to test and evaluate the principles of neuroplasticity. My own examination of psychobiology does not trace the lineage or originality of Meyer's conclusions about biological plasticity, nor do I claim here that his use of the term can be equated with current understandings of neuroplasticity. However, looking anew at the neurobiological foundations of his person-centered approach points to some intriguing directions for future exploration.[7]

Other key aspects of psychobiology are recognizable in scientific discussions today. In the field of experimental psychology, researchers are investigating a phenomenon labeled the "adaptive unconscious," reportedly responsible for the kind of instantaneous decision making often called "gut instinct" and powered by a different part of the brain than the one associated with conscious, thoughtful decisions. "The adaptive unconscious does an excellent job of sizing up the world," Timothy Wilson of the University of Virginia suggests, "warning people of danger, setting goals, and initiating action in a sophisticated and efficient manner." Wilson's description harmonizes with Meyer's understanding of common sense as an adaptive tool and the formation of psychobiological habits as an economizing measure for responding advantageously to the tacit stimuli of the common sense consensus. The popular science writer Malcolm Gladwell explores the benefits of the adaptive unconscious in his book *Blink: The Power of Thinking Without Thinking*, emphasizing that it should not be confused with the dark and murky place described by Sigmund Freud in which disturbing memories and fantasies are repressed. It is rather an "internal computer" that "quickly and quietly processes a lot of the data we need in order to keep functioning as human beings." Cognizant that readers may doubt the legitimacy of rapid cognition compared to rational decision making, Gladwell's stated goal is to demonstrate that "snap judgments and first impressions can offer a much better means of making sense of the world." What Meyer conceptualized using William James's instrumentalist terms as "common sense," these writers discuss in terms of a brain function that emerged during the course of human evolution and is used for "sizing up" or "making sense" of the world.[8]

Meyer's formulations have also become relevant to one of the most critical discourses within American psychiatry today. The American Psychiatric Association (APA) recently published a fifth edition of its influential *Diagnostic and Statistical Manual of Mental Disorders* (*DSM-5*). Years of preparation for the revisions were marked by controversy among some of the discipline's leading voices. There are those committed to diagnosing and classifying mental disorders on the disease model used by the rest of medicine, the dominant system since the third edition of the manual (*DSM-3*) was released in the 1980s. Another group of influential psychiatrists has been lobbying for the discipline to acknowledge the dimensional character of mental disorders, which they insist emerge on a spectrum of severity and lack clear boundaries with normality. Lobbyists for "dimensions" instead of "diseases" suggest reconsidering approaches that deemphasize the health-disease dichotomy. Meyer would feel right at home in these debates but surely dismayed that nosology continues to occupy the time and talent of psychiatrists. As he made clear in 1906, he considered his peers' enthusiasm for classification to be a diversion that risked undercutting the discipline's scientific credibility.

> Psychiatry has not reached, and probably never will reach, the stage where a small number of one-word diagnoses would be more than a formal index. It is, of course, convenient to have them, and there is a temptation to discuss the names and write papers about them; but as a rule that does not add much

to our knowledge in psychiatry, and only helps to undermine our standing beside other branches of medicine through appearance of continual vacillation of mere opinions; of aberration from real investigation of the nature of facts; from the normal instincts of the physician; and of something really being devoid of sufficient practical importance.[9]

Meyer urged a distinction between refining diagnostic practices and imposing a classification system based on etiological theories or descriptive categories that did not reflect observable clinical realities. Pressure from the American Bureau of the Census immediately after World War I had spurred the APA to produce a uniform nosology for statistical purposes. The result was a system reflecting Emil Kraepelin's theoretical disease concepts that was in use throughout the interwar period (ironically, devised decades earlier by Meyer for use in New York State). Following the Second World War, debates about classification were influenced by psychoanalysis and the dimensional models of maladjustment championed by Meyer's followers. The focus of clinical psychiatry subsequently shifted from incidents of severe mental illness in psychiatric hospitals to prevention of its early manifestations in the general population. Despite Meyer's longstanding dissent, his concept of reaction-types was mobilized in the production of the first *DSM* in 1952, two years after his death, to categorize and systematize the various forms of psychopathology collectively described in the manual as "diseases of the psychobiologic unit."[10]

In the end, the new *DSM-5* of 2013 included an alternative "hybrid" model for evaluating some disorders. It combines the categorical system and a new dimensional understanding of the personality. "We know that anxiety is often associated with depression, for example," explained the chair of the *DSM-5* task force, psychiatrist David Kupfer, in 2011, "but the current *DSM* doesn't have a good system for capturing symptoms that don't fit neatly into a single diagnosis." The rationale for the new hybrid model offered by John Oldham, a psychiatrist who was involved with the revisions, could have been cribbed from Meyer's psychobiology playbook: "Personality types, traits, and disorders are on a continuous spectrum, much like blood pressure and hypertension. Too much of a useful, adaptive trait may become a problem." A decade earlier, another psychiatrist urged, "To understand depression, look to psychobiology, not biopsychiatry."[11] The prospect of a resurgence of any approach resembling Meyer's unwieldy "person-centered"

model troubles other psychiatrists. Echoing several contemporary practitioners and historians, psychiatrist Nassir Ghaemi interprets Meyerian psychiatry as essentially eclectic—an approach in which all theories and methods are potentially correct and none definitively incorrect. He asserts that
Meyer's indecisiveness left postwar psychiatry prey to both dehumanized
biological extremism and the "intellectual anarchism" of bio-psycho-social
models.[12] With some psychiatrists lobbying for a reinstatement of Meyerian
principles and others insisting that nothing could be worse for the discipline, it seems that American psychiatrists are dragging Adolf Meyer back
onto the campaign trail.

During Meyer's tenure, between 1913 and 1941, the Phipps Clinic at Johns
Hopkins became an important center for clinical and laboratory research in
psychiatry that attracted investigators and trainees from around the world.
He thought that psychiatry must have strong ties to other medical departments and to the community it studied and served. He cultivated new ties
to sociology, neurology, mental hygiene, social work, psychology, law, pedagogy, and public health on behalf of psychiatry. The activities of those associated with the Phipps Clinic reflected this interdisciplinary philosophy.
For example, Phipps psychiatrist John Oliver became medical officer and
alienist to the Supreme Bench of Baltimore City. In 1919, he described how
Meyer had inspired him to apply psychiatry to work in the community, and
he encouraged other psychiatrists "to take up the work of a psychiatric missionary, not necessarily in the criminal courts, but in the schools, in the
families of some neighborhood, in the factories and the penitentiaries." Immediately after Stanley Cobb's specialty training at the Phipps, he joined the
faculty of Harvard Medical School, where he appeared to exemplify the
Meyerian approach: he conducted laboratory research with physiologist
Walter Cannon, created a multidisciplinary neurological unit at Boston
City Hospital, established a psychiatric service at Massachusetts General
Hospital, was active in the Boston Psychoanalytic Society, and treated private and hospital patients. He also made a practice of hiring psychiatrists,
social workers, and nurses who had trained at the Phipps Clinic. Psychiatrists Trigant Burrow and Paul Schilder were pioneers in developing social
psychiatry and group psychotherapy, which led to subsequent innovations
in therapeutics, such as milieu, art, and cognitive-behavior therapies. Influenced by both psychobiology and psychoanalysis, the work of Phipps
alumni, among them Phyllis Greenacre and Edward J. Kempf, contributed
significant new insights into specific forms of psychopathology. Curt Rich-

ter, director of the psychobiology laboratory in the Phipps Clinic for many decades, conducted diverse programs of original research, including his groundbreaking insights into circadian rhythmicity in animal behavior and the concept of the biological clock. Under Meyer's tutelage, Leo Kanner established the field of child psychiatry at Johns Hopkins, and Paul Lemkau founded the Department of Mental Hygiene in the university's School of Public Health. In the 1960s, the psychiatric epidemiologist Alexander Leighton instituted the multidecade Stirling County Study, based on Meyerian concepts and methods, to survey the distribution of mental illness in the general population. Paul McHugh, chief of psychiatry at Johns Hopkins from 1975 to 2001, explained recently that Meyer's direct contributions to American psychiatry during the first half of the twentieth century were obscured by the rise of psychoanalysis. During centenary celebrations for the Henry Phipps Psychiatric Clinic in 2013, he emphasized that Meyer's positive influence on modern practices was sustained through the work of his adherents and successors.[13]

Perhaps the best way to conclude this book, conceived as just a beginning, is to highlight two key insights that appear particularly valuable to the rediscovery of Meyerian psychiatry and the demystification of its originator, and to further exploration of the historical implications of both. First, Meyer viewed mental activity and brain activity as a single biological response. As I emphasized in the introduction to this book, to overlook this principle is to risk misconstruing Meyer's intentions, actions, and, especially, his abstruse published explanations of his ideas and methods. Second, in addition to recognizing this core concept of psychobiology, the key to unlocking his enigmatic formulations is to appreciate the essentially medical orientation of Meyer's thinking, practice, and teaching. Walter Cannon, well-known for his work on the physiological basis of emotional excitement and the so-called fight-or-flight response, remarked in 1931 that, as scientists, the only difference between himself and Meyer was Meyer's preoccupation with "practical treatment." Indeed, the importance of Meyer's neurobiological concepts and contributions to neuropathological research was readily acknowledged by his fellow neuroscientists throughout his career. Perhaps he might not have come under fire from his contemporary critics, or from psychiatrists and historians today, if he had remained a laboratory investigator. For a variety of reasons, some of which have been explored here, he ultimately embraced the role of physician and medical professor. "My entrance into psychiatry was through the autopsy room," he acknowledged in his Maudsley Lecture in 1933, "but with the temperament of the practitioner of medicine."[14] Unlike branches of science focused principally on discovery, medicine, Meyer emphasized, was always under a double obligation, not only to identify *disease* but also to prevent or treat *sickness*. As a clinical science, psychiatry bore the same responsibility to investigate and work to ameliorate the pathologies of the mind.

Abbreviations

AMC   Adolf Meyer Collection
Alan Mason Chesney Medical Archives, Johns Hopkins Medical Institutions, Baltimore, Maryland. Items cited are designated by a three-part catalogue code assigned by Guide to the Adolf Meyer Collection (medical-archives.jhmi.edu/sgml/amg-d.htm). The numbers identifying the item's location represent the "series" (topical grouping), the "unit" (section within the series), and the folder number.

CP   *Collected Papers*
The Collected Papers of Adolf Meyer, vols. 1–4, edited by Eunice E. Winters (Baltimore: Johns Hopkins University Press, 1950–1951).

EWC   Eunice E. Winters Collection
Alan Chesney Medical Archives, Johns Hopkins Medical Institutions, Baltimore, Maryland.

Note Concerning Medical Records

Each medical file is cited using an alphanumeric code, generated by the author-researcher and bearing no relation to the original record number. Scholars wishing to access these records should contact the Alan Chesney Medical Archives, Johns Hopkins Medical Institutions, Baltimore, Maryland.

Introduction

1. Franklin Ebaugh, "Adolf Meyer's Contribution to Psychiatric Education," *Bulletin of the Johns Hopkins Hospital* 89 (1951): 71; D. K. Henderson, Introduction to vol. 2 of *CP*, xx; Oskar Diethelm, "Adolf Meyer," *American Journal of Psychiatry* 107 (1950): 80; Leland Crafts, "Adolf Meyer," *American Journal of Psychology* 63 (1950): 620; Jerome Frank, "Adolf Meyer in Retrospect," unpublished address given 11 February 1980, Box 3, EWC; Paul McHugh, "Adolf Meyer: Achievement and Influence in Psychiatry," lecture delivered 30 January 2013, inaugural lecture of Centenary Lecture Series in honor of 100th anniversary of the Henry Phipps Psychiatric Clinic, Johns Hopkins Hospital,

Baltimore, MD. The special issue for the centenary of Meyer's birth is *American Journal of Psychiatry* 123/3 (1966).

2. Frank J. Sulloway, Freud, *Biologist of the Mind: Beyond the Psychoanalytic Legend* (Cambridge, MA: Harvard University Press, 1992).

3. Fredric Wertham, "Review of *The Collected Papers of Adolf Meyer*," *Journal of the History of Medicine and Allied Sciences* 8 (1953): 242. On the burlesque, see Edward Hanrahan to William Alanson White, 8 June 1925, Record Group 418, Entry 7, National Health Archives and Records Administration, Washington, DC; and Lawrence Kubie, "Problem of Instincts As It Looks to a Psychiatrist," in *Symposium: Genetic, Psychological, and Hormonal Factors, 17–19 November 1954*, typed transcript held by the University of Kansas Libraries (folio QP251.S96.1955). "Non-Stop Sentence Derby," *The New Yorker*, 16 November 1940, 40; Adolf Meyer, "Mental Health," *Science* 92 (27 September 1940): 271–276. On textbooks, see Wendell Muncie, *Psychobiology and Psychiatry: A Textbook of Normal and Abnormal Human Behavior* (St. Louis: C. V. Mosby Co., 1939); and Esther Richards, *Psychobiology and Psychiatry* (St. Louis: C. V. Mosby Co., 1941).

4. Alexander Leighton, Introduction to vol. 4 of *CP*, xiii; Adolf Meyer, "Critical Review of the Data and General Methods and Deductions of Modern Neurology: Part 2," *Journal of Comparative Neurology* 8/4 (1898): 287; J. Wortis, "Adolf Meyer: Some Recollections and Impressions," *British Journal of Psychiatry* 149 (1986): 677; Meyer, "Spontaneity," *CP*, vol. 4, 463.

5. Leo Kanner, unpublished autobiography, Part 6, Archives of the American Psychiatric Association, Arlington, VA.

6. Jerome Frank, "Impressions of Half a Century at Hopkins," *Maryland Psychiatrist Newsletter* 22/1 (1995).

7. Leighton, Introduction, xiii.

8. Wortis, "Meyer: Some Recollections," 678.

9. John MacCurdy to William Alanson White, 18 December 1928, Record Group 418, Entry 7, National Health Archives and Records Administration, Washington, DC. See also E. E. Southard, "On the Significance of Un—in the Term Unconscious," *Journal of Abnormal Psychology* 9 (1914): 355–360.

10. Robert Boakes, *From Darwin to Behaviorism: Psychology and the Minds of Animals* (Cambridge: Cambridge University Press, 1984), 168; Henri Ellenberger, *The Discovery of the Unconscious: The History and Evolution of Dynamic Psychiatry* (New York: Basic Books, 1970), 286; Richard W. Fox, *So Far Disordered in Mind: Insanity in California, 1870–1930* (Berkeley: University of California Press, 1978), 170; Nathan Hale, *Freud and the Americans: The Beginnings of Psychoanalysis in the United States, 1876–1917* (Oxford: Oxford University Press, 1995), 157; Ruth Leys, "Adolf Meyer: A Biographical Note," in *Defining American Psychology: The Correspondence between Adolf Meyer and Edward Bradford Titchener*, ed. Ruth Leys and Rand B. Evans (Baltimore: Johns Hopkins University Press, 1990), 54; Eva S. Moskowitz, *In Therapy We Trust: America's Obsession with Self-Fulfillment* (Baltimore: Johns Hopkins University Press, 2001), 41–44; Richard Noll, *American Madness: The Rise and Fall of Dementia Praecox* (Cambridge, MA: Harvard University Press, 2011), 36; Roy Porter, *A Social History of Madness: Stories of the Insane* (London: Phoenix, 1987), 197; Jack Pressman, *Last Resort: Psychosurgery and the Limits of Medicine* (Cambridge: Cambridge University Press, 1998), 431; Andrew Scull, *Madhouse: A Tragic Tale of Megalomania and*

Modern Medicine (New Haven, CT: Yale University Press, 2005), 161; Edward Shorter, *A History of Psychiatry: From the Era of the Asylum to the Age of Prozac* (New York: John Wiley and Sons, 1997), 101.

11. Scull, *Madhouse*, 239.

12. Shorter, *History of Psychiatry*, 111.

13. Noll, *American Madness*, 160.

14. Chris Feudtner, *Bittersweet: Diabetes, Insulin, and the Transformation of Illness* (Chapel Hill: University of North Carolina Press, 2003).

15. Eric J. Engstrom, *Clinical Psychiatry in Imperial Germany: A History of Psychiatric Practice* (Ithaca, NY: Cornell University Press, 2003), 199–201; Ruth Leys, "Types of One: Adolf Meyer's Life Chart and the Representation of Individuality," *Representations* 34 (1991): 26, n. 45.

16. See Gerald Grob, *The Inner World of American Psychiatry, 1890–1940: Selected Correspondence* (New Brunswick, NJ: Rutgers University Press, 1985), 199–211; Noll, *American Madness*, 179–182.

17. Charles Rosenberg, Foreword to *Mania: A Short History of Bipolar Disorder*, by David Healy (Baltimore: Johns Hopkins University Press, 2008), xiii.

18. Alan Chesney, "Adolf Meyer," *Transactions of the Association of American Physicians* 63 (1950): 9–10; Crafts, "Adolf Meyer" (1950): 620–622; Diethelm, "Adolf Meyer"; Franklin Ebaugh, "Memorial to Past President," *American Journal of Psychiatry* 107 (1950): 288–290 and "Adolf Meyer: A Tribute from Home," *American Journal of Psychiatry* 123 (1966): 334–336; Frank, "Impressions of Hopkins" and "Adolf Meyer in Retrospect"; D. K. Henderson, "Adolf Meyer: A Tribute from Abroad," *American Journal of Psychiatry* 123 (1966): 332–334; Kanner, autobiography, Part 6, 320; Thomas Rennie, "Adolf Meyer," *Psychosomatic Medicine* 12 (1950): 71–72; Walter Riese, "Adolf Meyer," *Journal of Nervous and Mental Disease* 113 (1951): 89–91; John C. Whitehorn, "Adolf Meyer," *Bulletin of the Johns Hopkins Hospital* 89 (1950): 53–55; Wortis, "Meyer: Recollections and Impressions."

19. Adolf Meyer, "My Experience with American Psychiatry," unpublished manuscript, 1898, X/1/27, AMC.

20. E. E. Southard, "Cross Sections of Mental Hygiene, 1844, 1869, 1894," *American Journal of Insanity* 76 (1919): 107–108.

21. See, for example, Ruth Leys, "Meyer's Dealings with Jones: A Chapter in the History of the American Response to Psychoanalysis," *Journal of the History of the Behavioral Sciences* 17 (1981): 445–465; Leys, "Meyer, Watson, and the Dangers of Behaviorism," *Journal of the History of the Behavioral Sciences* 20 (1984): 128–149; Scull, *Madhouse*.

22. S. Nassir Ghaemi, "Adolf Meyer: Psychiatric Anarchist," *Philosophy, Psychiatry, and Psychology* 14 (2007): 342; Andrew Scull, On-Line Forum for the television series *American Experience* on PBS, website accessed 20 July 2013, http://www.pbs.org /wgbh/americanexperience/features/general-article/lobotomist-online-forum.

23. Kanner, autobiography, 322.

24. Adolf Meyer to Hermann Meyer, 18 December 1892, IV/3/14, AMC; 28 June 1912, IV/3/142, AMC; and 5 November 1912, IV/3/145, AMC. See also Barbara Betz, "Adolf Meyer: Youth and Manhood, Part 3" *American Journal of Social Psychiatry* 1/2 (1981): 32–39; Eunice Winters, "Adolf Meyer's Two and Half Years at Kankakee," *Bulletin of the History of Medicine* 40 (1966): 455.

25. Adolf Meyer to Anna Meyer, 25 October 1892, IV/3/10, AMC; and November 1892, IV/3/12, AMC; Downey Harris, "Remarks," in *Contributions to Dr. Adolf Meyer by his Colleagues, Friends and Pupils*, ed. S. Katzenelbogen (Baltimore: Johns Hopkins Press, 1938), 25; and Winters, "Meyer's Years at Kankakee," 445; Meyer, "Critical Review of Data and Methods: Part 2," 299.

26. Gerald Grob, "Adolf Meyer on American Psychiatry in 1895," *American Journal of Psychiatry* 119 (1962–1963): 1136–1137; Daniel Rodgers, "In Search of Progressivism," *Reviews in American History* 10/4 (1982): 113–132.

27. A. McGehee Harvey et al., eds., *A Model of Its Kind* (Baltimore: Johns Hopkins University Press, 1989); Thomas Neville Bonner, *Becoming a Physician: Medical Education in Britain, France, Germany, and the United States, 1750–1945* (Baltimore: Johns Hopkins University Press, 1995); Stanley Joel Reiser, *Medicine and the Reign of Technology* (Cambridge: Cambridge University Press, 1978); Charles Rosenberg, *The Care of Strangers: The Rise of America's Hospital System* (Baltimore: Johns Hopkins University Press, 1987); John Harley Warner, *The Therapeutic Perspective: Medical Practice, Knowledge, and Identity in America, 1820–1885* (Princeton, NJ: Princeton University Press, 1997).

28. Roy Porter, *The Greatest Benefit to Mankind: A Medical History of Humanity* (New York: W. W. Norton, 1997), 306–313.

29. See Erwin Ackerknecht, *La médecine hospitalière à Paris (1794–1848)* (Paris: Payot, 1986); W. F. Bynum, *Science and the Practice of Medicine in the Nineteenth Century* (Cambridge: Cambridge University Press, 1994); Russell C. Maulitz, *Morbid Appearances: The Anatomy of Pathology in the Early Nineteenth Century* (Cambridge: Cambridge University Press, 1987/2002). John Harley Warner, *Against the Spirit of the System: The French Impulse in Nineteenth-Century American Medicine* (Baltimore: Johns Hopkins University Press, 1998).

30. Maulitz, *Morbid Appearances*, 9–19; Porter, *Greatest Benefit to Mankind*, 320–340.

31. Engstrom, *Clinical Psychiatry in Imperial Germany*, 88–120.

32. Meyer, "My Experience," 44–49.

33. Gerald Grob, *Mental Illness and American Society, 1875–1940* (Princeton: Princeton University Press, 1983), 7–29; Grob, *Inner World of American Psychiatry*, 1–18.

34. Quoted in Gerald Grob, *The Mad Among Us: A History of the Care of America's Mentally Ill* (New York: Free Press, 1994), 133.

35. Grob, *Inner World of American Psychiatry*, 1–18; Grob, *The Mad Among Us*, 129–140.

36. Gerald Grob, *The State and the Mentally Ill: A History of Worcester State Hospital in Massachusetts, 1830–1920* (Chapel Hill: University of North Carolina Press, 1966); Grob, *The Mad Among Us*; Grob, "Adolf Meyer on American Psychiatry," 1135–1142; Grob, *Inner World of American Psychiatry*.

37. Ruth Leys, "The Correspondence between Adolf Meyer and E. B. Titchener," in *Defining American Psychology: The Correspondence between Adolf Meyer and Edward Bradford Titchener*, ed. Ruth Leys and Rand B. Evans (Baltimore: Johns Hopkins University Press, 1990), 58–114; Leys, "Types of One."

38. Pressman, *Last Resort*, 18–46. See also John C. Burnham, "Psychiatry, Psychology and the Progressive Movement," *American Quarterly* 12/4 (1960): 457–465; Grob, *Mental Illness*, 113–118; Hale, *Freud and the Americans*, 116–249.

39. Michel Foucault, *Psychiatric Power: Lectures at the Collège de France, 1973–1974*, trans. Graham Burchell (New York: Arnold Davidson, 2008), 269–276.

40. Leys, "Types of One," 2.

41. Burnham, *Psychoanalysis and American Medicine, 1894–1918* (New York: International Universities Press, 1967), 18, 159–161; Hale, *Freud and the Americans*, 83–98, 151–173; Ruth Leys, "Meyer, Jung and the Limits of Association," *Bulletin of the History of Medicine* 59 (1985): 345–360; Leys, "Meyer's Dealings with Jones"; Moskowitz, *In Therapy We Trust*.

42. Adolf Meyer, "The Role of Mental Factors in Psychiatry," *American Journal of Insanity* 65 (1908): 41.

Chapter 1. Pathology as Method

1. Adolf Meyer, "A Few Demonstrations of Pathology of the Brain and Remarks on the Problems Connected With Them," *American Journal of Insanity* 52 (1895): 243–249.

2. Adolf Meyer, "The Integrative Function of a Hospital Laboratory" (1921), in *CP*, vol. 2, 79. Originally published in *State Hospital Quarterly* 6 (1920–1921): 445–451.

3. Gerald Grob, *The State and the Mentally Ill: A History of Worcester State Hospital in Massachusetts, 1830–1920* (Chapel Hill: University of North Carolina Press, 1966); Theodore Lidz, "Adolf Meyer and the Development of American Psychiatry," *American Journal of Psychiatry* 123 (1966): 320–332; Edward Shorter, *A History of Psychiatry: From the Era of the Asylum to the Age of Prozac* (New York: John Wiley and Sons, 1997), 109–112.

4. See Erwin Ackerknecht, *La médicine hospitalière à Paris (1794–1848)* (Paris: Payot, 1986); W. F. Bynum, *Science and the Practice of Medicine in the Nineteenth Century* (Cambridge: Cambridge University Press, 1994); John Harley Warner, *Against the Spirit of the System: The French Impulse in Nineteenth-Century American Medicine* (Baltimore: Johns Hopkins University Press, 1998).

5. Erwin Ackerknecht, *Rudolf Virchow: Doctor, Statesman, Anthropologist* (Madison: University of Wisconsin Press, 1953), 52–58; Eric J. Engstrom, *Clinical Psychiatry in Imperial Germany: A History of Psychiatric Practice* (Ithaca, NY: Cornell University Press, 2003), 88–89.

6. Michael Worboys, "Was There a Bacteriological Revolution in Late Nineteenth-Century Medicine?" *Studies in History and Philosophy of Science* 38 (2007): 20–42; John Harley Warner, *The Therapeutic Perspective: Medical Practice, Knowledge, and Identity in America, 1820–1885* (Princeton, NJ: Princeton University Press, 1997), 278–283.

7. Jan Goldstein, *Console and Classify: The French Psychiatric Profession in the Nineteenth Century* (Chicago: University of Chicago Press, 1987); Engstrom, *Clinical Psychiatry in Imperial Germany*, 51–120.

8. Henri Ellenberger, *The Discovery of the Unconscious: The History and Evolution of Dynamic Psychiatry* (New York: Basic Books, 1970), 434; Engstrom, *Clinical Psychiatry in Imperial Germany*, 123–126; Anne Harrington, *Reenchanted Science: Holism in German Culture from Wilhelm II to Hitler* (Princeton, NJ: Princeton University Press, 1996), 16–18; Otto Marx, "Nineteenth-century Medical Psychology: Theoretical Problems in the Work of Griesinger, Meynert and Wernicke," *Isis* 61 (1970): 355–370.

9. Gerald Grob, *The Mad Among Us: A History of the Care of America's Mentally Ill* (New York: Free Press, 1994), 129–140; Grob, *The Inner World of American Psychiatry, 1890–1940: Selected Correspondence* (New Brunswick, NJ: Rutgers University Press, 1985), 2–18.

10. Adolf Meyer, "A Personal History," n.d., Autobiographical Notes (1939–1944), Box 1, EWC; Meyer, Diary Entry, Autumn 1885, VI/7/4, AMC.

11. Adolf Meyer, Student Notebooks, VII/1/18-51, AMC.

12. Meyer, "My Experience with American Psychiatry," unpublished manuscript, 1898, X/1/27, 2–4, AMC.

13. Meyer, Travel Diary, 1891, VI/7/16, AMC; Barbara Betz, "Adolf Meyer: Youth and Young Manhood, 1866–1890, Part 2," *American Journal of Social Psychiatry* 1/1 (1981): 35–40; Adolf Meyer to Anna Meyer, 16 and 26 April 1890, IV/3/1, AMC. See also Meyer, "Contemporary Setting of the Pioneer," *Journal of Comparative Neurology* 74 (1941): 1–24.

14. Meyer, "My Experience," 2–16. See also Adolf Meyer, "British Influences in Psychiatry and Mental Hygiene," *Journal of Mental Science* 79 (1933): 435–464.

15. Meyer, "British Influences," 400; Meyer, "My Experience," 2–3.

16. For Forel, see August Forel, "Einege hirnanatomische Betrachtungen und Ergebnisse," *Archive für Psychiatrie und Nervenkrankenheit* 18 (1887): 162–198; Ellenberger, *History of the Unconscious*, 285–286; Adolf Meyer, "August Forel, 1848–1931," *Journal of Nervous and Mental Disease* 74 (1931): 785–787. For the Zurich School, see Angela Graf-Nold, "History of Psychiatry in Switzerland," *History of Psychiatry* 2 (1991): 321–328; Hans Walther, "Adolf Meyer: Student of the Zurich Psychiatric School," *Gesnerus* 41 (1984): 49–51.

17. Meyer, "My Experience," 8–11; Anna Meyer to Adolf Meyer, June 1890, quoted in Betz, "Meyer: Part 2," 38; Adolf Meyer to Hermann Meyer, 9 January 1893, IV/3/14, AMC. Meyer's doctoral thesis is reprinted in *CP* vol. 1, 13–69; for the original manuscript, see AMC, VII/3/1-22.

18. Meyer, "My Experience," 11–16.

19. Meyer, "My Experience," 12; Alfred Lief, *The Common Sense Psychiatry of Dr. Adolf Meyer: Fifty-two Selected Papers Edited With Biographical Narrative* (New York: McGraw Hill, 1948), 23.

20. Meyer, "My Experience," 21–24.

21. Meyer, "Contemporary Setting," 240.

22. Meyer, "My Experience," 22.

23. Meyer, "British Influences," 404; Meyer, "Contemporary Setting," 13–14; Meyer, "My Experience," 22.

24. Adolf Meyer to Anna Meyer, 25 October 1892, IV/3/10, AMC; and 13–17 November 1892, IV/3/11, AMC. See also James Gilbert, *Perfect Cities: Chicago's Utopias of 1893* (Chicago: Chicago University Press, 1991); Mina Carson, *Settlement Folk: Social Thought and the American Settlement Movement, 1885–1930* (Chicago: University of Chicago Press, 1990).

25. Adolf Meyer to Anna Meyer, 1 December 1892, IV/3/12, AMC; Meyer, "My Experience," 24; Adolf Meyer, "On Preserving Embryological Material," *Journal of the American Medical Association* 22 (1894): 251; Adolf to Anna, 29 September 1892, IV/3/10, AMC.

26. Ludwig Hektoen, Remarks at Twenty-fourth Anniversary of the Phipps Clinic, in *Contributions to Dr. Adolf Meyer by his Colleagues, Friends and Pupils,* ed. S. Katzenelbogen (Baltimore: Johns Hopkins Press, 1938), 23; Adolf Meyer to Anna Meyer, 2 February 1893, IV/3/14, AMC; and 1 December 1892, IV/3/12, AMC; Meyer, "My Experience," 31–32; Adolf Meyer to Hermann Meyer, 11 January 1893, IV/3/14, AMC; W. F. Windle, *The Pioneering Role of Clarence Luther Herrick in American Neurosciences* (New York: Exposition Press, 1979), 94.

27. Victor Robinson, *The Don Quixote of Psychiatry* (New York: Historico-Medical Press, 1919), 106–107; Meyer, "My Experience," 50; Grob, *The Mad Among Us,* 46–48.

28. Gerald Grob, *Mental Illness and American Society, 1875–1940* (Princeton, NJ: Princeton University Press, 1983), 30–45; Meyer, "My Experience," 70 and 35; Lief, *Common Sense Psychiatry,* 45.

29. Quoted in Eunice Winters, "Adolf Meyer's Two and a Half Years at Kankakee," *Bulletin of the History of Medicine* 40 (1966): 442. See also Meyer, "My Experience," 34.

30. Meyer, "My Experience," 36–40; for log books, see Grob, *The State and the Mentally Ill,* 295; Adolf Meyer to Stanley Hall, 7 December 1895, reprinted in Gerald Grob, "Adolf Meyer on American Psychiatry in 1895," *American Journal of Psychiatry* 119 (1963): 1135–1142.

31. Meyer, "My Experience," 35–43 (the other physician was William Stearn); Adolf Meyer to Hermann Meyer, 2 January 1895, cited in Winters, "Meyer's Years at Kankakee," 454.

32. "A New Departure," *Cincinnati Lancet-Clinic* 71 (9 June 1894), 633–634; Adolf Meyer to Stanley Hall, 7 December 1895, reprinted in Grob, "Adolf Meyer on American Psychiatry," 1139; Meyer, "My Experience," 44–46.

33. Winters, "Meyer's Years in Kankakee," 447–448; Meyer, "My Experience," 44–46. For brain modeling at Johns Hopkins, see Louis Hausman and Adolf Meyer, "A Reconstruction Course in the Functional Anatomy of the Nervous System," *Archives of Neurology and Psychiatry* 7 (1922): 287–310.

34. Adolf Meyer, Discussion Section, "Proceedings of the Fourteenth Annual Meeting of the American Psychological Association," *Psychological Bulletin* 3 (1906): 63; Meyer to Stanley Hall, 7 December 1895, reprinted in Grob, "Adolf Meyer on American Psychiatry," 1141.

35. Adolf Meyer to August Forel, 28 December 1893, Box 1, EWC; Meyer, "My Experience," 51–55; for Gapen, see Meyer to Stanley Hall, 7 December 1895, reprinted in Grob, "Adolf Meyer on American Psychiatry," 1139.

36. Emil Kraepelin, *Psychiatrie: Ein kurzes Lehrbuch für Studirende und Aertz,* 4th ed. (Leipzig: Ambr. Abel, 1893); Lief, *Common Sense Psychiatry,* 82.

37. Adolf Meyer to Hermann Meyer, 20 December 1894, quoted in Winters, "Meyer's Years at Kankakee," 454.

38. Adolf Meyer, "The Treatment of the Insane" (1894), in *CP,* vol. 2, 49; Meyer, "Considerations on the Findings in the Spinal Cord of Three General Paralytics," *American Journal of Insanity* 51 (1895): 378.

39. Adolf Meyer to Hermann Meyer, 28 December 1892, IV/3/14, AMC; Meyer, "Thirty-Five Years of Psychiatry in the United States and Our Present Outlook," *American Journal of Psychiatry* 85/1 (1928–1929): 15.

40. S. Weir Mitchell, "Address Before the Fiftieth Annual Meeting of the American Medico-Psychological Association" (1894), in *Proceedings of the American Medico-Psychological Association* (Utica, NY: American Medico-Psychological Association, 1895), 101–121; Meyer, "My Experience," 66; Meyer, "Thirty-five Years of Psychiatry," 2–5.

41. Meyer, "Integrative Function," 79; Meyer, "Special Reports of the Medical Department, Worcester State Lunatic Hospital" (1898), in *CP*, vol. 2, 62.

42. See Adolf Meyer to Stanley Hall, 7 December 1895, reprinted in Grob, "Adolf Meyer on American Psychiatry," 1135–1142; Grob, *The State and the Mentally Ill*, 294–296; Quinby's letter to Meyer extending and describing the job offer is reprinted in Meyer, "Integrative Function," 79; Meyer, "Thirty-five Years of Psychiatry," 7.

43. Engstrom, *Clinical Psychiatry in Imperial Germany*, 121–146.

44. Adolf Meyer, "A Few Trends in Modern Psychiatry," *Psychological Bulletin* 1 (1904): 222; Meyer, "Conditions for Psychiatric Research," *Medical News* 80 (1902): 465; Meyer, "My Experience," 90–91.

45. Meyer, "My Experience," 115–116.

46. Meyer, "Organization of the Medical Work in the Worcester State Hospital" (1912), in *CP*, vol. 2, 59.

47. Edward Cowles, "Progress in the Clinical Study of Psychiatry," *American Journal of Psychiatry* 56 (1899): 109–122; Adolf Meyer to Hermann Meyer, 29 May 1899, IV/3/72, AMC.

48. Adolf Meyer to Hosea Quinby (c. 1900), reprinted in Grob, *Inner World of American Psychiatry*, 64; Meyer, "Conditions for Psychiatric Research," 465.

49. Adolf Meyer to Stanley Hall, 7 December 1895, reprinted in Grob, "Adolf Meyer on American Psychiatry," 1141; Meyer, "My Experience," 78; Grob, *The State and the Mentally Ill*, 294–296.

50. Adolf Meyer, "Aims and Plans of the Pathological Institute for the New York State Hospitals" (1902), in *CP*, vol. 2, 95; Meyer, "My Experience," 97; Meyer, "Special Reports, Worcester Hospital," 62.

51. Adolf Meyer to Stanley Hall, 7 December 1895, reprinted in Grob, "Adolf Meyer on American Psychiatry," 1141.

52. Adolf Meyer, "Twenty-fourth Anniversary of the Henry Phipps Psychiatric Clinic," in *Contributions to Dr. Adolf Meyer by his Colleagues, Friends and Pupils*, ed. S. Katzenelbogen, 47; Grob, *The State and the Mentally Ill*, 294–298; Meyer, "Organization of Work in Worcester Hospital," 59.

53. "A New University Course in Psychiatry," *Worcester Daily Spy*, 23 June 1896 (likely written by Meyer or based on an interview with him); for "morbid exhibits," see Lief, *Common Sense Psychiatry*, 78; Meyer, "My Experience," 75; Meyer, Clark University Lectures, X/2/1-36, AMC; Nathan Hale, *Freud and the Americans: The Beginnings of Psychoanalysis in the United States, 1876–1917* (Oxford: Oxford University Press, 1995), 100.

54. Meyer, "Integrative Function," 80.

55. Adolf Meyer to Hosea Quinby (c. 1900), reprinted in Grob, *Inner World of American Psychiatry*, 65–69; Meyer, "My Experience," 99 and 115; Adolf Meyer to Anna Meyer, 15 March 1902, IV/3/91, AMC.

56. Adolf Meyer to Hermann Meyer, 4 February 1900, IV/3/76. See also Ian Dowbiggin, *Keeping America Sane: Psychiatry and Eugenics in the United States and*

Canada (Ithaca, NY: Cornell University Press, 1997), 53–54; Grob, *The Mad Among Us*, 145–146.

57. Meyer, "Special Reports, Worcester Hospital," 75; Grob, *The State and the Mentally Ill*, 294.

58. For a list of physicians trained by Meyer at Worcester, see Meyer, "Organization of Work in Worcester Hospital," 60–62.

59. Meyer, "Findings in the Spinal Cord," 378; Henry Hurd, *The Institutional Care of the Insane in the United States and Canada*, vol. 2 (Baltimore: Johns Hopkins Press, 1916), 407.

60. Meyer, "Aims and Plans of the Pathological Institute," 90; Meyer, "A Few Remarks Concerning the Organization of the Medical Work in Large Hospitals for the Insane" (1902), in *CP*, vol. 2, 89–90.

61. For an overview of hospital tours, see Meyer, "Aims and Plans of the Pathological Institute"; for Meyer's recollection, see Lief, *Common Sense Psychiatry*, 98–102; for intra-hospital activities, see "Dr. Ferris Defends Psychiatric Work," *New York Times*, 18 September 1910; and Frederick Peterson, "What the State of New York is Doing for the Insane," *Medical News* 86 (1905): 733.

62. D. K. Henderson, "Adolf Meyer: A Tribute from Abroad," *American Journal of Psychiatry* 123 (1966): 334.

63. A. A. Brill, *Psychanalysis: Its Theory and Practical Application* (Philadelphia: W. B. Saunders, 1912), 1; Alexander MacDonald, "Presidential Address," *American Journal of Insanity* 61 (1905): 569–580.

64. Adolf Meyer, "Discussion of Cooperation Between the State Hospitals and the Institute" (1904), in *CP*, vol. 2, 129. Originally published in *16th Annual Report of the State Commission in Lunacy*, New York, 1904.

65. Meyer, excerpt from the *21st Annual Report of the State Commission in Lunacy*, 30 September 1909, New York, in *CP*, vol. 2, 158.

Chapter 2. Mind as Biology

1. Adolf Meyer, "The Role of the Mental Factors in Psychiatry," *American Journal of Insanity* 65 (1908): 51–52.

2. Meyer, "Mental Factors in Psychiatry," 52.

3. Adolf Meyer, "A Short Sketch of the Problems of Psychiatry," *American Journal of Insanity* 53 (1897): 539.

4. Meyer, "Problems of Psychiatry," 549; Meyer, "The Problems of Mental Reaction-Types, Mental Causes and Diseases," *Psychological Bulletin* 5 (1908): 254–255.

5. Adolf Meyer, "Discussion of Cooperation Between the State Hospitals and the Institute" (1904), in *CP*, vol. 2, 126. Originally published in *16th Annual Report of the State Commission in Lunacy*, New York, 1904.

6. Adolf Meyer, "My Experience With American Psychiatry," unpublished manuscript, 1898, X/1/27, AMC: 120.

7. Eric Bredo, "Evolution, Psychology, and John Dewey's Critique of the Reflex Arc Concept," *Elementary School Journal* 98 (1998): 447–466.

8. Garland Allen, "Mechanism, Vitalism and Organicism in Late Nineteenth and Early Twentieth-century Biology: the Importance of Historical Context," *Studies in*

the History and Philosophy of Biological and Biomedical Sciences 36 (2005): 261–283; Anne Harrington, *Reenchanted Science: Holism in German Culture from Wilhelm II to Hitler* (Princeton, NJ: Princeton University Press, 1996), 7–14.

9. Eric J. Engstrom, *Clinical Psychiatry in Imperial Germany: A History of Psychiatric Practice* (Ithaca, NY: Cornell University Press, 2003), 123–126; Harrington, *Reenchanted Science*, 15–18; Franz Seitelberger, "Theodor Meynert: Pioneer and Visionary of Brain Research," *Journal of the History of the Neurosciences* 6 (1997): 264–274; David Steinberg, "Cerebral Localization in the Nineteenth Century," *Journal of the History of the Neurosciences* 18 (2009): 254–261.

10. Peter J. Bowler, *The Eclipse of Darwinism: Anti-Darwinian Evolution Theories in the Decades around 1900* (Baltimore: Johns Hopkins University Press, 1992); Scott F. Gilbert and Sahotra Sarkar, "Embracing Complexity: Organicism for the Twenty-first Century," *Developmental Dynamics* 219 (2000): 1–9; Viktor Hamburger, "Wilhelm Roux: Visionary with a Blind Spot," *Journal of the History of Biology* 30 (1997): 229–238; Harrington, *Reenchanted Science*, 7–14.

11. Maurice Mandelbaum, *Man, History and Reason: A Study of Nineteenth-century Thought* (Baltimore: Johns Hopkins Press, 1971); Gerald Grob, *Mental Illness and American Society, 1875–1940* (Princeton. NJ: Princeton University Press, 1983), 39–41; Harrington, *Reenchanted Science*, 12–30.

12. Henri Ellenberger, *The Discovery of the Unconscious: The History and Evolution of Dynamic Psychiatry* (New York: Basic Books, 1970), 434.

13. Engstrom, *Clinical Psychiatry in Imperial Germany*, 123–126; Roy Porter, *The Greatest Benefit to Mankind: A Medical History of Humanity* (New York: W. W. Norton, 1997), 340.

14. John C. Burnham, *Psychoanalysis and American Medicine, 1894–1918* (New York: International Universities Press, 1967), 55; Daniel Rodgers, "In Search of Progressivism," *Reviews in American History* 10/4 (1982): 113–132.

15. Adolf Meyer, "Critical Review of the Data and General Methods and Deductions of Modern Neurology: Part 1," *Journal of Comparative Neurology* 8/3 (1898): 134.

16. Ellenberger, *Discovery of the Unconscious*, 290–290; E. Stengel, "Hughlings Jackson's Influence in Psychiatry," *British Journal of Psychiatry* 109 (1963): 348–355; Frank J. Sulloway, *Freud, Biologist of the Mind: Beyond the Psychoanalytic Legend* (Cambridge, MA: Harvard University Press, 1992), 271–273.

17. Louis Hausman, *Atlases of the Spinal Cord and Brainstem and the Forebrain* (Springfield, IL: Charles C. Thomas, 1951), 5. Hausman collaborated with Meyer at the Phipps Clinic in the 1920s to develop courses in brain anatomy based on this concept. He produced this textbook to accompany their teaching method.

18. John Hughlings Jackson, "Evolution and Dissolution of the Nervous System," *British Medical Journal* 1/1214 (1884): 660–663. See also Robert Richards, *Darwin and the Emergence of Evolutionary Theories of Mind and Behavior* (Chicago: University of Chicago Press, 1987), 280–287.

19. Jackson, "Evolution and Dissolution," 660; Adolf Meyer, "Etiological, Clinical and Pathological Factors in Diagnosis and Rational Classification of Infectious, Toxic and Asthenic Disease of the Peripheral Nerves, Spinal Cord, and Brain" (1896), in *CP*, vol. 1, 323–333. Originally published in *Medicine* 2 (1896): 639–652. See also G. E. Berrios, "The Factors of Insanities," *History of Psychiatry* 12 (2001): 353–373.

20. Stewart Paton, "The Development of the Higher Brain Centres," *American Journal of Insanity* 54 (1897): 167–179; E. E. Southard, "On the Mechanism of Gliosis in Acquired Epilepsy," *American Journal of Insanity* 64 (1908): 607–641.

21. Adolf Meyer, Private Notes, VII/2/7, AMC; Meyer, "British Influences in Psychiatry and Mental Hygiene," *Journal of Mental Science* 79 (1933): 445 (emphasis in original).

22. John Hughlings Jackson, "On Epilepsies and the After-effects of Epileptic Discharge," *West Riding Lunatic Asylum Medical Report* 6 (1876): 266–309. See also Walther Riese, "The Sources of Hughlings Jackson's View on Aphasia," *Brain* 88 (1965): 811–822; George K. York and David A. Steinberg, "The Philosophy of Hughlings Jackson," *Journal of the Royal Society of Medicine* 95 (2002): 314–318; Robert Young, *Mind, Brain and Adaptation in the Nineteenth Century* (Oxford: Oxford University Press, 1990), 197–223.

23. Edwin Clarke and L. S. Jacyna, *Nineteenth-century Origins of Neuroscientific Concepts* (Berkeley: University of California Press, 1987), 58–100; E. Pannese, "The Gogli Stain: Invention, Diffusion, and Impact on Neurosciences," *Journal of the History of the Neurosciences* 8 (1999): 132–140.

24. C. U. M. Smith, "Sherrington's Legacy: Evolution of the Synapse Concept, 1890s–1990s," *Journal of the History of the Neurosciences* 5 (1996): 43–55.

25. Adolf Meyer, "Neurologists and Neurological Laboratories, Part IV: Neurological Work at Zurich," *Journal of Comparative Neurology* 3 (1893): 114.

26. Meyer, "Etiological, Clinical and Pathological Factors," 324.

27. Meyer, "Critical Review of Data and Methods: Part 1": 113–147; and Meyer, "Critical Review of the Data and General Methods and Deductions of Modern Neurology: Part 2," *Journal of Comparative Neurology* 8/4 (1898): 249–313.

28. Adolf Meyer, "Considerations on the Findings in the Spinal Cord of Three General Paralytics," *American Journal of Insanity* 51 (1895): 375. For Jackson, see Berrios, "Factors of Insanities."

29. Meyer, "Critical Review of Data and Methods: Part 2," 312.

30. Cheryl Logan, "The Legacy of Adolf Meyer's Comparative Approach: Worcester Rats and the Strange Birth of the Animal Model," *Integrative Physiological and Behavioral Science* 40 (2005): 169–181.

31. Meyer, "Critical Review of Data and Methods: Part 1," 146. For Sherrington and Horsley, see Meyer, "British Influences." For Roux, see Meyer, "Remarks on Habit Disorganizations in the Essential Deteriorations, and the Relation of Deterioration to the Psychasthenic, Neurasthenic, Hysterical and Other Constitutions," in *Studies in Psychiatry* 1 (1912): 104.

32. Meyer, "Critical Review on Data and Methods: Part 2," 291 and 296.

33. Meyer, "Critical Review on Data and Methods: Part 2," 294–296 (emphasis in original).

34. Louis Hausman, Introduction to vol. 1 of *CP*; Jerzy Rose, "Adolf Meyer's Contributions to Neuroanatomy," *Bulletin of the Johns Hopkins Hospital* 89 (1951): 56–59.

35. Meyer, "Problems of Mental Reaction-Types," 255.

36. Meyer, "Critical Review of Data and Methods: Part 2," 311 (emphasis in original).

37. Adolf Meyer, "Relationship of Hysteria, Psychasthenia, and Dementia Praecox," in *Studies in Psychiatry* 1 (1912): 157.

38. Adolf Meyer, "Anatomical Facts and Clinical Varieties of Traumatic Insanity," *American Journal of Insanity* 60 (1904): 373.

39. Adolf Meyer, "The Aims of a Psychiatric Clinic" (1912), in *Proceedings of the Mental Hygiene Conference, New York, 1912* (New York: Committee on Mental Hygiene of the State Charities Aid Association, 1912): 119.

40. Meyer, "Problems of Mental Reaction-Types," 258.

41. Meyer, "Remarks on Habit Disorganizations," 104.

42. Meyer, "Mental Factors in Psychiatry," 44.

43. Meyer, "Critical Review of Data and Methods: Part 2," 296.

44. Meyer, "A Few Demonstrations of Pathology of the Brain and Remarks on the Problems Connected With Them," *American Journal of Insanity* 52 (1895): 248.

45. Meyer, "Critical Review of Data and Methods: Part 1," 114.

46. Meyer, "Insanity: General Pathology," in *Reference Handbook of the Medical Sciences*, 2nd ed., vol. 5, ed. A. H. Buck (New York: William Wood, 1902): 40; Meyer, "Review of Recent Text-books of Anatomy and Pathology of the Nervous System: Third Article," *Journal of Comparative Neurology* 11 (1901): xliii.

47. Ruth Leys, "The Correspondence between Adolf Meyer and E. B. Titchener," in *Defining American Psychology: The Correspondence between Adolf Meyer and Edward Bradford Titchener*, ed. Ruth Leys and Rand B. Evans (Baltimore: Johns Hopkins University Press, 1990), 87–89; Meyer, "Critical Review of Data and Methods: Part 2," 290.

48. Meyer, "Psychopathology" (1899), in *CP* vol. 2, 283. Originally published in *Clark University, 1889–1899: Decennial Celebration Volume*, 1899.

49. Meyer, "Insanity: General Pathology," 40.

50. Adolf Meyer, "The Mental Factor in Medicine," *Journal of Comparative Neurology* 12 (1902): xvii; Meyer, "Problems of Mental Reaction-Types," 259–260; Meyer, "Insanity: General Pathology," 38.

51. Meyer, "My Experience," 75; Meyer, Clark University Lectures, X/2/1-36, AMC.

52. William James to Clifford Beers, 22 September 1909, and Adolf Meyer to Mary Potter Brooks Meyer, 8 May 1902, both quoted in Ruth Leys, "Types of One: Adolf Meyer's Life Chart and the Representation of Individuality," *Representations* 34 (1991): 22, n. 5.

53. Adolf Meyer, "Hypnotism and Meyer's Views of Mental Dynamics," c. 1923, Box 4, EWC. See also Leys, "Adolf Meyer: A Biographical Note," in *Defining American Psychology: The Correspondence between Adolf Meyer and Edward Bradford Titchener*, ed. Ruth Leys and Rand B. Evans (Baltimore: Johns Hopkins University Press, 1990), 43.

54. Richards, *Darwin and Theories of Mind*, 430–435.

55. William James, "Are We Automata?" *Mind* 4 (1879): 1–22; James, *Principles of Psychology*, vol. 1 (New York: Henry Holt, 1890), 138–144. See also Richards, *Darwin and Theories of Mind*, 430–435.

56. James, *Principles of Psychology*, vol. 1, 487; Wayne Viney, "The Radical Empiricism of William James and Philosophy of History," *History of Psychology* 4 (2001): 211–227.

57. Meyer, "Critical Review of Data and Methods: Part 1," 146.

58. Meyer, "Critical Review of Data and Methods: Part 2," 299. See also Adolf Meyer, "Misconceptions at the *Bottom of Hopelessness of all Psychology* by P. J. Möbius," *Psychological Bulletin* 4/6 (1907): 170–179.

59. Leo Kanner, unpublished autobiography, Part 6, Archives of the American Psychiatric Association, Arlington, VA. See also Kanner, "The Significance of a Pluralistic Attitude in the Study of Human Behavior," *Journal of Abnormal and Social Psychology* 28 (1933): 30–41.

60. Meyer, "Critical Review of Data and Methods: Part 2," 299; Dugald Stewart, "Of the Fundamental Laws of Human Belief," in *Collected Works of Dugald Stewart*, vol. 3, ed. William Hamilton (Edinburgh: Clark, 1877), 45; William James, "A World of Pure Experience" (1904), in *James and Dewey on Belief and Experience,* ed. John M. Capps and Donald Capps (Chicago: University of Chicago Press, 2005), 144–161; Charlene Haddock Seigfried, "The Philosopher's 'License': William James and Common Sense," *Transactions of the Peirce Society* 9 (1983): 273–290.

61. Charles Darwin, *Journal and Remarks: 1832–1836* (London: Colburn, 1839), 237; George John Romanes, *Mental Evolution in Animals* (New York: Appleton, 1884), 38 and 192. See also Richards, *Darwin and Theories of Mind*, 365–368.

62. James, *Principles of Psychology*, vol. 1, 76–113, quote on 113; Meyer, "Critical Review of Data and Methods: Part 2," 293.

63. Bredo, "Evolution and Dewey's Critique," 452–453; John Dewey, "The Postulate of Immediate Empiricism," in *The Influence of Darwin on Philosophy* (New York: Henry Holt, 1910), 227; Andrew Backe, "John Dewey and Early Chicago Functionalism," *History of Psychology* 4 (2001): 323–340.

64. Thomas C. Dalton, *Becoming John Dewey: Dilemmas of a Philosopher and Naturalist* (Bloomington: Indiana University Press, 2002), 11, 65–66, 85.

65. Meyer, "Critical Review of Data and Methods: Part 1," 146; John Dewey, "Educational Psychology: Syllabus of a Course of Twelve Lecture Studies" (1896), in *Early Works of John Dewey, 1882–1898,* ed. Jo Anne Boydston (Carbondale: Southern Illinois University Press, 1972): 304; John Dewey, "The Reflex Arc Concept in Psychology," *Psychological Review* 3/4 (1896): 358; Meyer, "Etiological, Clinical and Pathological Factors," 325. Notably, the ideas of Charles Sanders Peirce influenced those of Dewey and Meyer.

66. Dewey, "Educational Psychology," 314; Dalton, *Becoming John Dewey,* 80; Meyer, "Problems of Psychiatry," 541.

67. Alfred Lief, *The Common Sense Psychiatry of Dr. Adolf Meyer: Fifty-two Selected Papers Edited With Biographical Narrative* (New York: McGraw Hill, 1948), 152.

68. Adolf Meyer, "Objective Psychology or Psychobiology with Subordination of the Medically Useless Contrast of Mental and Physical," *Journal of the American Medical Association* 65 (1915): 862.

69. Adolf Meyer, "Discontent—A Psychobiological Problem of Hygiene" (1919), in *CP,* vol. 4, 384; Meyer, "Treatment of Paranoic and Paranoid States," in *The Modern Treatment of Nervous and Mental Diseases*, vol. 1, ed. W. A. White and S. E. Jelliffe (Philadelphia: Lea and Febiger, 1913), 660.

70. Adolf Meyer, "Organization of the Work of the Henry Phipps Psychiatric Clinic, Johns Hopkins Hospital, With Special Reference to the First Year's Work," *Transactions of the American Medico-Psychological Association 70th Annual Meeting* 21 (1914): 402; Meyer, "Aims of a Psychiatric Clinic" (1912), 119.

71. Adolf Meyer, "Mental Factors in Psychiatry," 44; Meyer, "Fundamental Conceptions of Dementia Praecox," *British Medical Journal* 2387 (29 September 1906): 757–758.

72. Private Notes, 8–10 April 1912, and Meyer to Anna Meyer, 14 April 1912, IV/3/142, AMC; Meyer, "Mental Factors in Psychiatry," 47.

73. Thomas H. Huxley, "On the Educational Value of the Natural History Sciences" (1854), in *Science and Education*, vol. 3 of *Essays by Thomas H. Huxley* (New York: Appleton, 1897), 45–46; Adolf Meyer, "A Few Remarks Concerning the Organization of the Medical Work in Large Hospitals for the Insane" (1902), in *CP*, vol. 2, 89.

74. Meyer, "Misconceptions," 172–173 (emphasis in original); Meyer, "Objective Psychology or Psychobiology," 860. See also Leys, "Correspondence between Meyer and Titchener," 84.

75. Meyer, "Objective Psychology or Psychobiology," 860 (emphasis in original); Meyer, "Problems of Psychiatry," 538 n.1; Meyer, "Misconceptions," 171; Adolf Meyer, "Conditions for a Home of Psychology in the Medical Education," *Journal of Abnormal Psychology* 7 (1912–1913): 315.

76. Adolf Meyer, "Freedom and Discipline" (1928), in *CP*, vol. 4, 417. Originally published in *Progressive Education* 5 (1928): 205–210.

77. Meyer, "Mental Factors in Psychiatry," 48.

78. Adolf Meyer, "The Aims of a Psychiatric Clinic" (1913), in *CP* vol. 2, 200. Originally published in *Transactions of the 17th International Congress of Medicine in London*, Section 12, Part 1 (1913): 1–11.

79. Adolf Meyer, "A Few Remarks Concerning the Organization of the Medical Work in Large Hospitals for the Insane" (1902), in *CP*, vol. 2, 89–90. Originally printed privately.

80. Meyer, "Problems in Psychiatry," 543; Meyer, "The Relation of Psychogenic Disorders to Deterioration," *Journal of Nervous and Mental Disease* 34/6 (1907): 404.

81. Meyer, "Insanity: General Pathology," 37–38; Meyer, "Aims of a Psychiatric Clinic" (1912), 120–121.

82. Meyer, "Organization of Work of Phipps Clinic," 402.

83. Meyer, "Habit Disorganizations," 104.

84. Meyer, "Aftercare and Prophylaxis" (1911), in *CP*, vol. 4, 215.

85. Meyer, "Insanity: General Pathology," 40.

86. Meyer, "Aims of A Psychiatric Clinic" (1912), 119.

87. Case DZE-923. See also Meyer, "Problems of Mental Reaction-Types."

88. Meyer, "Mental Factors in Psychiatry," 40.

89. Meyer, "Problems of Mental Reaction-Types," 252.

90. Meyer, "Problems of Mental Reaction-Types," 258; Meyer, "Mental Factors in Psychiatry," 47.

91. Adolf Meyer, "On the Observation of Abnormalities of Children" (1895), in *CP*, vol. 4, 327. Originally published in *Child-Study Monthly* 1 (1895): 1–12. See also Meyer, "Psychogenic Disorders."

92. Case CGV-482.

93. Meyer, "Fundamental Conceptions of Dementia Praecox," 759; Meyer, "Hysteria, Psychasthenia, and Dementia Praecox," 160.

94. Case LQT-895; Meyer, "Habit Disorganizations," 104.

95. Adolf Meyer, "The Right to Marry" (1916), in *CP*, vol. 4, 295. Originally published in *Survey* 36 (1916): 243–246.

96. Meyer, "Mental Factors in Psychiatry," 40–41.

97. Case CGV-482.

98. Adolf Meyer, "A Review of the Signs of Degeneration and of Methods of Registration," *American Journal of Insanity* 52 (1895–1896): 345. See also Robert Nye, "Sociology and Degeneration: The Irony of Progress," in *Degeneration: The Dark Side of Progress*, ed. J. Edward Chamberlain and Sander Gilman (New York: Columbia University Press, 1985): 49–71.

99. Adolf Meyer, "An Attempt at Analysis of the Neurotic Constitution," *American Journal of Psychology* 14 (1903): 95.

100. Adolf Meyer, "Organization of Eugenic Investigation" (1917), in *CP*, vol. 4, 304.

101. Meyer, "Abnormalities of Children," 326.

102. Meyer, "Fundamental Conceptions of Dementia Praecox," 759.

103. Meyer, "The Problem of the State in the Care of the Insane," *American Journal of Insanity* 65 (1909): 689.

104. Nathan Hale, *Freud and the Americans: The Beginnings of Psychoanalysis in the United States, 1876–1917* (Oxford: Oxford University Press, 1995), 279–280; Stanley Abbot, "Meyer's Theory of the Psychogenic Origin of Dementia Praecox: A Criticism," *American Journal of Psychiatry* 68 (1911): 15–22.

105. Richards, *Darwin and Theories of Mind*, 234.

106. See Adolf Meyer, "Constructive Formulation of Schizophrenia," *American Journal of Psychiatry* 78 (1921): 355–362.

107. Meyer, "Mental Factors in Psychiatry," 47.

108. Meyer, "Psychogenic Disorders," 403; Meyer, "Mental Factors in Psychiatry," 54; Meyer, "Objective Psychology or Psychobiology," 863; Adolf Meyer, *Outlines of Examinations* (New York: Bloomingdale Hospital Press, 1918), 13.

109. Meyer, "Review of *A Mind That Found Itself* by Clifford Beers," *North American Review* 187, no. 629 (1908): 611; Meyer, "The Psychiatric Clinic, Its Aims (Educational and Therapeutic), and the Results Obtained in Respect to Promotion of Recovery" (1913), in *CP*, vol. 2, 203. Originally published in *Transactions of the 17th International Congress of Medicine in London*, Section 12, Part 2 (1913): 9–11.

110. Thomas Rennie, "Adolf Meyer and Psychobiology," in *Papers from the Second American Congress on General Semantics*, ed. M. Kendig (Chicago: Institute of General Semantics, 1943): 165 (emphasis added for clarity).

111. Jerome Frank, "Adolf Meyer in Retrospect," unpublished address given 11 February 1980, Box 3, EWC.

112. Lawrence Davidson, "The Strange Disappearance of Adolf Meyer," *Orthomolecular Psychiatry* 9 (1980): 135–143.

Chapter 3. Unique Soil in Baltimore

1. Adolf Meyer, "The Purpose of the Psychiatric Clinic," *American Journal of Insanity* 69 Special Issue (1913): 857; William Osler, "The Fixed Period" (1905), in William Osler, *Aequanimitas, With Other Addresses to Medical Students, Nurses and Practitioners of Medicine*, 2nd ed. (London: Lewis, 1906): 389–411; "Human Interest Strongly in Evidence at Opening," *Baltimore Sun*, 17 April 1913, 6; Osler, "Specialism in the General Hospital," *British Medical Journal* (17 May 1913): 1055–1056.

2. See Gerald Grob, "Adolf Meyer on American Psychiatry in 1895," *American Journal of Psychiatry* 119 (1962–1963): 1135–1142; Grob, *The Inner World of American Psychiatry, 1890–1940: Selected Correspondence* (New Brunswick, NJ: Rutgers University Press, 1985), 19–25; Grob, *Mental Illness and American Society, 1875–1940* (Princeton, NJ: Princeton University Press, 1983), 46–71, 108–123; Edward Shorter, *A History of Psychiatry: From the Era of the Asylum to the Age of Prozac* (New York: John Wiley and Sons, 1997), 65–98.

3. Emil Kraepelin, *Psychiatrie: Ein Lehrbuch für Studirende und Aerzte* 5th ed. (Leipzig: J. A. Barth, 1896).

4. Adolf Meyer to Hermann Meyer, 29 May 1899, IV/3/72, AMC; and 14 July 1912, IV/3/143, AMC; "Phipps Psychiatric Clinic at the Johns Hopkins Hospital," *American Journal of Insanity* 65 (1908): 186–203.

5. Adolf Meyer, "Review of Recent Text-books of Anatomy and Pathology of the Nervous System," *Journal of Comparative Neurology* 11 (1901): xliii.

6. Gerald Grob, *The Mad Among Us: A History of the Care of America's Mentally Ill* (New York: Free Press, 1994), 142; Jack Pressman, *Last Resort: Psychosurgery and the Limits of Medicine* (Cambridge: Cambridge University Press, 1998), 20–21.

7. Nathan Hale, *Freud and the Americans: The Beginnings of Psychoanalysis in the United States, 1876–1917* (Oxford: Oxford University Press, 1995), 151–173; Eva S. Moskowitz, *In Therapy We Trust: America's Obsession with Self-Fulfillment* (Baltimore: Johns Hopkins University Press, 2001).

8. "Phipps Gives $500,000 to Johns Hopkins," *New York Times*, 15 June 1908, 1; "Splendid Gift to Advance Psychiatry," *Journal of the American Medical Association* 51 (1908): 43–44; Notes section, *Boston Medical and Surgical Journal* 158 (1908): 948.

9. W. F. Bynum, "The Rise of Science in Medicine" in *The Western Medical Tradition, 1800–2000*, ed. W. F. Bynum et al. (Cambridge: Cambridge University Press, 2006): 149.

10. A. McGehee Harvey et al., eds., *A Model of Its Kind* (Baltimore: Johns Hopkins University Press, 1989), 1–22; William Osler to Henry Phipps, 8 July 1908, CUS417/114.135, Harvey Cushing Fonds (P417), Osler Library Archive Collections, Osler Library of the History of Medicine, McGill University, Montreal, Canada.

11. "New York Alienist Called to Direct New Hospital for the Curable Insane," *New York Times*, 23 June 1908, 1. See also, Christopher Lawrence, "Anaesthesia in the Age of Reform," *History of Anaesthesia Proceedings* 20 (1997): 11–16; Owsei Temkin, "The Role of Surgery in the Rise of Modern Medical Thought," *Bulletin of the History of Medicine* 25 (1951): 248–259.

12. Henry Phipps to William Welch, 18 May 1908, quoted in Alan Chesney, *The Johns Hopkins Hospital and The Johns Hopkins School of Medicine: A Chronicle*, vol. 3 (Baltimore: Johns Hopkins University Press, 1963), 64–65; "Phipps Interested in Thaw," *New York Times*, 16 June 1908, 16; Osler, "The Fixed Period," 409; William Welch to Adolf Meyer, 15 July 1908, I/3988/2, AMC. See also Clifford Beers, *A Mind That Found Itself* (New York: Longmans-Green, 1907); Eunice Winters, "Adolf Meyer and Clifford Beers, 1907–1910," *Bulletin of the History of Medicine* 43 (1969): 414–443.

13. Henry Hurd to the Executive Committee of the Board of Trustees of The Johns Hopkins Hospital, 1 June 1908, quoted in Chesney, *Hopkins Hospital and School of*

Medicine, 66–67; William Welch to Henry Hurd, 12 June 1908, I/3988/1, AMC; Henry Phipps to William Welch, 13 June 1908, quoted in Chesney, ibid., 67–68.

14. Adolf Meyer to Hermann Meyer, 12 June 1908, IV/3/122, AMC; Adolf Meyer, "My Experience With American Psychiatry," unpublished manuscript, 1898, X/1/27, AMC: 85.

15. Meyer, "Twenty-fourth Anniversary of the Henry Phipps Psychiatric Clinic," in *Contributions to Dr. Adolf Meyer by his Colleagues, Friends and Pupils*, ed. S. Katzenelbogen (Baltimore: Johns Hopkins Press, 1938), 51; Meyer to Hermann Meyer, 12 June 1908, IV/3/122, AMC.

16. Drafts of letters, Adolf Meyer to William Welch, n.d., I/3988/1, AMC; Welch to Meyer, 15 and 19 June 1908, I/3988/2, AMC.

17. "Dr. Meyer The Man," *Baltimore Sun*, 23 June 1908, 12; "The Phipps Psychiatric Clinic in Connection with the Johns Hopkins Hospital," *Johns Hopkins Hospital Bulletin* 19/209 (1908): 241–243; "Phipps Psychiatric Clinic," *American Journal of Insanity* 65 (1908): 186–203; "Dr. Meyer The Man," *Baltimore Sun*, 23 June 1908, 12.

18. Draft of letter, Meyer to William Welch, n.d., I/3988/1, AMC.

19. Osler, "The Fixed Period," 409; D. K. Henderson, Introduction to vol. 2 of *CP*, xvi–xvii; *Twentieth-fifth Report of the Superintendent of The Johns Hopkins Hospital, for the year ending January 31, 1914* (Baltimore: Johns Hopkins Press, 1914): 32–33.

20. Jay Schulkin, *Curt Richter: A Life in the Laboratory* (Baltimore: Johns Hopkins University Press, 2005), 16.

21. Meyer rarely commented publicly on religion or politics, keeping his personal views on these subjects close to the vest. When he emigrated to the United States in 1892, he joined the Swiss Socialist Society in Chicago (see Adolf Meyer to Anna Meyer, 13–17 November 1892, IV/3/11, AMC). In the 1920s, he was a private supporter of the political activist Elisabeth Gilman, who ran for governor, senator, and mayor of Baltimore on the Socialist Party ticket (Adolf Meyer to Elisabeth Gilman, Meyer's calling card with handwritten note dated 1926, Elisabeth Gilman Papers Ms. 235, Box 2, Special Collections, Milton S. Eisenhower Library, Johns Hopkins University).

22. Adolf Meyer to Hermann Meyer, 14 July 1912, IV/3/143, AMC; Adolf Meyer to Grosvenor Atterbury, 25 July 1910, I/114/7, AMC.

23. Eric J. Engstrom, *Clinical Psychiatry in Imperial Germany: A History of Psychiatric Practice* (Ithaca, NY: Cornell University Press, 2003), 144–145.

24. Adolf Meyer, "The Psychiatric Clinic, Its Aims (Educational and Therapeutic), and the Results Obtained in Respect to Promotion of Recovery" (1913), in *CP*, vol. 2, 204. Originally published in *Transactions of the 17th International Congress of Medicine in London*, Section 12, Part 2 (1913): 9–11.

25. Adolf Meyer, "Where Should We Attack the Problem of the Prevention of Mental Defect and Mental Disease?" (1915), *CP*, vol. 4, 197. Address to the 42nd Annual Session of the National Conference of Charities and Correction, May 1915. Published in its *Proceedings* 42 (1915): 298–307.

26. John Oliver, "Experience of a Psychiatric Missionary in the Criminal Courts," *Journal of the American Institute of Criminal Law and Criminology* 9/4 (1919): 559; Private Correspondence, 19 December 1913, XV/Box B4, AMC.

27. Adolf Meyer to Hosea Quinby (c. 1900), reprinted in Grob, *Inner World of American Psychiatry*, 64.

28. Adolf Meyer, "Aftercare and Prophylaxis" (1911), in *CP*, vol. 4, 212.

29. "A Brief History of the Henry Phipps Psychiatric Clinic," typed draft by unknown author with Meyer's marginalia, c. 1909, XII/1/374, AMC.

30. Adolf Meyer, "Plans for Work in the Phipps Psychiatric Clinic" (1913), in *CP*, vol. 2, 186. Originally published in *Modern Hospital* 1 (1913–1914): 69–76.

31. Adolf Meyer, "Treatment of Paranoic and Paranoid States," in *The Modern Treatment of Nervous and Mental Diseases*, vol. 1, ed. W. A. White and S. E. Jelliffe (Philadelphia: Lea and Febiger, 1913): 643.

32. Adolf Meyer, "The Aims of a Psychiatric Clinic" (1913), in *CP*, vol. 2, 202. Originally published in *Transactions of the 17th International Congress of Medicine in London*, Section 12, Part 1 (1913): 1–11.

33. Henry Hurd, *The Institutional Care of the Insane in the United States and Canada*, vol. 2 (Baltimore: Johns Hopkins Press, 1916), 571 (Meyer wrote the entry for the Phipps Clinic); Meyer, "Aims of a Psychiatric Clinic" (1913), 199.

34. Adolf Meyer, "Some Common Misunderstandings About State Hospitals and the Way to Make Them Unnecessary" (1913), in *CP*, vol. 4, 181. Address to North Carolina's first Conference on Mental Hygiene, 1913.

35. Meyer, "Plans for Phipps Clinic," 190; D. K. Henderson, "Remarks on Cases Received in the Henry Phipps Psychiatric Clinic," *Bulletin of the Johns Hopkins Hospital* 25/277 (1914): 71; "To Cure Insanity Fear," *Baltimore Sun*, 28 July 1912, 8. For quantitative analyses, see Susan D. Lamb, "Pathologist of the Mind: Adolf Meyer, Psychobiology and the Phipps Psychiatric Clinic at The Johns Hopkins Hospital, 1908–1917" (Ph.D. diss., Johns Hopkins University, 2010). ProQuest, UMI Dissertations Publishing, 3440753.

36. Meyer, "Aims of a Psychiatric Clinic" (1913), 195.

37. Meyer, "Where Should We Attack?" 197; Trigant Burrow to his mother, 22 January 1910, in *A Search for Man's Sanity: The Selected Letters of Trigant Burrow*, ed. William Galt (New York: Oxford University Press, 1958), 29 (emphasis in original); C. Macfie Campbell, "The Mental Health of the Community and the Work of the Psychiatric Dispensary," *Mental Hygiene* 1 (1917): 572.

38. Meyer, "Aims of a Psychiatric Clinic" (1913), 194 and 201 (emphasis in original). See also, Adolf Meyer, "Organization of the Work of the Henry Phipps Psychiatric Clinic" (1914), in *CP*, vol. 2, 205. Originally published in *Transactions of the American Medico-Psychological Association 70th Annual Meeting* 21 (1914): 397–403.

39. Meyer, "Organization of Work of Phipps Clinic," 205. There is a copy of the agreement in the record of Case WSR-288.

40. Meyer, "Aims of a Psychiatric Clinic" (1913), 198.

41. Adolf Meyer, "The Aims of a Psychiatric Clinic" (1912), in *Proceedings of the Mental Hygiene Conference, New York, 1912* (New York: Committee on Mental Hygiene of the State Charities Aid Association, 1912): 119.

42. Meyer, Working Notes, XII/1/353, AMC.

43. Meyer, "Plans for Phipps Clinic," 191; Adolf Meyer, "The Problems of the Physician Concerning the Criminal Insane and Borderline Cases," *Journal of American Medicine* 54 (1910): 1931.

44. Private Correspondence, 23 July 1912, 25 June 1912, 5 August 1912, and 6 August 1912, in XV/A5, AMC.

45. Meyer, "Organization of Work of Phipps Clinic," 208–209.

46. Effie J. Taylor, "Nursing in the Henry Phipps Psychiatric Clinic," *Johns Hopkins Nurses Alumnae Magazine* 13/4 (1914): 235; "Phipps Clinic Ready Today," *Baltimore Sun*, 1 May 1913, 4.

47. Private Correspondence, 5 January 1913, Box A2, Series XV, AMC.

48. "6,818 Patients in Year," *Baltimore Sun*, 27 June 1914, 4; *Twenty-Seventh Report of the Superintendent of the Johns Hopkins Hospital for the year ending January 31, 1916* (Baltimore: Johns Hopkins Press, 1916).

49. Adolf Meyer, "Extra-Institutional Responsibilities of State Hospitals for Mental Diseases" (1916), in *CP*, vol. 4, 228–263.

50. See Susan Lamb, "Social, Motivational, and Symptomatic Diversity: Analysis of the Patient Population of the Phipps Psychiatric Clinic, 1913–1917," *Canadian Bulletin for Medical History* 29/2 (2012): 243–263.

51. Private Correspondence, XV/A2, AMC; Case RZG-112; Case NSU-863.

52. Case CGV-482.

53. Meyer, "Organization of Work of Phipps Clinic," 207; Henderson, "Remarks on Cases in Phipps Clinic," 69. The letter of instruction from Johns Hopkins regarding his bequest is reprinted in Harvey et al., *A Model of Its Kind*, 9.

54. Meyer, "Organization of Work of Phipps Clinic," 205; Walter O. Jahrreiss, *History of Mount Hope Retreat* (Baltimore: Thompsen-Ellis-Hutton, 1940).

55. Private Correspondence, XV/B4, AMC.

56. Meyer, "Plans for Phipps Clinic," 186.

57. Case DAU-952; Case SGM-137.

58. Case JGC-374; Case RZG-112; Case WSR-288.

59. Case ZWF-362; Case NUY-166; Case ARD-545.

60. Case WTQ-149.

61. Case OLR-652.

62. Case LFM-844; Case RCZ-156.

63. Case DVE-705.

64. For example, Philippe Pinel, Benjamin Rush, and Thomas Story Kirkbride.

65. Meyer, "Aims of a Psychiatric Clinic" (1913), 195; Adolf Meyer, *Outlines of Examinations* (New York: Bloomingdale Hospital Press, 1918), 14; Meyer to Charles P. Emerson, 2 January 1914, quoted in Grob, *Inner World of American Psychiatry*, 82.

66. Meyer, "Plans for Phipps Clinic," 188.

67. Case USN-176; Case FYN-934.

68. Correspondence between Adolf Meyer and Effie Taylor, 1912–1919, I/3770/1-5, AMC. See also Effie J. Taylor, "Nursing in the Phipps Clinic" and Mame Warren, ed., *Our Shared Legacy: Nursing Education at Johns Hopkins, 1889–2006* (Baltimore: Johns Hopkins University Press, 2006).

69. Case CGV-482. For Kraepelin, see Engstrom, *Clinical Psychiatry in Imperial Germany*, 133.

70. Meyer, "Organization of Work of Phipps Clinic," 208 (emphasis in original).

71. Henry Phipps Psychiatric Clinic Nursing Manual, circa 1916, unpublished instruction manual consisting of sixty-eight typewritten pages bound with cloth tape and organized alphabetically by topic, XII/24/41, AMC (hereafter, Phipps Nursing Manual), 8; Case ACM-910; Case PTY-140.

72. Case RDL-968; Phipps Nursing Manual, 7.

73. Meyer, *Outlines of Examinations*, 11. See also Adolf Meyer, "The Complaint as the Center of Genetic-Dynamic and Nosological Teaching in Psychiatry," *New England Journal of Medicine* 199 (1928): 360–370.

74. Meyer, *Outlines of Examinations*, 10–16; Franklin Ebaugh, Introduction to vol. 3 of *CP*, ix.

75. Case HTR-132; Case LHM-638.

Chapter 4. The Baptismal Child of American Psychiatry

1. Adolf Meyer, "Discussion of Cooperation Between the State Hospitals and the Institute" (1903), in *CP*, vol. 2, 124 (emphasis in original). Originally published in *16th Annual Report of the State Commission in Lunacy, New York, 30 September 1904*.

2. Adolf Meyer, "Remarks on Habit Disorganizations in the Essential Deteriorations, and the Relation of Deterioration to the Psychasthenic, Neurasthenic, Hysterical and Other Constitutions," in *Studies in Psychiatry* 1 (1912): 109.

3. Adolf Meyer, "Critical Review of the Data and General Methods and Deductions of Modern Neurology: Part 2," *Journal of Comparative Neurology* 8/4 (1898): 309.

4. Joel Braslow, *Mental Ills, Bodily Cures: Psychiatric Treatment in the First Half of the Twentieth Century* (Berkeley: University of California Press, 1997), 76–78.

5. Adolf Meyer, "The Extra-Institutional Responsibilities of State Hospitals for Mental Diseases" (1916), in *CP*, vol. 4, 228; Meyer, "Considerations on Psychiatry or Ergasiatrics as an Essential and Natural Part of All Medical Training and Practice" (1940), in *CP*, vol. 3, 462.

6. Xavier Bichat, *Anatomie générale appliquée à la physiologie et à la médecine*, nouvelle edition (Paris: J. S. Chaud, 1830), xciv.

7. Eric J. Engstrom, *Clinical Psychiatry in Imperial Germany: A History of Psychiatric Practice* (Ithaca, NY: Cornell University Press, 2003), 125–143; Matthias M. Weber and Eric J. Engstrom, "Kraepelin's 'Diagnostic Cards': The Confluence of Clinical Research and Preconceived Categories," *History of Psychiatry* 8 (1997): 375–385.

8. Adolf Meyer, "Review of *Psychiatrie: Ein Lehrbuch für Studirende und Aertze* by Emil Kraepelin," *American Journal of Psychiatry* 53 (1896): 302.

9. Adolf Meyer, "Thirty-Five Years of Psychiatry in the United States and Our Present Outlook," *American Journal of Psychiatry* 85/1 (1928–1929): 1–31.

10. Engstrom, *Clinical Psychiatry in Imperial Germany*, 143.

11. Adolf Meyer, "Excerpt from the *67th Annual Report of the Trustees of the Worcester Insane Hospital, 30 September 1899*," in *CP*, vol. 2, 68.

12. Meyer, "Cooperation Between Hospitals and the Institute," 121.

13. Meyer, "Review of *Psychiatrie*," 300.

14. Meyer, *Outlines of Examinations* (New York: Bloomingdale Hospital Press, 1918), 6.

15. For Henle, see Edwin Clarke and L. S. Jacyna, eds., *Nineteenth-century Origins of Neuroscientific Concepts* (Berkeley: University of California, 1987), 22; for Wernicke, see Engstrom, *Clinical Psychiatry in Imperial Germany*, 101; for Jackson, see L. S. Jacyna, *Lost Words: Narratives of Language and the Brain, 1825–1926* (Princeton, NJ: Princeton University Press, 2000), 127.

16. Adolf Meyer, "Dynamic Interpretation of Dementia Praecox," *American Journal of Psychology* 21 (1910): 402.

17. Meyer, *Outlines of Examinations*, 13; Adolf Meyer, "Aims and Meaning of Psychiatric Diagnosis," *American Journal of Insanity* 74 (1917): 166–167.

18. Adolf Meyer, "Principles in Grouping the Facts in Psychiatry" (1905), in *CP*, vol. 2, 138.

19. Adolf Meyer, Scientific Notes and Records, n.d.. XI/1/2, AMC; Meyer, "Insanity: General Pathology," in *Reference Handbook of the Medical Sciences*, vol. 5, ed. A. H. Buck (New York: William Wood, 1902): 36.

20. Adolf Meyer, discussion section, *Journal of Philosophy, Psychology, and Scientific Methods* 9 (1912): 179; Meyer, "Principles in Grouping Facts," 138.

21. Adolf Meyer, "The Problems of Mental Reaction-Types, Mental Causes and Diseases." *Psychological Bulletin* 5 (1908): 253; Meyer, "Objective Psychology or Psychobiology with Subordination of the Medically Useless Contrast of Mental and Physical," *Journal of the American Medical Association* 65 (1915): 861; Meyer, "Insanity: General Pathology," 36.

22. Adolf Meyer, "The Role of the Mental Factors in Psychiatry," *American Journal of Insanity* 65 (1908): 41; Meyer, *Outlines of Examinations*, 5.

23. Meyer, *Outlines of Examinations*, 4–11.

24. Meyer, "Extra-Institutional Responsibilities," 228–229; Meyer, *Outlines of Examinations*, 13.

25. Case ARD-545.

26. Case ZWF-362; Case QES-174; Case ARD-545.

27. Meyer, Private Correspondence, 1914, Series XV, Box B4, AMC.

28. Meyer, *Outlines of Examinations*, 3.

29. Meyer, *Outlines of Examinations*, 2–5 (emphasis in original); Case SGE-889.

30. Meyer, *Outlines of Examinations*, 10.

31. Meyer, *Outlines of Examinations*, 6.

32. Edward Shorter, *A History of Psychiatry: From the Era of the Asylum to the Age of Prozac* (New York: John Wiley and Sons, 1997), 129.

33. This is a composite description based on the case histories for the years 1913 to 1917.

34. Meyer, *Outlines of Examinations*, 6; Institutional Records, "Routine Laboratory Examinations To Be Made On All Patients," c. 1914, XII/24/27, AMC; Case CPE-165.

35. Case ARD-545.

36. Meyer, *Outlines of Examinations*, 16–20.

37. Elizabeth Lunbeck, *The Psychiatric Persuasion: Knowledge, Gender, and Power in Modern America* (Princeton, NJ: Princeton University Press, 1994), 54–61.

38. Case SDZ-157; Meyer, *Outlines of Examinations*, 20; Case LFM-844.

39. Meyer, *Outlines of Examinations*, 13; Case RUR-949; Case NNT-197.

40. Meyer, *Outlines of Examinations*, 9 and 24.

41. Scientific Notes, n.d., XII/1/779, AMC; Jerome Frank, "Adolf Meyer in Retrospect," 11 February 1980, Box 3, EWC.

42. Case FYN-934 (emphasis added).

43. Gerald Grob, *Mad Among Us: A History of the Care of America's Mentally Ill* (New York: Free Press, 1994), 143; Richard Noll, *American Madness: The Rise and Fall*

of Dementia Praecox (Cambridge, MA: Harvard University Press, 2011), 158–159; Shorter, *History of Psychiatry*, 112; Andrew Scull, *Madhouse: A Tragic Tale of Megalomania and Modern Medicine* (New Haven, CT: Yale University Press, 2005), 239.

44. Ruth Leys, "The Correspondence between Adolf Meyer and E. B. Titchener," in *Defining American Psychology: The Correspondence between Adolf Meyer and Edward Bradford Titchener*, ed. Ruth Leys and Rand B. Evans (Baltimore: Johns Hopkins University Press, 1990), 83–88.

45. Adolf Meyer, "Excerpt from the *18th Annual Report of the State Commission in Lunacy, New York, 1906*," in *CP*, vol. 2, 149–150; Meyer to Edward Titchener, 28 September 1909, quoted in Leys and Rand, *Defining American Psychology*, 145. For phenomenology, see Stephen Kern, *The Culture of Time and Space 1880–1918* (Cambridge, MA: Harvard University Press, 2003; orig. published 1983), x–xi.

46. Jackson is quoted in Jacyna, *Lost Words*, 128; Adolf Meyer, "The 'Complaint' as the Center of Genetic-Dynamic and Nosological Teaching in Psychiatry" (1928), in *CP*, vol. 3, 15. See also Volker Hess and J. Andrew Mendelsohn, "Case and Series: Medical Knowledge and Paper Technology, 1600–1900," *History of Science* 48 (2010): 287–314.

47. Quoted in Grob, *Mad Among Us*, 146.

48. John Oliver, "Experience of a Psychiatric Missionary in the Criminal Courts," *Journal of the American Institute of Criminal Law and Criminology* 9/4 (1919): 569.

49. Franklin Ebaugh, Introduction to vol. 3 of *CP*, xi.

50. Jacob Conn, "Adolf Meyer Discusses the Pathology of Dementia Praecox," *American Journal of Psychiatry* 112/5 (1955): 366; Adolf Meyer, "Suggestions Concerning a Grouping of Facts According to Cases" (1905), in *CP*, vol. 2, 145.

51. Guenter Risse and John Harley Warner, "Reconstructing Clinical Activities: Patient Records in Medical History," *Social History of Medicine* 5/22 (1992): 189.

52. Meyer, *Outlines of Examinations*, 10 and 24; Meyer, "Insanity: General Pathology," 39.

53. Joel Howell, *Technology in the Hospital: Transforming Patient Care in the Early Twentieth Century* (Baltimore: Johns Hopkins University Press, 1995), 32. For the Life Chart, see Ruth Leys, "Types of One: Adolf Meyer's Life Chart and the Representation of Individuality," *Representations* 34 (1991): 1–28; Meyer, *Outlines of Examinations*, 23–24; Adolf Meyer, "The Life Chart and the Obligation of Specifying Positive Data in Psychopathological Diagnosis," in *Contributions to Medical and Biological Research, dedicated to Sir William Osler, in Honour of his Seventieth Birthday*, vol. 2, ed. Charles L. Dana et al. (New York: Hoeber, 1919), 1128–1133.

54. Effie Taylor (Phipps Clinic Head Nurse), 13 October 1915, *Questionnaire Relating to Hospitals for the Insane in the Public and Private Institutions in the Various States*, solicited by Senate Committee on Civil Service, Albany, New York, I/3770/1, AMC.

55. Case WSR-288; Case LAR-823; Case LFM-844.

56. Meyer, *Outlines of Examinations*, 12.

57. Meyer, *Outlines of Examinations*, 23; Adolf Meyer, "Notes of Clinics in Psychopathology" (1908), in *CP*, vol. 1, 45.

58. Gerald Grob, *Mental Illness and American Society, 1875–1940* (Princeton, NJ: Princeton University Press, 1983), 42–43 and 118–120; Engstrom, *Clinical Psychiatry in Imperial Germany*, 26–30 and 121–146.

59. Meyer, "Suggestions concerning a Grouping of Facts," 146.

60. Adolf Meyer to Hermann Meyer, 28 January 1912, IV/3/142, AMC; Adolf Meyer, "Ninth Lecture: Dementia Praecox," unpublished lecture, 1905, Series XV, Box 1, AMC.

61. Adolf Meyer, "The Psychiatric Clinic, Its Aims (Educational and Therapeutic), and the Results Obtained in Respect to Promotion of Recovery" (1913), in *CP*, vol. 2, 204. Originally published in *Transactions of the 17th International Congress of Medicine in London*, Section 12, Part 2 (1913): 9–11.

62. Adolf Meyer, "Scope of Psychopathology" (1916), *CP*, vol. 2, 622. Presidential address delivered to American Psychopathological Association. Originally published in *Psychiatric Bulletin of the New York State Hospitals* 1 (1916): 297–305.

63. Adolf Meyer, "An Attempt at Analysis of the Neurotic Constitution," *American Journal of Psychology* 14 (1903): 91; Meyer, "Trends in Modern Psychiatry," *Psychological Bulletin* 1/7-8 (1904): 218.

64. See Adolf Meyer, "Movement for a Change in Statistics" (1905), in *CP*, vol. 2, 132–146. Originally published in the *17th Annual Report of the State Commission in Lunacy, New York, 1905*.

65. Adolf Meyer to Hermann Meyer, 14 March 1914, IV/3/151, AMC.

66. See Horatio Pollock, "The Classification of Mental Diseases in New York," *Publications of the American Statistical Association* 15/117 (1917): 502–510.

67. Meyer, "Aims and Meaning of Diagnosis," 168 (both comments appear in the Discussion Section of Meyer's paper); Meyer, "Progress in Teaching Psychiatry," *Journal of the American Medical Association* 69 (1917): 861–863. See also Gerald Grob, "Origins of DSM-I: A Study in Appearance and Reality," *American Journal of Psychiatry* 148 (1991): 421–431.

68. Meyer, "Neurotic Constitution," 92.

69. Meyer, "Insanity: General Pathology," 38; Adolf Meyer, "The Anatomical Facts and Clinical Varieties of Traumatic Insanity," *American Journal of Insanity* 60/3 (1904): 429.

70. Adolf Meyer, "Fundamental Conceptions of Dementia Praecox," *British Medical Journal* 2387 (29 September 1906): 757.

71. Meyer was influenced especially by the work of Frédéric Paulhan. See Adolf Meyer, "Neurotic Constitution," 93–94; and Leys, "Types of One," 23 n. 11.

72. Meyer, "Mental Reaction-Types," 255.

73. Meyer, "Fundamental Conceptions of Dementia Praecox," 757; *Diagnostic and Statistical Manual: Mental Disorders*, prepared by the Committee on Nomenclature and Statistics of the American Psychiatric Association (Washington, DC: American Psychiatric Association, 1952).

74. Meyer, "Mental Reaction-Types," 257.

75. Charles Thompson, *Mental Disorders Briefly Described and Classified* (Baltimore: Warwick and York, 1920), 5–12.

76. See Meyer, "Habit Disorganizations"; Adolf Meyer, "The Relation of Psychogenic Disorders to Deterioration," *Journal of Nervous and Mental Disease* 34 (1907): 401–405; Adolf Meyer, "Review of *Studies in Psychopathology* by Boris Sidis," *Journal of Philosophy, Psychology and Scientific Methods* 4/23 (1907): 633–639; Meyer, "A Discussion of Some Fundamental Issues in Freud's Psychoanalysis" (1909), in *CP*, vol. 2, 604–617.

77. Adolf Meyer, "Moral Insanity," in *Dictionary of Philosophy and Psychology,* vol. 3, ed. James Mark Baldwin (New York: Macmillan, 1905), 104; Thompson, *Mental Disorders Classified,* 11.

78. Adolf Meyer, "The Aims of A Psychiatric Clinic" (1912), in *Proceedings of the Mental Hygiene Conference, New York, 1912* (New York: Committee on Mental Hygiene of the State Charities Aid Association, 1912): 121.

79. Adolf Meyer, "Evolution of the Dementia Praecox Concept" (1928), in *CP,* vol. 2, 485. Originally published in *Schizophrenia (Dementia Praecox)* (New York: Hoeber, 1928): 3–15. Meyer cites Kraepelin's 1919 work *Erscheinungsformen des Irreseins.*

80. *Johns Hopkins Circular* (Baltimore: Johns Hopkins Press), 1913, 1914, 1915, 1916, 1917.

81. Adolf Meyer, "The Aims of a Psychiatric Clinic" (1913), in *CP,* vol. 2, 193 and 200. Originally published in *Transactions of the 17th International Congress of Medicine in London,* Section 12, Part 1 (1913): 1–11.

82. Meyer, "My Experience," 88; Meyer, "Considerations on Psychiatry," 461.

83. Case FYN-934.

Chapter 5. *A Wonderful Center for Mental Orthopedics*

1. Adolf Meyer, "The Purpose of the Psychiatric Clinic," *American Journal of Insanity* 69 Special Issue (1913): 858.

2. Adolf Meyer, "Organization of the Work of the Henry Phipps Psychiatric Clinic" (1914), in *CP,* vol. 2, 207. Originally published in *Transactions of the American Medico-Psychological Association 70th Annual Meeting* 21 (1914): 397–403.

3. Meyer, "Purpose of the Psychiatric Clinic," 858 (emphasis added); Adolf Meyer, "What is the Safest Psychology for a Nurse?" (1916), in *CP,* vol. 4, 84; Meyer, Private Notes, n.d., XII/1/242, AMC (emphasis in original).

4. Adolf Meyer, *Outlines of Examinations* (New York: Bloomingdale Hospital Press, 1918), 23.

5. Adolf Meyer, "The Philosophy of Occupation Therapy" (1921), in *CP,* vol. 4, 89. Originally published in *Archives of Occupation Therapy* 1 (1922): 1–10.

6. Adolf Meyer to Hermann Meyer, 21 May 1913, IV/3/149, AMC.

7. Adolf Meyer to Effie Taylor, n.d. notes from organizational meeting and Taylor's response to *Questionnaire Relating to Hospitals for the Insane in the Public and Private Institutions in the Various States,* 13 October 1915, Senate Committee on Civil Service, New York, I/3770/1, AMC. See also *Report of the Superintendent of The Johns Hopkins Hospital,* nos. 25 through 30. For New York State, see Gerald Grob, *Mental Illness and American Society, 1875–1940* (Princeton, NJ: Princeton University Press, 1983), 19.

8. D. K. Henderson, Introduction to vol. 2 of *CP,* xvi; Case FDS-732.

9. Meyer, "Plans for Work in the Phipps Psychiatric Clinic" (1914), in *CP,* vol. 2, 188; Meyer, "Organization of Work of Phipps Clinic," 206.

10. Meyer, "Plans for Phipps Clinic," 188.

11. Case WJX-145.

12. Case EWB-937; Case LFM-844.

13. Case LFM-844; Case ARD-545.

14. Case NNT-917.

15. Case MDU-173.

16. Henry Phipps Psychiatric Clinic Nursing Manual, circa 1916, unpublished instruction manual consisting of sixty-eight typewritten pages bound with cloth tape and organized alphabetically by topic, XII/24/41, AMC (hereafter, Phipps Nursing Manual), 11 and 52.

17. Effie Taylor to Winford Smith, 22 August 1919, I/3770/2, AMC.

18. Jennifer Laws, "Crackpots and Basket-cases: A History of Therapeutic Work and Occupation," *History of the Human Sciences* 24 (2011): 65–81.

19. David G. Schuster, *Neurasthenic Nation: America's Search for Health, Happiness, and Comfort, 1869–1920* (New Brunswick, NJ: Rutgers University Press, 2011), 129.

20. *Maryland Psychiatric Quarterly*, Volumes 1–8 (1911–1917). For the history of occupational therapy, including the contributions of Slagle, Dunton, and Meyer, see Don M. Gordon, "Therapeutics and Science in the History of Occupational Therapy" (Ph.D. diss. University of Southern California, 2002), and Virginia Quiroga, *Occupational Therapy: The First Thirty Years* (Bethesda, MD: American Occupational Therapy Association, 1995).

21. Meyer, "What is the Safest Psychology?" 83–84.

22. Case RAB-261; Meyer, "Philosophy of Occupation Therapy," 90.

23. Meyer, "Philosophy of Occupation Therapy," 89–90.

24. Case EAF-170; Case AGH-133.

25. Case RAB-261; Meyer, "Philosophy of Occupation Therapy," 87.

26. Case RAB-261; Meyer, "Philosophy of Occupation Therapy," 87; Case ZWF-362.

27. Phipps Nursing Manual, 55.

28. Case AEN-477.

29. Phipps Nursing Manual (emphasis in original), 55.

30. See Meyer, "Philosophy of Occupation Therapy."

31. Case GFG-104; Case EVC-283; Phipps Nursing Manual, 2.

32. Case KLE-646

33. Case KLE-646.

34. Phipps Nursing Manual, 49.

35. Wendell Muncie, *Psychobiology and Psychiatry: A Textbook of Normal and Abnormal Human Behavior* (St. Louis: C. V. Mosby, 1939), 490.

36. Meyer, "Philosophy of Occupation Therapy," 86 (emphasis in original); Geoffrey Reaume, *Remembrance of Patients Past: Patient Life at the Toronto Hospital for the Insane, 1870–1940* (Oxford: Oxford University Press, 2000); Adolf Meyer to Effie Taylor, notes from an organizational meeting, c. 1913, I/3770/1, AMC.

37. Private Correspondence, Series XV, Medical Records Box 2, AMC; Case VMP-108; Case EAF-170.

38. Case EWB-937.

39. Case FYN-934; Case NSU-863.

40. Case KLE-646.

41. Case UBT-144; J. P. Müller, *My System: 15 Minutes' Work a Day for Health's Sake!* (Copenhagen: Tillge's Boghandel, 1904).

42. William James, *Principles of Psychology*, vol. 1 (New York: Henry Holt, 1890), 126; Adolf Meyer to Hermann Meyer, 14 March 1914, IV/3/151, AMC.

43. Howard Jones, personal telephone conversation with author, 17 March 2008. The pioneering research of Drs. Howard and Georgeanna Jones at Johns Hopkins Hospital in the 1960s was instrumental in the development of in vitro fertilization. When I spoke with Dr. Jones in 2008, he was the director of the Jones Institute for Reproductive Medicine at Eastern Virginia Medical School.

44. Adolf Meyer to Edward Kempf, 29 November 1912, I/2059/2, AMC.

45. Rebekah Wright, "Hydrotherapy for the Insane," c. 1908, Medical Staff Manual, Boston State Hospital, XII/4/10, AMC.

46. *Black's Medical Dictionary*, ed. John D. Comrie (New York: Macmillan, 1914), 845; Phipps Nursing Manual, 60.

47. "Bathtub the New Insanity Cure," *New York Herald*, 6 December 1903, I/114/2, AMC. See also Grob, *Mental Illness*, 17–19.

48. See, Susan Cayliff, *Wash and Be Healed: The Water-Cure Movement and Women's Health* (Philadelphia: Temple University Press, 1987); Roy Porter, ed., *The Medical History of Waters and Spas* (*Medical History*, Supplement 10 [1990]); Müller, *My System*, 68–69.

49. Joel Braslow, *Mental Ills, Bodily Cures: Psychiatric Treatment in the First Half of the Twentieth Century* (Berkeley: University of California Press, 1997), 37; Gerald Grob, *The Mad Among Us: A History of the Care of America's Mentally Ill* (New York: Free Press, 1994), 149. For theories see, for example, Simon Baruch, *The Principles and Practice of Hydrotherapy* (New York: William Wood, 1897), which was added to the Johns Hopkins Medical Library in 1908, and L. Hill and M. Flack, "The Influence of Hot Baths on Pulse Frequency, Blood Pressure, Body Temperature, Breathing Volume and Alveolar Tensions of Man," *Journal of Physiology* 38 (1909): 57–61.

50. Adolf Meyer, "A Few Remarks Concerning the Organization of the Medical Work in Large Hospitals for the Insane" (1902), *CP*, vol. 2, 96; Adolf Meyer, "My Experience With American Psychiatry," unpublished manuscript, 1898, X/1/27, AMC: 47.

51. Adolf Meyer to Rebekah Wright, 26 May 1913, I/4117/1, AMC.

52. Lloyd Felton, "Hydrotherapy as Applied to Psychiatric Patients," *Maryland Psychiatric Quarterly* 5 (1915): 33–43.

53. Case BSG-152; Case HTR-976; Case EVC-283.

54. Case RZG-112; Case TDA-469.

55. Phipps Nursing Manual, 60; Eric J. Engstrom, *Clinical Psychiatry in Imperial Germany: A History of Psychiatric Practice* (Ithaca, NY: Cornell University Press, 2003), 133–134; George Kirby, "The Psychiatric Clinic at Munich," *Medical Record* 70 (1906): 191; Felton, "Hydrotherapy," 39; William A. White, "Dangers of the Continuous Bath," *American Journal of Insanity* 72 (1916): 481–484.

56. Phipps Nursing Manual, 61; Case LFM-844; Case OLR-652.

57. Phipps Nursing Manual, 60–61; Case KLE-646.

58. Phipps Nursing Manual, 62.

59. Andrea Tone, *The Age of Anxiety: A History of America's Turbulent Affair with Tranquilizers* (New York: Basic Books, 2009), 9.

60. Case TDA-469.

61. Case EWB-937.

62. Adolf Meyer, "The Treatment of the Insane" (1894), in *CP*, vol. 2, 46; Meyer, "Die Irrenpflege in Schottland" (1893), in *CP*, vol. 2, 27–36; Meyer, "Plans for Phipps Clinic," 192–194.

63. Adolf Meyer, Discussion Section, *Transactions of the 17th International Congress of Medicine in London*, Section 12, Part 2 (1913): 9–11.

64. More likely a reference to William Jennings Bryan, the high-profile Democratic presidential candidate and secretary of state, than to the Democratic Florida senator William James Bryan.

65. Case NSU-863 (emphasis in original).

66. Case RZE-927.

67. Case LFM-844; Case EWB-937.

68. Case ARD-545.

69. Case KLE-646.

70. Case KLE-646.

71. Case EWB-937; Phipps Nursing Manual, 63.

72. Adolf Meyer, "Dynamic Interpretation of Dementia Praecox," *British Medical Journal* 2387 (29 September 1906): 757–758.

73. Charles Thompson, *Mental Disorders Briefly Described and Classified* (Baltimore: Warwick and York, 1920), 29–31; Case OLR-652.

74. Case ARD-545.

75. Case EWB-937; Case ARD-545.

76. Case WJX-145; Adolf Meyer, "Some Common Misunderstandings About State Hospitals and the Way to Make Them Unnecessary" (1913), in *CP*, vol. 4, 181. Address to North Carolina's first Conference on Mental Hygiene, 1913.

77. Case ARD-545.

78. Case VMR-108.

79. Case CNK-117; Phipps Nursing Manual, 51.

80. Case RZG-112.

81. Adolf Meyer, "The Purpose of the Psychiatric Clinic," 860; Meyer, "The Relation of Psychogenic Disorders to Deterioration," *Journal of Nervous and Mental Disease* 34 (1907): 404.

Chapter 6. *Subconscious Adaptation*

1. Adolf Meyer, "Review of *Psychotherapy* by Hugo Münsterberg," *Science*, new series 30/761 (1909): 150; Sonu Shamdasani, "Psychotherapy, 1909: Notes on a Vintage," in *After Freud Left: A Century of Psychoanalysis in America*, ed. John C. Burnham (Chicago: University of Chicago Press, 2012): 31–47.

2. Sigmund Freud, *On the History of the Psychoanalytic Movement* (1914), in *The Standard Edition of the Complete Psychological Works of Sigmund Freud*, ed. and trans. James Strachey (London: Hogarth), vol. 14, 30; Richard Skues, "Clark Revisited: Reappraising Freud in America" in Burnham, *After Freud Left* (see n. 1), 55. See also John C. Burnham, *Psychoanalysis and American Medicine, 1894–1918* (New York: International Universities Press, 1967), 13–32; and Nathan Hale, *Freud and the Americans: The Beginnings of Psychoanalysis in the United States, 1876–1917* (Oxford: Oxford University Press, 1995), 3–23.

3. Adolf Meyer, "Aftercare and Prophylaxis" (1911), in *CP*, vol. 4, 214.

4. Adolf Meyer, "Problems of Mental Reaction-Types, Mental Causes and Diseases," *Psychological Bulletin* 5 (1908): 255; Meyer, "Remarks on Habit Disorganizations in the Essential Deteriorations, and the Relation of Deterioration to the Psychasthenic, Neurasthenic, Hysterical and Other Constitutions," in *Studies in Psychiatry* 1 (1912): 95–109.

5. See Marijke Gijswijt-Hofstra and Roy Porter, eds., *Cultures of Neurasthenia from Beard to the First World War* (Amsterdam: Rodopi, 2001); Sander L. Gilman et al., eds., *Hysteria Beyond Freud* (Berkeley: University of California Press, 1993); Laura Hirshbein, *American Melancholy: Constructions of Depression in the Twentieth Century* (New Brunswick, NJ: Rutgers University Press, 2009); Mark Micale, *Approaching Hysteria: Disease and Its Interpretations* (Princeton, NJ: Princeton University Press, 1995); Richard Noll, *American Madness: The Rise and Fall of Dementia Praecox* (Cambridge, MA: Harvard University Press, 2011); David G. Schuster, *Neurasthenic Nation: America's Search for Health, Happiness, and Comfort, 1869–1920* (New Brunswick, NJ: Rutgers University Press, 2011).

6. Pierre Janet, "The Major Symptoms of Hysteria," *Boston Medical and Surgical Journal* 155/23 (1906): 667–668. Janet's clinical and theoretical contributions were comprehensive and multifaceted. See Janet, *L'Automatisme Psychologique* (Paris: Felix Akan, 1889); Janet, *The Major Symptoms of Hysteria* (New York: Macmillan, 1907); Henri Ellenberger, *The Discovery of the Unconscious: The History and Evolution of Dynamic Psychiatry* (New York: Basic Books, 1970), 331–417; Hale, *Freud and the Americans*, 126–128; Otto van der Hart and Barbara Freidman, "A Reader's Guide to Pierre Janet on Dissociation," *Dissociation* 2/1 (1989): 3–16.

7. Ellenberger, *Discovery of the Unconscious*, 311–318; Hans Walther, "Adolf Meyer—Student of the Zurich Psychiatric School," *Gesnerus* 41 (1984): 49–51.

8. Burnham, *Psychoanalysis and American Medicine*, 458–465; Hale, *Freud and the Americans*, 116–249; Eva S. Moskowitz, *In Therapy We Trust: America's Obsession with Self Fulfillment* (Baltimore: Johns Hopkins University Press, 2001).

9. Sigmund Freud, "Origin and Development of Psychoanalysis," *American Journal of Psychology* 21 (1910): 194. Lectures delivered at Clark University in September 1909 and translated from German by Harry W. Chase. To guard against anachronistic comparisons of psychoanalysis and psychobiology, I have drawn primarily on this English translation of the five lectures, delivered in German by Freud (and heard by Meyer), at Clark University in September 1909. For Freud's and Meyer's respective uses of the term *dynamic*, see Ellenberger, *Discovery of the Unconscious*, 289–291. For comparative analyses of Meyerian and Freudian ideas, see the works of Ruth Leys: "Meyer's Dealings With Jones: A Chapter in the History of the American Response to Psychoanalysis," *Journal of the History of the Behavioral Sciences* 17 (1981): 445–465; "Meyer, Jung and the Limits of Association," *Bulletin of the History of Medicine* 59 (1985): 345–360; "Adolf Meyer: A Biographical Note" and "Correspondence between Adolf Meyer and E. B. Titchener," in *Defining American Psychology: The Correspondence between Adolf Meyer and Edward Bradford Titchener*, ed. Ruth Leys and Rand B. Evans (Baltimore: Johns Hopkins University Press, 1990): 39–114; "Types of One: Adolf Meyer's Life Chart and the Representation of Individuality," *Representations* 34

(1991): 1–28. For the reception and interpretation of psychoanalytic concepts in the United States in this period, see Burnham, *Psychoanalysis and American Medicine*; and Hale, *Freud and the Americans*. For insightful historical interpretations of Freud's early ideas, see George Makari, *Revolution in Mind: The Creation of Psychoanalysis* (New York: Harper Perennial, 2009); and Frank Sulloway, *Freud, Biologist of the Mind: Beyond the Psychoanalytic Legend* (Cambridge, MA: Harvard University Press, 1992).

10. Freud, "Origin and Development of Psychoanalysis," 197; Sigmund Freud and Josef Breuer, *Studien über Hysterie* (Leipzig: Deuticke, 1895).

11. Freud, "Origin and Development of Psychoanalysis," 215.

12. C. G. Jung, "Die Psychopathologische Bedeutung des Assoziationsexperimentes," *Archiv für Kriminal-Anthropologie und Kriminalistik* 22 (1906): 145–162; Frederick Peterson and C. G. Jung, "Psycho-Physical Investigations with the Galvanometer and Pneumograph in Normal and Insane Individuals," *Brain* 30 (1907): 153–218. See also Ellenberger, *Discovery of the Unconscious*, 657–748; Leys, "Meyer, Jung and Limits of Association," 345.

13. Adolf Meyer, "Normal and Abnormal Association," *Psychological Bulletin* 2 (1905): 242–259; Meyer, "Application of Association Studies," *Psychological Bulletin* 3 (1906): 275–280.

14. Leys, "Meyer, Jung and Limits of Association," 347–351; C. G. Jung to Sigmund Freud, 21 August 1908 (Letter 107J), and Jung to Freud, 9 September 1908 (Letter 108J), in *The Jung/Freud Letters*, ed. William McGuire, trans. Ralph Manheim and R. F. C. Hull (Princeton, NJ: Princeton University Press, 1974); Meyer, "Problems of Mental Reaction-Types," 258.

15. Adolf Meyer, "Interpretation of Obsessions," *Psychological Bulletin* 3 (1906): 280–283; Burnham, *Psychoanalysis and American Medicine*, 19–20, 128–133, 158–161; Hale, *Freud and the Americans*, 157–164; Leys, "Meyer, Jung and Limits of Association," 347.

16. Adolf Meyer, "The Relationship of Hysteria, Psychasthenia, and Dementia Praecox," in *Studies in Psychiatry* 1 (1912): 157.

17. Adolf Meyer, "The Aims of a Psychiatric Clinic" (1912), in *Proceedings of the First Mental Hygiene Conference, New York, 1912* (New York: Committee on Mental Hygiene, 1912): 119.

18. Meyer, "Hysteria, Psychasthenia, and Dementia Praecox"; Adolf Meyer, "Critical Review of the Data and General Methods and Deductions of Modern Neurology: Part 2," *Journal of Comparative Neurology* 8 (1898): 293.

19. Adolf Meyer, "Fundamental Conceptions of Dementia Praecox," *British Medical Journal* 2387 (29 September 1906): 757–758.

20. Adolf Meyer, "The Mental Factor in Medicine," *Journal of Comparative Neurology* 12 (1902): xviii; Meyer, "Normal and Abnormal Association," 258; Adolf Meyer to August Hoch, 30 December 1904, quoted in Gerald Grob, *The Inner World of American Psychiatry, 1890–1940: Selected Correspondence* (New Brunswick, NJ: Rutgers University Press, 1985), 23–24.

21. Adolf Meyer, "A Discussion of Some Fundamental Issues in Freud's Psychoanalysis" (1909), in *CP*, vol. 2, 615. See also Meyer, "Fundamental Conceptions of Dementia Praecox."

22. Meyer, "Hysteria, Psychasthenia, and Dementia Praecox," 155–156; Meyer, Private Correspondence, n.d., Series XV, Box A4, AMC.

23. Diary of William Lyon Mackenzie King, 1 November 1916, Mackenzie King Papers, Library and Archives Canada. W. L. Mackenzie King was Prime Minister of Canada during the interwar period. Complaining of chronic nervousness, he was referred by his friend John D. Rockefeller, Jr., to the neurologist and chief of medicine at Johns Hopkins, Lewellys Barker, in 1916. Barker, in turn, referred him to Meyer. For a detailed account, see Paul Rozen, *Canada's King: An Essay in Political Psychology* (Oakville, ON: Mosaic Press, 1998).

24. Case CAL-942.

25. Meyer, "Fundamental Issues in Psychoanalysis," 615 and 604.

26. Meyer, "Interpretation of Obsessions," 282.

27. Meyer, "Application of Association Studies," 277; Meyer, "Normal and Abnormal Association," 256.

28. Meyer, "Normal and Abnormal Association," 256 (emphasis added). Meyer performed similar maneuvers in "A Discussion of Some Fundamental Issues in Freud's Psychoanalysis," published in 1909.

29. Sigmund Freud to C. G. Jung, 2 February 1910 (Letter 177F), in *The Jung/Freud Letters*. See also Hale, *Freud and the Americans*, 198 and Leys, "Meyer, Jung and Limits of Association," 345.

30. August Hoch to Adolf Meyer, 22 November 1909, I/1725/24, AMC; Macfie Campbell to Meyer, 2 July 1908, I/595/3, AMC; Trigant Burrow to Meyer, 12 February 1909 and 9 September 1909, I/565/3, AMC.

31. For Brill, see Leys, "Meyer, Jung and Limits of Association," 348; for Jones, see Leys, "Meyer's Dealings with Jones," 458–459; Adolf Meyer to C. G. Jung, 21 November 1911, I/1987/1, AMC.

32. Case SGE-889.

33. Adolf Meyer, *Outlines of Examinations* (New York: Bloomingdale Hospital Press, 1918), 5–11.

34. Case AGH-133; Meyer, Private Correspondence, 1913, Series XV, Box B4, AMC.

35. Case FYN-934; Meyer, *Outlines of Examinations,* 14; Case AGH-133.

36. Case YDA-919.

37. Meyer, Scientific Notes, n.d., XII/1/779, AMC; D. K. Henderson, Introduction to vol. 2 of *CP*, xiii; Stanley Cobb, "Adolf Meyer," *Archives of Neurology and Psychiatry* 64 (1950): 879; Thomas Rennie, "Adolf Meyer and Psychobiology: The Man, His Methodology and Its Relation to Therapy," *American Congress on General Semantics*, vol. 2 (1943): 160.

38. Diary of Mackenzie King, 27 October 1916, Archives of Canada.

39. Case GFG-104.

40. Case EWB-937.

41. Adolf Meyer, "Conditions for a Home of Psychology in the Medical Curriculum," *Journal of Abnormal Psychology* 7 (1912–1913): 325.

42. Case LFP-831.

43. Case BBW-164; Case CAL-942.

44. Meyer, "Fundamental Issues in Psychoanalysis," 605 and 614; Adolf Meyer to Morton Prince, 4 November 1909, quoted in Hale, *Freud and the Americans*, 18.

45. "Psychanalysis a New Science," *Chicago Daily Tribune*, 20 August 1911, E2; "By Dual Personality New Theory of Nervous Disease to be Tested Here," *Baltimore Sun*, 11 January 1911, 14.

46. Private Correspondence, 6 April 1916, Series XV, Box A13, AMC; Case FYN-934.

47. Case KCM-729 (emphasis added for clarity).

48. Meyer, "Normal and Abnormal Association," 258; Case CAL-942; Private Correspondence, 1914, Series XV, Box B4, AMC.

49. Case SSC-355.

50. Case PTT-389.

51. Meyer, "Fundamental Issues in Psychoanalysis," 605–606.

52. Meyer, "Fundamental Issues in Psychoanalysis," 614; Leys, "Types of One," 26 n. 45; Leys, "Meyer, Jung and Limits of Association," 358; Meyer, "Problems of Mental Reaction-Types," 260.

53. Private Correspondence, 1911, Series XV, Box A3, AMC; Meyer, Scientific Notes, n.d., XII/1/780, AMC; Private Correspondence, n.d., Series XV, Box A4, AMC.

54. Case SGE-889; Meyer, "Mental Factors in Psychiatry," 586; Case YDA-919.

55. Meyer, "Aims of a Psychiatric Clinic" (1912), 198; Private Correspondence, 1914, Series XV, Box B4, AMC; Meyer, "Mental Factors in Psychiatry," 586.

56. Case CNK-117 (Henrietta).

57. Case SGE-889 (Irving). Shire and Schmitt are also pseudonyms.

58. The name of the psychoanalyst is withheld to protect the patient's anonymity. It can be found in the case history of Irving.

59. Jung to Freud, 19 January 1909 (Letter 126J), in *The Jung/Freud Letters*.

60. "Has the Psycho-Analyst Fad Struck You Yet?" *New York Tribune*, 16 May 1915, B11.

61. Freud, "Origin and Development of Psychoanalysis," 216; Sulloway, *Freud, Biologist of the Mind*, 272.

62. Adolf Meyer, "Preparation for Psychiatry," *Archives of Neurology and Psychiatry* 30 (1933): 1123.

63. Leys, "Types of One," 16.

64. Adolf Meyer, *Psychobiology: A Science of Man* (Springfield, IL: Charles Thomas, 1957), 165. Originally presented as the inaugural Salmon Lectures in April 1932 in New York City.

Conclusion

1. Adolf Meyer, "The Opening of the Henry Phipps Psychiatric Clinic" (1913), in *CP*, vol. 2, 185. Originally published in *American Journal of Insanity* 69 (1913): 1079–1086; Adolf Meyer to Hermann Meyer, 12 June 1908. IV/3/122, AMC.

2. Richard Noll, *American Madness: The Rise and Fall of Dementia Praecox* (Cambridge, MA: Harvard University Press, 2011), 278–280; D. B. Double, "What Would Adolf Meyer Have Thought of the Neo-Kraepelinian Approach?" *Psychiatric Bulletin* 14 (1990): 472–474; Jack Pressman, *Last Resort: Psychosurgery and the Limits of Medicine* (Cambridge: Cambridge University Press, 1998), 18–46.

3. Eric R. Kandel et al., *Principles of Neural Science*, 5th ed. (New York: McGraw-Hill, 2012).

4. Adolf Meyer, "A Few Demonstrations of Pathology of the Brain and Remarks on the Problems Connected With Them," *American Journal of Insanity* 52 (1895): 248.

5. Adolf Meyer, "Insanity: General Pathology," in *Reference Handbook of the Medical Sciences*, 2nd ed., vol. 5, ed. A. H. Buck (New York: William Wood, 1902): 40.

6. Michael Merzenich, "Growing Evidence of Brain Plasticity," lecture delivered at TED2004 Conference, February 2004, web video retrieved 3 August 2013: http://www.ted.com/talks/michael_merzenich_on_the_elastic_brain.html. Michael Merzenich is Professor Emeritus in Neuroscience at University of California, San Francisco. For an overview of his academic research, see Dean B. Buonomano and Michael Merzenich, "Cortical Plasticity: From Synapses to Maps," *Annual Review of Neuroscience* 21 (1998): 149–186. His 2007 Public Television Broadcast *The Brain Fitness Program* promotes and sells "brain-training" exercises, developed by his company PositScience, based on the idea that active learning causes the brain to remodel itself. Alison Gopnik, Andrew N. Meltzoff, Patricia K. Kuhl, *The Scientist in the Crib: What Early Learning Tells Us About the Mind* (New York: Harper-Perennial, 2002), 184.

7. Louis Hausman, "Review of *The Collected Papers of Adolf Meyer*, vol. 1, Neurology," *Psychosomatic Medicine* 14 (1952): 422. The following publications are good examples of recent experimental and clinical applications of neuroimaging techniques to the study of neuroplasticity: Christy L. Ludlow et al., "Translating Principles of Neuroplasticity into Research on Speech Motor Control Recovery and Rehabilitation," *Journal of Speech, Language, and Hearing Research* 51/1 (2008): S240–S258 and C. L. Kipps et al., "Understanding Social Dysfunction in the Behavioural Variant of Frontotemporal Dementia: The Role of Emotion and Sarcasm Processing," *Brain* 132 (2009): 592–603.

8. Timothy Wilson, *Strangers to Ourselves: Discovering the Adaptive Unconscious* (Cambridge, MA: Harvard University Press, 2002), 6–7; Malcolm Gladwell, *Blink: The Power of Thinking Without Thinking* (New York: Back Bay Books, 2005), 11–14.

9. Adolf Meyer, "Fundamental Conceptions of Dementia Praecox," *British Medical Journal* 2387 (29 September 1906): 758. I have added punctuation to facilitate comprehension.

10. Gerald Grob, "Origins of DSM-I: A Study in Appearance and Reality," *American Journal of Psychiatry* 148 (1991): 421–431; *Diagnostic and Statistical Manual: Mental Disorders*, prepared by the Committee on Nomenclature and Statistics of the American Psychiatric Association (Washington, DC: American Psychiatric Association, 1952): 1.

11. For Kupfer, see Rebecca A. Clay, "Revising the DSM," *Monitor on Psychology* 42/1 (2011): 54; for Oldham, see Mark Moran, "DSM Section Contains Alternative Model for PD," *Psychiatric News* 48/9 (2013): 11; René J. Muller, "To Understand Depression, Look to Psychobiology, Not Biopsychiatry," *Psychiatric Times* 20/8 (2003): 46.

12. S. Nassir Ghaemi, "Adolf Meyer: Psychiatric Anarchist," *Philosophy, Psychiatry, and Psychology* 14 (2007): 343.

13. John Oliver, "Experience of a Psychiatric Missionary in the Criminal Courts," *Journal of the American Institute of Criminal Law and Criminology* 9/4 (1919): 570; Eugene Taylor, *The Mystery of a Personality: History of Psychodynamic Theories* (New York: Springer, 2009), 190–193; William Galt, ed., *A Search for Man's Sanity: The Selected Letters of Trigant Burrow* (New York: Oxford University Press, 1958); Jay Schulkin, *Curt*

Richter: A Life in the Laboratory (Baltimore: Johns Hopkins University Press, 2005); Marc-Adélard Tremblay, "Alexander H. Leighton's and Jane Murphy's Scientific Contributions in Psychiatric Epidemiology," *Transcultural Psychiatry* 43 (2006): 7–20; for McHugh, see Aaron Levin, "Despite Critics, Adolf Meyer's Influence Still Very Much Alive," *Psychiatric News* 48/6 (2013): 17–29.

14. Cannon is quoted in Gerald Grob, *Mental Illness and American Society* (Princeton, NJ: Princeton University Press, 1983), 117; Adolf Meyer, "Fourteenth Maudsley Lecture: British Influences in Psychiatry and Mental Hygiene," *Journal of Mental Science* 79/326 (1933): 445.

INDEX